WATER IN R(

MW01670623

Dear MyCopy Customer,

This Springer book is a monochrome print version of the eBook to which your library gives you access via SpringerLink. It is available to you as a subsidized price since your library subscribes to at least one Springer eBook subject collection.

Please note that MyCopy books are only offered to library patrons with access to at least one Springer eBook subject collection. MyCopy books are strictly for individual use only.

You may cite this book by referencing the bibliographic data and/or the DOI (Digital Object Identifier) found in the front matter. This book is an exact but monochrome copy of the print version of the eBook on SpringerLink.

GEOTECHNICAL, GEOLOGICAL AND EARTHQUAKE ENGINEERING

Volume 5

Series Editor

Atilla Ansal, *Kandilli Observatory and Earthquake Research Institute, Boğaziçi University, Istanbul, Turkey*

Editorial Advisory Board

Julian Bommer, *Imperial College London, U.K.*
Jonathan D. Bray, *University of California, Berkeley, U.S.A.*
Kyriazis Pitilakis, *Aristotle University of Thessaloniki, Greece*
Susumu Yasuda, *Tokyo Denki University, Japan*

For other titles published in this series, go to
www.springer.com/series/6011

Water in Road Structures

Movement, Drainage and Effects

edited by

ANDREW DAWSON

Nottingham Transportation Engineering Centre
University of Nottingham, UK

ESF provides the COST Office through an EC contract

COST is supported by the EU RTD Framework programme

Springer

Written by members of the WATMOVE team,

Edited by Andrew Dawson
Nottingham Transportation Engineering Centre,
University of Nottingham, UK
Andrew.Dawson@nottingham.ac.uk

COST Action No. 351
October 2008

"This publication is based on a study entitled 'Water Movements in Road Pavements and Embankments' ('*WATMOVE*') which was supported by COST. More details of COST can be found at the end of this book.

Legal notice

Neither the ESF-COST Office nor any person acting on its behalf is responsible for the use which might be made of the information contained in this publication. The views expressed in this publication do not necessarily reflect the views of the ESF-COST Office. The ESF-COST Office is not responsible for the external websites referred to in this publication.

DOI 10.1007/978-1-4020-8562-8

© Springer Science+Business Media B.V. 2009
No part of this work may be reproduced, stored in a retrieval system, or transmitted
in any form or by any means, electronic, mechanical, photocopying, microfilming, recording
or otherwise, without written permission from the Publisher, with the exception
of any material supplied specifically for the purpose of being entered
and executed on a computer system, for exclusive use by the purchaser of the work.

Printed on acid-free paper

9 8 7 6 5 4 3 2 1

springer.com/mycopy

Preface

Background

This book is the most obvious outcome of the "COST 351, WATMOVE" project (see www.watmove.org). For most readers these terms probably mean little so some explanation is called for. In 2001/2002 a small group led by Kent Gustafson of the Swedish Road and Traffic Institute (VTI) made a proposal to the European programme on "Co-operation in Science and Technology" (COST). They proposed that a pan-European team be set up to study the issue of "Water Movements in Road Pavements and Embankments" (acronym = WATMOVE). The COST organisation agreed the proposal and the study formally began in December 2003 with the support of COST. Due to ill-health, Kent was not able to lead the project and I was asked by the Management Committee to chair the project team.

Scope of the Book

This book is NOT about "Water and Roads", nor on "Water on Roads". There are other books which deal with surface water drainage in great detail and there are other source materials that deal with the impact of roads on water in the general environment. To cover every aspect of the interaction between water and highways would have required a much greater effort and a much thicker book. So this book seeks to limit itself to:

(i) Water inside the road construction and the underlying subgrade soils and rocks;
(ii) Water from the surface down to the phreatic surface,[1] and
(iii) Water in the road and ground between the fence-lines of the highway.

Sometimes these boundaries to the book's scope were a little bit too limiting so, from time-to-time the book wanders somewhat further. For example, it proved impossible (and undesirable) to discuss water in the road construction without

[1] Annex C contains a Glossary of terms that may be unfamiliar to some readers.

mentioning from where it comes. So, there is some consideration of run-off in so far as it is the major contributor to sub-surface road water, but readers will not find a full description of surface water drainage systems. Within these self-set limits the authors have tried to address most conceivable topics in some detail, bringing together both established theory and practice and some of the latest developments.

Acknowledgements

The job of chairman is not always easy, but in the case of WATMOVE it has been a privilege and a pleasure to work with a wonderful team of experts drawn from engineering, environmental and geological backgrounds across 18 European countries. Their hard work can be seen in the pages that follow. Every member of the WATMOVE project has contributed in some way or another. Most have authored and/or edited parts of the text and you will find the names of the chapter co-ordinator and the contributing authors at the head of each chapter. Many have provided raw information. Some have contributed nationally developed research findings. All have entered into developing a mutual understanding and appreciation of each other's viewpoints. By this method this book is more than a summary of individual contributions, it is truly a state-of-the-art-and-practice on subsurface road drainage. Mention should also be made of the many scientists and engineers who contributed indirectly by providing information and, in the case of a few, even contributed text.

The team, listed at page xiii, wish to thank their employers for actively supporting their contributions and the European Science Foundation's COST office in Brussels for all their support and funding. In particular we would like to single out the COST office's scientific and administrative secretaries, Jan Spousta, Marcu Zisenis, Thierry Groger, Isabel Silva and Carmencita Malimban for their understanding and responsive assistance in so many matters, great and small. This publication has been supported by COST.

It's their overall aim that the relevant and useful aspects of modern (and not-so-modern) research be implemented into practice. So this book is directed towards practitioners – engineers, environmentalists and hydrogeologists who have to provide for pavement and earthworks drainage – and those who will soon become practitioners – students taking advanced courses in pavement engineering, hydrogeology and geo-environmental engineering. Inevitably some will find some sections more pertinent and accessible than others .... but we hope that readers will find this a valuable resource from which to learn and to which they will often turn for reference. If you still want more then you may find some other resources at our web-site – www.watmove.org

So, finally, a really big "thank you" to everyone in the WATMOVE team and also to you, the reader. If the contents of this book prove useful to you, the the efforts of the team will certainly have been worthwhile!

Nottingham, UK
October 2008

Andrew Dawson

Contents

Contributors

The following have contributed to this book. Where a name is shown in bold, he or she is a contributor to one of the main chapters. Where a name is shown in bold and italics, the person was a contributor to one of the main chapters but not a member of the COST 351 project. Special thanks is due to these authors for being external contributors to the book. The others listed were members of the COST 351 project team but their contribution has not been separately identified. This doesn't mean that it was an unimportant part .... in several cases these people have made major contributions in editing, providing material, organizing the appendices, etc. Some of those listed only participated in the COST 351 Action for a short period. Particular recognition is due to those who helped establish the direction of the study but were then unable to continue to the final stages of which this book is the principal result.

Jostein Aksnes
Vegvesen/Norwegian Road Administration

Radu Andrei
Technical University "Gh. Asachi", Iaşi, Romania

Torleif Bækken
Norsk Institutt for Vannforskning/Norwegian Institute for Water Research (NIVA), e-mail: torleif.backken@niva.no

José Baena
Centro de Estudios y Experimentación de Obras Públicas/Research and Experimentation Centre for Public Works (CEDEX), Spain, e-mail: jose.m.baena@cedex.es

Susanne Baltzer
Vejteknisk Institut/Danish Road Institute, e-mail: sub@vd.dk

Gunnar Bjarnason
Vegagerdin/Icelandic Road Administration, e-mail: gbj@vegag.is

Virgil Breaban
Ovidius University of Constanţa, Romania

Mihael Brenčič
Geoloski Zavod Slovenije/Geological Survey of Slovenia,
e-mail: mihael.brencic@geo-zs.si

Cane Cekerevac
Stucky SA, Switzerland, e-mail: ccekerevac@stucky.ch

Robert Charlier
University of Liège, Belgium (vice chairman), e-mail: robert.charlier@ulg.ac.be

Cyrille Chazallon
University of Limoges/National Institute of Applied Sciences, Strasbourg, France,
e-mail: cyrille.chazallon@insa-strasbourg.fr

Frédéric Collin
University of Liège, Belgium, e-mail: f.collin@ulg.ac.be

Andrew Dawson
University of Nottingham, UK (chairman),
e-mail: andrew.dawson@nottingham.ac.uk

Sigurður Erlingsson
Háskóli Íslands/University of Iceland & Statens väg- och transportforskningsinstitut/Swedish National Road and Transport Research Institute (VTI) (vice chairman), e-mail: sigger@hi.is

José Santinho Faísca
Estradas de Portugal/Roads of Portugal, e-mail: jose.faisca@estradasdeportugal.pt

Lennart Folkeson
Statens väg- och transportforskningsinstitut/Swedish National Road and Transport Research Institute (VTI), e-mail: lennart.folkeson@vti.se

Denis François
Laboratoire Central des Ponts et Chaussées/Central Bridges and Highways Laboratory (LCPC), France, e-mail: denis.francois@lcpc.fr

Beata Gajewska
Instytut Badawczy Dróg i Mostów/Road and Bridge Research Institute (IBDiM), Poland, e-mail: bgajewska@ibdim.edu.pl

Klas Hansson
(formerly) Uppsala University, Sweden, e-mail: klas.hansson@vti.se

Diane Harrower
Scott Wilson, UK

Åke Hermansson
Statens väg- och transportforskningsinstitut/Swedish National Road and Transport Research Institute (VTI), e-mail: ake.hermansson@vti.se

Pierre Hornych
Laboratoire Central des Ponts et Chaussées/Central Bridges and Highways Laboratory (LCPC), France, e-mail: pierre.hornych@lcpc.fr

Aleksandar Kostadinović
Institut za Ispitivanje Materijala/Institute for Testing Materials (IMS)

Cezary Kraszewski
Instytut Badawczy Dróg i Mostów/Road and Bridge Research Institute (IBDiM), Poland, e-mail: ckraszewski@ibdim.edu.pl

Niki Kringos
Delft University of Technology, Netherlands, e-mail: n.kringos@tudelft.nl

Petra Kuřímská
Centrum Dopavniho Vyzkumu/Transport Research Centre (CDV), Czech Republic

Rainer Laaksonen
VTT/Technical Research Centre of Finland (VTT),
e-mail: petra.kurimska@cdv.cz

Michael Larsen
Vejteknisk Institut/Danish Road Institute

Lyesse Laloui
Ecole Polytechnique Fédéral de Lausanne/Swiss Federal Institute of Technology, e-mail: lyesse.laloui@epfl.ch

Teresa Leitão
Laboratório Nacional de Engenharia Civil/National Laboratory for Civil Engineering (LNEC), Portugal, e-mail: tleitao@lnec.pt

Roman Ličbinský
Centrum Dopavniho Vyzkumu/Transport Research Centre (CDV), Czech Republic, e-mail: roman.licbinsky@cdv.cz

Wyn Lloyd
Highways Agency, UK

Angel Mateos Moreno
Centro de Estudios y Experimentación de Obras Públicas / Research and Experimentation Centre for Public Works (CEDEX), Spain

Nikolaos Michas
Οδομελ/ Greek Road Federation

Goran Mladenović
University of Belgrade, Serbia

Antero Nousiainen
Destia Ltd., Finland, e-mail: antero.nousiainen@destia.fi

Primož Pavšič
Zavod za Gradbeništvo Slovenije/Slovenian National Building and Civil Engineering Institute (ZAG), e-mail: primoz.pavsic@zag.si

Karel Pospíšil
Centrum Dopavniho Vyzkumu/Transport Research Centre (CDV), Czech Republic

Paulică Răileanu
Technical University "Gh. Asachi", Iaşi, Romania

Tom Scarpas
Delft University of Technology, Netherlands, e-mail: a.scarpas@tudelft.nl

Mate Sršen
Institut Građevinarstva Hrvatske/Civil Engineering Institute of Croatia (IGH), e-mail: mate.srsen@igh.hr

Eric van den Kerkhof
Formerly Centre de Recherches Routières/Opzoekingscentrum voor de Wegenbouw/Belgian Road Research Centre (BRRC)

Martin Vojtěšek
Centrum Dopavniho Vyzkumu/Transport Research Centre (CDV), Czech Republic, e-mail: martin.vojtesek@cdv.cz

List of Figures

List of Tables

Chapter 1
Introduction

Andrew Dawson

Abstract This introduction provides a brief review of the history of highway subdrainage before setting out the aims and organisation of the book of which it forms the first chapter. It gives an overview of the subjects to be covered in the following chapters, introduces the key topics including definitions of subgrade and pavement layers, their classification from a drainage point-of-view together with a brief coverage of the principle of effective stress, suction, leaching and water movement due to evaporation and frost-heave. It outlines the way in which pavements and the hydrological environment interact before introducing the reader to the varieties of climate in which highways and pavements have to operate – a task that is likely to become more onerous in the light of climate change effects.

Keywords Introduction · history · definition · drainage classification of pavements · alternative materials · drainage systems · climate

1.1 Some History

In Europe road construction may date back as far as 3,500 years ago. These early roads were probably largely for ceremonial purposes, over short distances, and may have carried little, if any, wheeled traffic. It was not until the growth of the Roman Empire that a large network of engineered pavements was first constructed (Fig. 1.1). Such was the desire to secure the Empire against enemies and to enhance trade that, at the peak time, about 0.5 km was being built *daily*. Although foot and hoof traffic probably predominated, those roads were certainly used for wheeled vehicles too.

The engineers responsible for these pavements understood some important truths about pavement drainage – truths which, in practice, sometimes are still not recognised today. Figure 1.2 shows a cross section of a high quality Roman road. It illustrates that its designer:

✉ A.R. Dawson
University of Nottingham, Nottingham, UK
e-mail: andrew.dawson@nottingham.ac.uk

A.R. Dawson (ed.), *Water in Road Structures*, DOI 10.1007/978-1-4020-8562-8_1,

© Springer Science+Business Media B.V. 2009

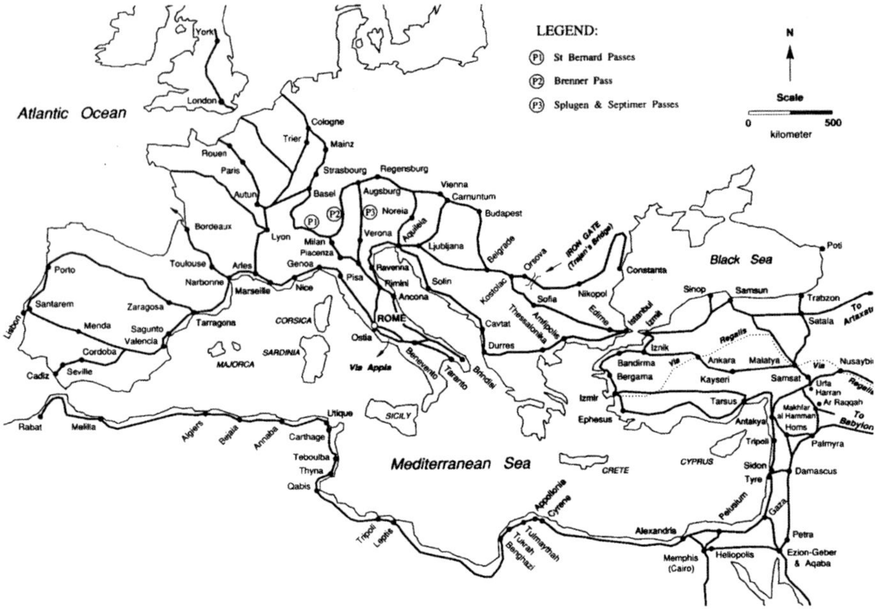

Fig. 1.1 Roman principal road network (Lay, 1992). Reproduced by permission of M.G. Lay

- provided cross-fall to help shed surface water to the margins rather than to soak in,
- raised the pavement well above the groundwater level of the surrounding ground so as to keep the amount of water in the embankment low and the effective stress (and, hence, soil strength) high,
- provided lateral ditches to prevent water table rise in wet periods and to convey draining water away from the construction.

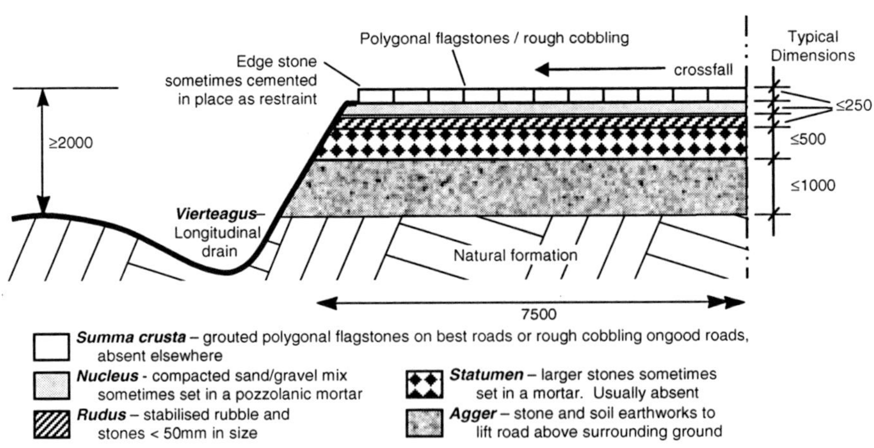

Fig. 1.2 Cross-section of a high quality Roman pavement (dimensions in mm)

Only a small proportion of pavements were built as illustrated, more generally a two layer pavement, comprising *nucleus* and *rudus* as explained in the notes to the Figure, was constructed on an embankment (*agger* – see Fig. 1.2). Despite the economizing on materials, the aim of keeping the construction well-drained remained unaltered.

It was not until the late 1700s and early 1800s that a similar understanding was once again developed. Figure 1.3 shows a cross-section of a main coach pavement as designed by Thomas Telford, the Scottish engineer who worked in the UK and other European countries between about 1800 and 1830. In this design an effort has been made to provide a relatively impermeable surface to prevent water infiltration and a drainable foundation, but the route to lateral drains from these is not well developed.

Despite these early evidences of the understanding that drainage is needed for a well-functioning road, the lesson hasn't always been fully appreciated. Arthur Cedergren, the American engineer, famously said that "there are three things that a road requires – drainage, drainage and more drainage" (Cedergren, 1974, 1994). He said this many years ago yet, despite many advances in the subject and a huge rise in environmental concerns since he was active, little further has been published in the area. This book is our attempt to redress that omission.

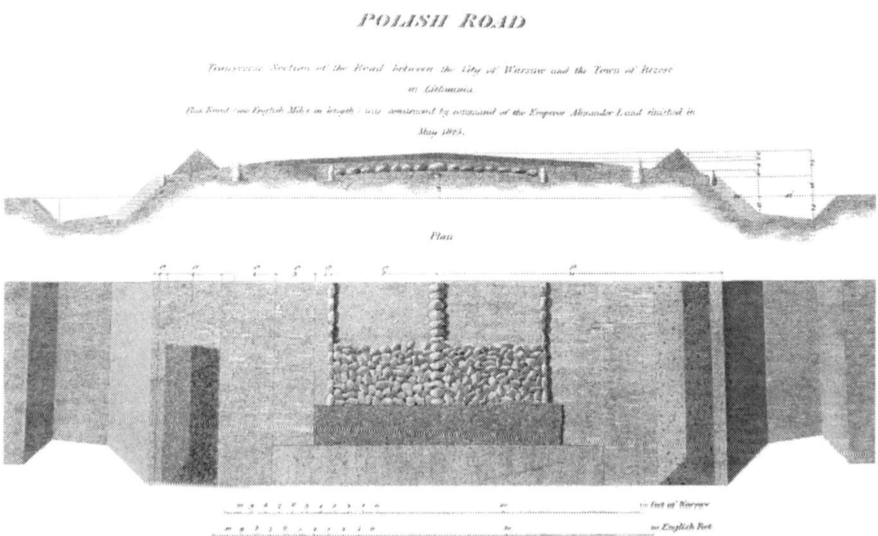

Fig. 1.3 Facsimile of Telford's design for a road from Warsaw to Brzseć[1] (Telford, 1838)

[1] The text at the top of the figure reads "*POLISH ROAD / Transverse Section of the Road between the City of Warsaw and the Town of Brzesć in Lithuania. / This Road (100 English Miles) was constructed by command of the Emperor Alexander I. and finished in May 1825.*" Nowadays, Brzesć is known as Brèst, and forms the border crossing town between Poland and present day Belarus. It lies about 215 km East of Warsaw (Warszawa) not "100 English miles" $\approx$ 160 km as indicated in the figure.

In the following sections the modern manifestation of the same objectives – to keep pavements dry by limiting ingress and assisting drainage – are introduced. Alongside this updating of the age-old principle of drainage, the modern pavement engineer or geo-environmentalist has to consider the quality of the draining water. What chemical components does the water contain? Is that a problem? Where will they go? The following sections seek to introduce this modern concern as well.

1.2 Aims and Objectives

The main aim of this book is to increase the knowledge about water in the sub-surface road environment so as to improve highway performance and minimise the leaching of contaminants from roads. Improvement of pavement performance will lead to less road closures, better use of the road network, longer service life and more effective transportation of goods and people.

This aim can be further divided into the following four secondary objectives:

- to describe the most up-to-date understanding of water movements and moisture conditions in unbound pavement layers and subgrades for different types of road constructions in various climatic conditions,
- to explain the relationship between the mechanical behaviour of materials/soils and their permeability[2] and moisture condition,
- to report on advanced modelling of water movement and condition in the sub-surface pavement environment developed from laboratory analysis and field studies,
- to inform about the identification, investigation and control of contaminants leaching from soils, natural aggregates and by-products in the sub-surface layers.

It is important to add that this book is NOT about road surface drainage. There are many books that address this topic, e.g. Ksaibati & Kolkman (2006), whereas this book aims to concentrate on sub-surface water. Inevitably, there is some overlap between the two aspects but, in this book, the information on runoff is only that necessary to complete a coverage of the sub-surface condition. Regarding contamination, traffic is, of course, a significant source too (see Chapter 6, especially Fig. 6.1) but the reduction in contaminants from that source is beyond the scope of this book. Rather, the aim is to understand what happens, or should happen, to contaminants in the highway environment.

[2] The terms permeability and hydraulic conductivity are both used to mean the ease with which water travels through saturated, porous media. In this book the term permeability is used preferentially. In particular the 'coefficient of hydraulic conductivity' and the term 'coefficient of permeability' are identical and given the symbol K.

1.3 Organisation of Book

The book covers both theory and practice and addresses both sub-surface drainage and water quality issues. Chapter 2 deals with basic flow and suction theory and Chapter 4 with heat transfer, which is implicated in driving water movement by phase change mechanisms (freezing, thawing, evaporation). Chapter 6 establishes a basis for discussing environmental aspects, introducing contaminant transport processes. Chapters 3 and 7 give an overview of the available techniques to monitor water flow, suction pressures and contaminants in water. After this introductory and descriptive section of the book, Chapters 8–11 aim to explain interaction between water in soil and aggregate and the mechanical response of such materials. This explanation covers theory, field behaviour, laboratory testing and theoretical and numerical modelling. The book ends with two chapters that aim to guide readers to improve the environmental condition, to reduce water pressure and to reduce the volume of water in the pavement and adjacent soils. Aspects only of relevance to environmental matters are covered in Chapter 12 while the much longer Chapter 13 describes the many techniques that can be used to manage water in the pavement and near-pavement soils. Often such management has benefits to both the mechanical and environmental performance of the pavement structure and earthworks.

1.4 Pavements and Earthworks

1.4.1 Definitions

In today's world almost all traffic runs on an improved surface and not on the natural soil profile . . . and most users give such improved surfaces little thought. This improved surface is known as a pavement and may be a simple layer of imported aggregate or the structure that comprises a modern expressway. In fact, all but the very simplest pavements comprise a series of carefully designed layers of imported, selected, construction materials placed on the "subgrade" (Fig. 1.4). The subgrade

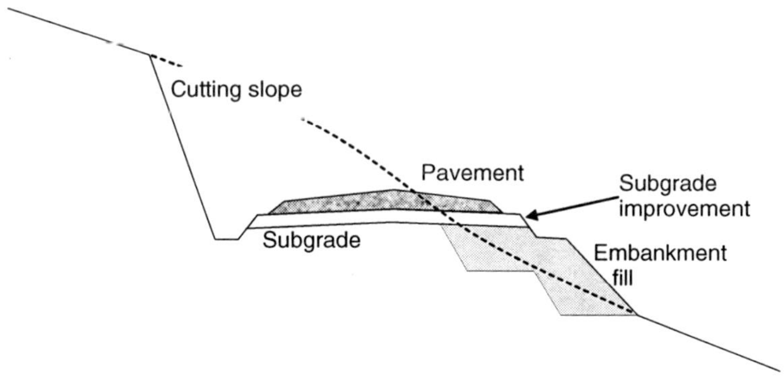

Fig. 1.4 Indicative highway cross section

may either be the natural soil profile encountered at the site (e.g. when the pavement construction is built at the same level as the surrounding ground, or when it is built in a cutting) or it may be a "fill" material that has been imported to create an embankment.

This fill is often subgrade soil from some other location, but might also be a by-product or waste material. Even in a cutting, some material may have been imported to form the subgrade as it can aid drainage – and thus improve the performance of the pavement. Material specifically designed to provide an improved subgrade is known as "capping". It may either be imported or may be achieved by some improvement technique applied to the natural subgrade in-situ (Fig. 1.4).

It is not only the pavement that needs draining; drainage of embankments and cuttings is also important to give soil-based slope stability and longevity of performance.

Finally, what about "moisture" and "water"? Both terms are used, more or less synonymously in this book. In general the direct term "water" is preferred to the more indirect term "moisture", although "moisture" is often used to describe water in the unsaturated parts of the sub-surface.

1.4.2 Pavements and Their Construction

Modern pavements normally comprise one or more bound layers overlying one or more unbound aggregate layers which, in turn, rest on the subgrade. In almost all cases the uppermost layers are bound by bitumen or cement. In the case of an embankment the subgrade is comprised of imported fill. In the case of a cutting it will often be the natural rock or soil at that location. Figure 1.5 provides two typical pavement profiles. Considering these from the bottom upwards, the following layers are, typically encountered:

- The **pavement foundation** consists of the natural ground (subgrade), and often a capping layer, the role of which is to improve the levelling, homogeneity and bearing capacity of the subgrade, and often also to ensure frost protection.
- The **sub-base** layer is normally comprised of an aggregate layer which acts as a platform for construction and compaction of the higher pavement layers and continues to function during the life of the pavement as an intermediate distributor

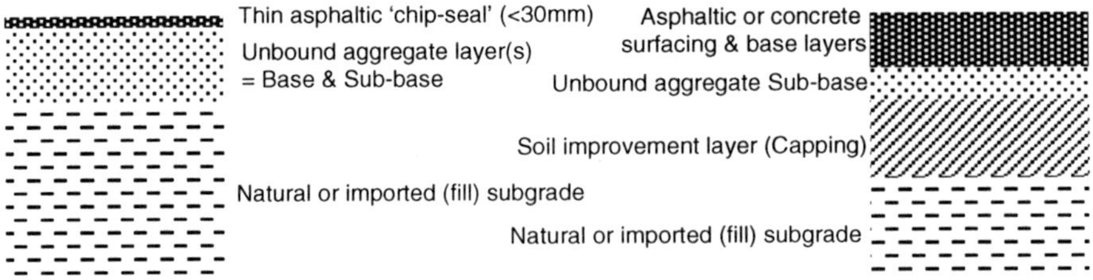

Fig. 1.5 Some typical pavement profiles (*left* = low volume, *right* = higher volume)

of stress from the higher layers of the pavement down to the foundation. It may also have a frost protection role.

- The **pavement base** is usually comprised of treated materials in high-traffic pavements or may be untreated in low traffic pavements. These layers provide the pavement with the mechanical strength to withstand the loads due to traffic and distribute these loads to the weaker lower pavement layers.
- The **surface course** (and possibly a binder course below) is the top layer of the pavement, exposed to the effects of traffic and climate. It must resist traffic wear and also protect the structural layers, in particular against infiltration of water.

These are described more fully in Chapter 8. The higher volume cross-section is typical of those found in major highways. It is critical that it continues to function well, so drainage of these pavements is provided, principally, to maximise the life of the pavement structure and thus to minimise the cost. Although longevity is also an issue for the low volume pavement, it also needs drainage to perform at critical times – e.g. after heavy rain storms or during spring-thaw (see Section 1.8) – when it would otherwise become impassable.

The pavement construction is there to provide an almost fixed, plane surface on which tyred vehicles may pass without difficulty. To meet this requirement the surface:

- must not deflect much transiently – otherwise vehicles will be travelling in a depression of their own making and using excess fuel in a vain attempt to climb out of it;
- must not deform plastically – otherwise ruts will form, hindering steerage, leading to increased fuel and tyre costs due to a greater contact with the tyre and tending to feed rainwater to the wheel path thereby promoting aquaplaning;
- must provide adequate skidding resistance – to enhance safety; and
- must continue to meet these requirements for a long time so that the pavement is economic and so that users are not unduly affected by pavement rehabilitation needs.

As far as the lower unbound layers and subgrades are concerned, they have to provide the necessary support to the upper layers so that those layers do not flex too much under trafficking as this could lead to those upper, bound, layers failing prematurely by fatigue. The upper layers need to be thick enough so that they spread the traffic loading so that the lower layers are not over-stressed and can provide their function successfully (Fig. 1.6). Successful pavement design is all about satisfying

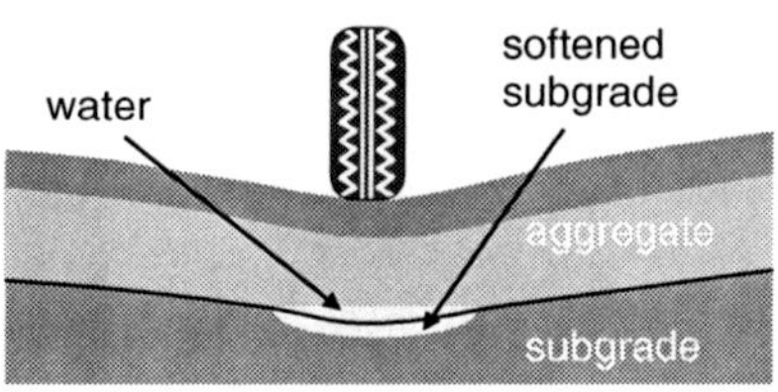

Fig. 1.6 Malfunction of the lower pavement layers. A depression in an impermeable sub-grade allows water in the aggregate layer to collect there, leading to subgrade softening and consequent rutting of the whole pavement

these two needs in the most efficient manner given the properties of the available materials.

1.4.3 A Drainage Classification of Pavements

It is conventional to classify pavements according to their construction – flexible (i.e. principally made of asphalt or only granular), rigid (i.e. concrete) and semi-rigid (i.e. made of both concrete and asphalt layers). From the point of view of water movements these classifications are not very relevant. Instead, pavements may be classified by the way in which water enters and moves in the pavement. On this basis the following classification is more appropriate:

A. Impermeable throughout the construction
B. Impermeable surface and structural layers
C. Permeable surface over impermeable structural layers
D. Permeable throughout with water storage capacity within the structure
E. Permeable throughout without water storage capacity
F. Cracked or jointed surface layers over permeable lower layers

Each type of pavement can be constructed on pervious or impervious ground and the ground- (or surface-) water level could be below or above the bottom of the construction. Thus each of the above 6 classes could, in principle, be sub-divided according to these conditions. In fact, except in limited circumstances, only a few of the classes and their sub-divisions have meaning in situations that are at all frequent. Thus circumstances in which the surrounding water is above the base of the construction are rare. Normally drainage (e.g. in the form of lateral drains) is provided to avoid this possibility.

Class A constructions are relatively rare. Full-depth asphaltic construction has been used in a few situations, most often in city streets, but its adoption has not been widespread. In this case significant water flows to the subgrade through the pavement layers are not expected but if water does become trapped at the subgrade surface (e.g. where the subgrade is impermeable), then it may be difficult to get it out of the pavement.

Class B pavements are probably the most common in developed countries, typically comprising asphalt or Portland concrete over a granular base or sub-base. The granular layers can act as drainage layers if they are permeable enough and have appropriate falls and outlets.

Class C constructions have become relatively common in recent years. The most common form of these is a flexible asphalt pavement with a porous asphalt surfacing. Rain water infiltrates the surface and then runs sub-horizontally within the asphalt to a drain that must be provided (some more details are included in Chapter 5).

Class D pavements seem, at first an undesirable concept. Storing water in the pavement will be likely to reduce the structural capacity of the construction. However, the chief motivation for this is to reduce runoff rates to surface water bodies (streams, rivers, lakes). With increasing urbanisation and areas of "hard" surfaces, rainfall arrives more rapidly at receiving watercourses than it does in "green" environments where vegetation and partial sorption into soil delay the arrival. The consequence is that river hydrographs become more "peaky" and flooding more common. Therefore, the provision of water storage within the pavement reverses this trend, delaying arrival of rain to the watercourse. Furthermore, the slowing of water as it percolates through the storage area means that it drops particulates. Also, some sorption of contaminants from the percolating water is achieved. Thus, the water arriving at the water body is also cleaner than it would otherwise have been. They are discussed a little more in Chapter 13.

Class E pavements are usually those with no sealed surface. They are common in parts of Scandinavia and form the minor road networks in many countries. Although unsealed, a well compacted surface of material with sufficient fine particles to block the pores and without potholes and ruts can shed a large proportion of the rain that falls on it. Conversely, distressed pavements of this type rather easily take in water and then tend to rapidly deteriorate further.

In countries with a network of jointed concrete pavements – Class F pavements – entry of water through joints can be significant, especially as the pavement ages and the joint fill compounds become less effective at keeping the water out. Asphaltic pavements that have suffered significant cracking could also be placed in Class F.

Where water does enter the pavement through the upper layers (Classes D, E and F) then the type of subgrade is likely to have more significance than in other pavements. Impermeable subgrades will necessitate horizontal or sub-horizontal egress. Permeable subgrades will allow vertical drainage towards the water table. Impermeable pavement subgrades are typically comprised of clay. When water reaches these it can cause softening and deterioration of the mechanical behaviour of the pavement (see Fig. 1.6).

Porous pavements that are designed to beneficially carry water through their layers (Classes C and D) are liable to deteriorate in their ability to do so, with time, as solid particles block the pore spaces. Porous surfacings are prone to ravelling as direct trafficking and the induced water pressure pulses in the pores and micro-cracks between particles tends to cause particles to separate one from another. This process, known as "stripping", is common to all asphaltic mixtures to some degree, but is more prevalent in porous asphalts where, of necessity, particles are less firmly fixed together than is conventional in densely-graded materials (see Chapter 5).

1.5 Pavement Materials – Geotechnical Behaviour

The upper, bound layers in a pavement are little affected by pavement moisture. "Stripping" can occur in repeated wet weather when the traffic loading causes pulses of pressure of water which has seeped into cracks in the bound materials. Exploiting micro-cracks this water can separate the binder from the aggregate it is supposed to bind, leading to ravelling of the bound material. Similarly, water may cause delamination of one bound layer from another, thereby reducing pavement load-carrying capacity, by exploiting inter-layer cracks.

A whole book could be written on this aspect alone, but this volume only devotes part of Chapter 5 to this topic as its focus is on the lower unbound and subgrade soil layers. These are geotechnical materials and behave according to the basic principles of soil mechanics as described in many standard text books on the subject. As explained further in Chapter 9 (Section 9.2) mechanical performance depends, largely, on the frictional interaction developed between one particle and the next. When the grains in an aggregate or soil are pushed together, greater friction is developed between the grains. The greater friction leads to improved strength of the subgrade soil or unbound granular pavement material, greater stiffness and greater resistance to rutting.

These inter-particle forces, considered over a large volume of particles, can be treated as a stress, known as the effective stress, σ', which is defined as:

$$\sigma' = \sigma - u \tag{1.1}$$

where σ is the stress applied externally to the volume of particles and u is the pressure of water in the soil pores which may be trying to push the particles apart. Thus, well-drained pavements lead to higher values of σ' which means more friction which, in turn, yields a material (and thus a road) that lasts longer and/or is more economic to construct and maintain. For this reason it is the road engineer's task to keep the effective stress high and, from Eq. 1.1, it can be seen that this condition is achieved when the pore pressure is smallest. This is the underlying reason why drainage is so important for efficient pavement and earthworks structures.

Nevertheless, even if it were possible, a completely dry geotechnical material is not wanted, instead a partially-saturated condition is often desired. When soil or aggregate is kept relatively (but not totally) dry, matric suctions will develop in the pores due to meniscus effects at the water-air interfaces. This suction would be represented in Eq. 1.1 by a negative value of u such that the effective stress, σ', increases as the suction develops additional inter-particle stresses by pulling the soil grains together. The topic of suction is discussed in more detail in Chapter 2.

For these reasons the pavement engineer wants to stop surface water (i.e. rain) from entering the pavement and wants to help any water that is in the pavement to leave as quickly as possible. Sealed layers and sealed lateral trenches may be used as barriers to prevent water from entering into the pavement or earthworks although, in practice, barriers are often not very effective due to defects or flow routes around them. Thus, drains to aid water egress are the primary weapon in the highway engineer's fight against water-induced deterioration. Although there are other techniques than drains that may be employed to stop ingress and aid drainage (discussed further in Chapter 13), for now it is sufficient to mention drains as interceptors that both cut-off the arrival of groundwater at the pavement and that provide an exit route for water already in the pavement and earthworks. The scope for drainage of pavements is somewhat limited by the need to keep the pavement trafficable – thus steep longitudinal or cross-carriageway slopes cannot be used. For this reason drainage gradients are, typically, small ($\ll$5%) necessitating that highly permeable materials are used that exhibit low suction potential.

High permeability materials are, characteristically, those with open pore structures. In geotechnical terms, the permeability is described using the coefficient of permeability, *K*, such that

$$q = -AKi = -AK\frac{dh}{dl} \tag{1.2}$$

where q is the volume of water flowing in unit time through an area, A, under a hydraulic gradient, i, and i is defined as the change in head, dh, over a small distance, dl. The negative sign is a mathematical indication that water flows down the hydraulic gradient. Inspecting Eq. 1.2 it is apparent that more effective drainage can be achieved by:

- increasing the area of flow intercepted – e.g. by providing drains with greater face area;
- increasing the hydraulic gradient – e.g. by installing deeper drains or drainage layers with steeper cross-falls; and
- increasing the coefficient of permeability – e.g. by selecting a more open-graded drainage material.

1.6 Interaction Between Percolating Water and the Pavement

In the road environment, contaminants are almost entirely moved by water-based or air-based processes. Air-based processes are, principally, by dust and spray, but these are not discussed further, being beyond the scope of this book. In water-based processes, contaminants are carried in and/or by the water through soil or aggregate pores, over the top of the pavement and through drainage systems. As water moves through the sub-surface, water that is fairly pure will have the ability to pick-up chemicals from the soil through which is flowing and to carry these elsewhere while runoff water that arrives from the pavement surface may percolate into the construction carrying impurities with it which are then "dropped" by one mechanism or another into the layer. The interaction between water and contaminants is the subject of Chapter 6 with methods of measurement being described in Chapter 7.

Movement of contaminants is often considered using a "Source – Pathway – Receptor" framework (Fig. 1.7). In the context of the highway, the "source" would probably be the surface runoff water (although its antecedents of vehicle cargoes, atmospheric pollution, etc. could also be considered as the source) or the pavement

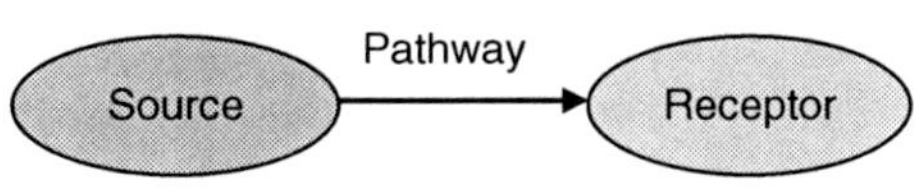

Fig. 1.7 "Source – Pathway – Receptor" framework

construction. The contaminated water then moves through the pavement and the drainage system which provides the "pathway" for the contaminant to move. Eventually it arrives at a place where it has a potentially deleterious impact – the "receptor". In an ideal understanding the receptor is a human, animal or plant that is affected by the contaminant. In practice, policing of impacts on humans would be almost impossible to monitor and, anyway, too late to change an undesirable response. So, instead, it is normal, in most practical circumstances, to treat either the surface water body (river, lake, etc.) or the ultimate groundwater body (e.g. drinking water aquifer) as the receptor. Monitoring their quality is the subject of Chapter 7.

Chapter 12 gives information that, with the help of Chapter 13, can be used to mitigate problems due to contaminant movement in and with the water percolating in the near-pavement environment. Usually, these involve either interruption of the pathway or removal of the source or target.

1.7 Water and Alternative Materials

Recycled and alternative materials are finding increased use in the construction of road pavements and embankments. Environmental concerns are leading to constraints on quarrying of the materials that have, conventionally, been used while tax incentives and legislative limitations are encouraging the uptake of wastes, by-products and recycled elements in their place. These materials do not, necessarily, behave in the same manner in the presence of water as many conventional materials do. Self-cementation due to pozzolanic activity may help to stabilise some, others may exhibit undesirable leaching, yet others may have much higher permeability than the material they replace. As a consequence, it is important for the road designer or manager to understand these characteristics and their implication for the hydrological and environmental performance of a highway that incorporates such materials. Relying on experience alone is likely to be insufficient.

It is very common for regulators and potential users to express concern about leaching when construction with such materials is proposed – yet this is a concern that often has no basis in fact! Many alternative materials come from an industrial process in which some chemical has been involved which no-one would wish to become widely distributed in the environment. Thus, environmental regulators often show particular concerns about materials deriving from metal processing industries – e.g. slags, foundry sands, etc. However, their real impact depends not on the actual content of the chemical of concern in the solid but on its availability to pore fluids, its solubility and its transportability. Many alternative materials have been through a hot process which vitrifies the solids making it extremely difficult for chemicals, now held in a glass-like matrix, to leave the solid phase. Alternatively, the pH level in-situ may render the contaminant essentially non-soluble. Chapters 6 and 12 discuss these issues further.

1.8 The Effect of Temperature

Pore suctions have the effect of "pulling" the saturated zone nearer the ground surface than it would otherwise have been from where evaporation becomes possible. When evaporation is significant then upward water flow takes place to replace the water being evaporated. Evapotranspiration by vegetation also introduces an upward water flow towards roots in a similar manner. In hot climates, evaporation can lead to upward moving water tens of metres above the phreatic surface and it can also lead to salts being lifted to the surface where they precipitate out in the soil pores forming calcretes and silcretes (Sabkha soils are an example of this).

Another cause of suction is seasonal ground freezing in high latitudes or at high altitudes. For reasons explained in Chapter 4, considerable ice wedges may form at the boundary between freezing caused by cold road surface temperatures and underlying non-frozen soils (Fig. 1.8). By this means, soils may heave by hundreds of millimetres in a winter season.

In spring, this ice melts, but the warmth comes from the surface so that the ice nearest the surface melts first. As there is still ice below it (and probably in the margin of the roads where the surrounding ground is covered by snow and ice cleared from the road pavement), the water from the melted ice has nowhere to drain and extremely weak conditions can result (Fig. 1.9).

Just as evaporation, evapotranspiration and freezing are non-constant processes drawing water up from the groundwater zone, so rainfall is an intermittent supplier of water at, or near (via drains or soakaways) the surface causing a downward flow. A small amount of rain very rapidly cancels a high suction (thus suddenly reducing the effective stress and the frictional strength of a soil – see Section 1.4.3). It is partly for this reason that wet weather is so often associated with occurrence of slope distress in earthworks and in deformation of pavements.

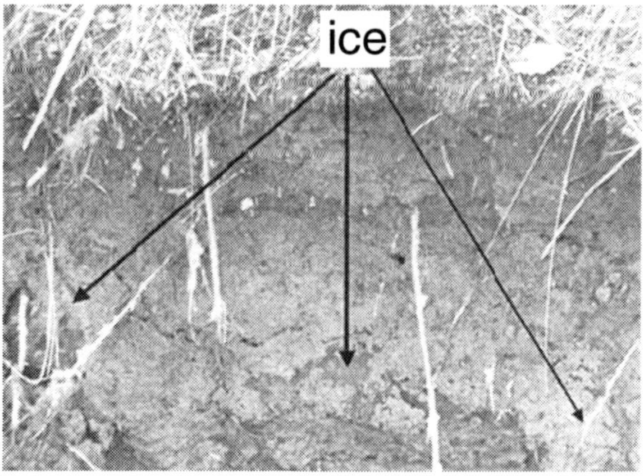

Fig. 1.8 Ice lenses in a silty subgrade, Kuorevesi, Finland, Spring 2003. Reproduced by permission of N. Vuorimies

Fig. 1.9 Thin asphaltic surfacing of a road lifted by traffic-induced water pressure during spring-thaw in Northern-Karelia, Finland. Reproduced by permission of M. Leppänen

1.9 Runoff

Runoff derives, principally, from rainfall falling on the pavement and surrounding ground. Although surface water drainage falls outside the scope of this book, runoff becomes of interest as some soaks in through cracks or through pervious surfacings. The proportion soaking-in will vary depending on the rainfall pattern, road surface quality and the permeability of the road's margins and surrounding earthworks. At the margins the water should be routed into a drainage system. If a positive drainage system is provided then kerbs or gullies will intercept the surface flow and feed it to gulley pots and/or a piped drainage system. From there, water may be fed to some disposal system – this may be a soakaway to the ground or it may be to a surface water course. Normally road runoff will be given some treatment before it is disposed. Treatment will usually include solids settlement and oil separation.

In areas where land is available, the runoff may be fed into an open, vegetated, lateral ditch known as a swale (see Figs. 1.10 and 1.11). These can form part of "SUDS" (Sustainable Urban Drainage Systems). Together with filter strips, infiltration trenches and basins, porous pavement surfaces, constructed wetlands (e.g. reedbeds) and detention and retention ponds, they tend to act as natural attenuators of contaminants that will be sorbed into the ditch lining and taken up into the vegetation which, periodically, can be cut and removed. They also act as sediment traps, removing suspended solids. Excess water that does eventually arrive at a surface water course, or that soaks down to the water table, will usually be relatively

Fig. 1.10 A well designed and maintained roadside swale operating correctly in heavy rain. Reproduced by permission of Chris Jefferies

Note: The grass-lined swale has a drainage grill set 300 mm above the base to encourage storage and infiltration of water into the ground.

clean due to the natural attenuation processes encountered *en route*. However, this means of treating and disposing of water that potentially contains contaminants cannot be used unthinkingly. There are many natural environments that are "highly vulnerable". High vulnerability exists when a near-highway environment can be easily polluted by runoff water (for example, where seeping water provides drinking water or sustains a quality ecosystem). The rare, but critical, occurrence of an accidental spillage, e.g. following a tanker crash, can cause acute effects to down-gradient waters.

1.10 Drainage Systems

Drainage systems come in many shapes and forms (see Chapter 13) but they also share many common features – they are placed lower than the section of road or earthworks they are intended to drain and they comprise materials (and/or pipes) that

Fig. 1.11 A gravel-lined swale with planted reeds to function as a soakaway. Reproduced by permission of VicUrban

Note: Infiltration is encouraged by providing a porous surfacing.

are more permeable than the surrounding materials. Broadly, they may be classified as follows:

i) Horizontal (or sub-horizontal) drainage layers.

   - When placed in, or more usually at the bottom of, some imported soil used for earthworks, these are termed blanket drains. Then they are used to isolate earthworks from underlying groundwaters, allowing any up-flowing water to be intercepted before it causes deterioration of earthworks and to catch water draining down from higher layers.
   - Drainage layers may be provided only to carry small seepage flows consequent upon leakages in the otherwise impermeable pavement surface. Typically these are provided as an integral function of one of the road's construction layers.

ii) Vertical, in soil, drainage trenches.

   - Some are intended to provide drainage of earthworks structures. The amount of water to be carried (and, hence, the drain's design) will depend on the permeability of the ground to be drained and on the height of the natural water table.

- Pavement median or edge drains are usually installed at the edge of pavements, often extending down into the underlying natural soil or into the earthworks. Depending on the arrangements in force these may be expected to handle runoff water arriving from a pavement surface as well as from seepages carried to the pavement edge by a drainage layer (ib. above).

iii) Drains for structures. Drainage systems are usually installed behind constructed walls and bridge abutments so as to reduce the lateral water pressures on these structures. They are not considered in this book.

Conventional drains are provided by aggregate, with a low proportion of fine sizes, placed in or under the zone to be drained. When the natural ground, imported earthworks, or pavement layer that is to be drained is particularly fine graded, it may be necessary to place a filter layer between the natural soil and the drainage system element.

Nowadays, alternatives to aggregates are available to effect drainage. Except where large flows are expected, geosynthetic fin drains comprising an exterior "filter fabric" and an interior highly permeable core are generally accepted as suitable substitutes for drainage ditches. They can be installed very rapidly and avoid expensive quarrying and associated transport of large volumes of dense materials. The "filter fabric" layer of the composite geosynthetic will normally be a felt-like layer around 1mm thick having a fairly small pore size. Although its pores will be too large to prevent every grain of the surrounding soil from going through into the interior core, they will halt somewhat larger particles that will, in their own turn, then allow layers of progressively finer particles to block the gaps between them. In this way a natural bridging layer will develop such that an effective filter zone will be catalysed by the "filter fabric".

Where low flows are expected, the water may be carried within the drainage medium (stone or geosynthetic), but where moderate or high flows can be anticipated it becomes necessary to install a pipe at the bottom of the trench or fin. This has to be permeable in some way (slotted, holed or integral with the fin drain) so that the water collected by the drain may be fed to the pipe and thereby carried away.

1.11 Climate and Climate Change

The ultimate reason for having drainage is because of rain! Therefore the drainage needs and solutions will be heavily influenced by the climate where the road is built. Broadly, climate may be divided by temperature and by rain/snowfall. Typically climatologists further differentiate on temperature variation across the year and on rainfall distribution. A very commonly used classification that takes this approach is that due to Koeppen (McKnight and Hess, 2000). This divides the world into 5 major zones, each with subdivisions:

- Tropical – subdivided into Rain Forest, Monsoon and Savannah;
- Dry – subdivided into Desert and Steppe;
- Temperate – subdivided into Mediterranean, Sub-tropical, Maritime and Maritime Sub-polar;
- Continental – subdivided into Hot Summer, Warm Summer and Sub-arctic; and
- Polar – subdivided into Tundra and Ice-cap. Alpine climates can be grouped here, too, although their climate results from elevation, not latitude.

The characteristics of each climate will have major effects on the water in the road and adjacent ground. In particular, where the potential for evaporation is

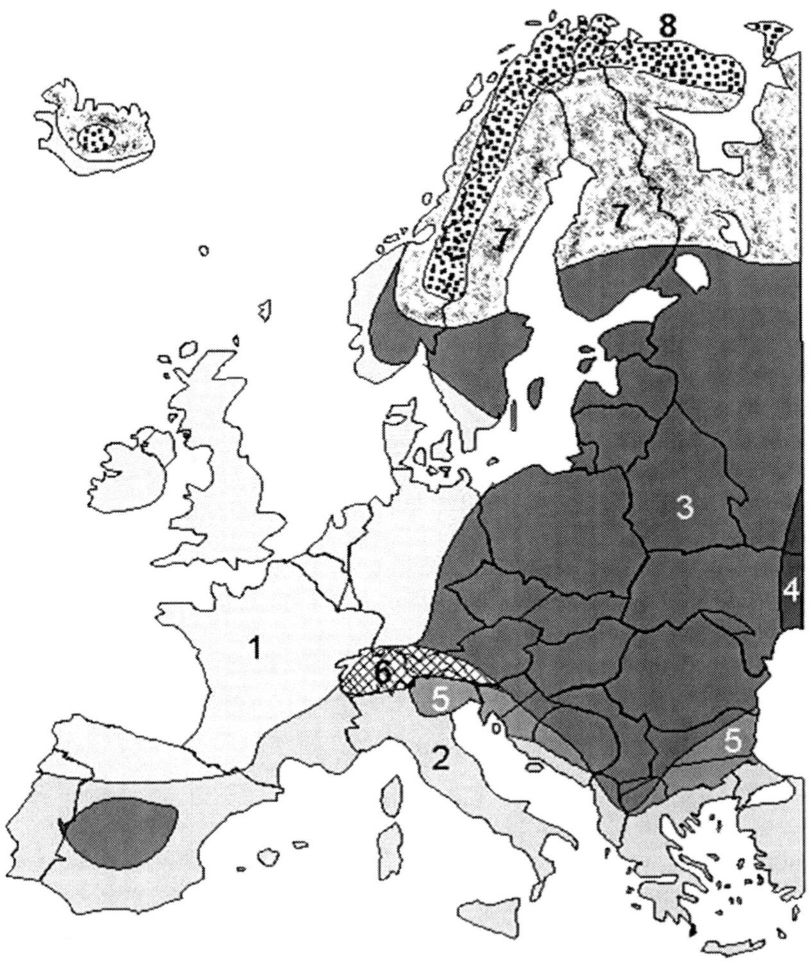

Fig. 1.12 Simplified climatic zone map of Europe. 1 – Temperate, maritime; 2 – Temperate, Mediterranean; 3 – Continental, warm summer; 4 – Dry, steppe; 5 – Temperate, sub-tropical; 6 – Alpine; 7 – Continental, sub-arctic; 8 – Polar, undifferentiated

significantly greater than rainfall, the effects of drainage may be less noticeable. However, this would need to be true throughout the year. Thus in tropical monsoon climates it may be true averaged over the year, but it is certainly not true during the wet season. It is at such a time that effective drainage may make the difference between survival of the road and its rapid deterioration.

Figure 1.12 shows a schematic division of Europe into the appropriate climatic zones.

In recent years the topic of climate change has become a consideration for almost everyone and road builders and operators are no exception. Temperature rise itself is unlikely to make a lot of difference to water in road structures in the warmer temperate areas, but in areas with seasonal freezing it could make a big difference. If the seasonal freeze period is reduced in length or lost or occurs repeatedly with intermittent thaw periods, then much longer periods of wet, non-frozen conditions (as currently experienced for shorter periods in the Autumn and Spring) can be expected, necessitating more stringent drainage requirements and causing much more frequent thaw-weakening problems. However, the shorter frozen period and the shallower penetration of the freezing front into the ground means that drainage trenches should more easily continue to operate year round.

Of likely importance to almost everyone is the increased rainfall that can be expected. Warmer ocean temperatures will lead to higher amounts of water in the atmosphere and, thus, more rain and snow. Whether this water will come regularly or in a few, but more severe, storms is less certain and the answer may vary by locality. The exact result may be difficult to predict, but the urgency for keeping water out of the pavement and associated highway earthworks can only increase. With greater runoff volumes anticipated, the need to provide positive drainage increases commensurately. Below ground it would be wise to anticipate greater volumetric flow rates of longer duration than previously experienced.

1.12 Legal Considerations

1.12.1 National and Trans-National

The demands of legislation can greatly influence the design and management of a road in order to control its influence on the water environment. General legislation, such as European directives (e.g. the Water Framework Directive), define water protection in a general manner with universal requirements that no pollutant input is allowed and that high cleanliness of water bodies should be established. However valid legislation at a national level, or even locally valid ordinances, can precisely define such requirements in terms of the level and manner of these protection measures. The level of the protective prescriptions depends on the legal system. In some countries the valid legislation is very general repeating obligations from the European directives and the proper implementation and then the functioning of protection measures are the responsibility of the road owner, operator and the designer. In some other cases, protection is very precisely defined by technical legislation that has been mandated centrally. In these kinds of document all the technical details can be found.

1.12.2 Local

During the development of a road scheme, existing hydrological and hydrogeological zones, crossed by road, are classified on the basis of legally existing water protection zones (e.g. drinking water source protection zones, Natura 2000 zones) and natural field conditions encountered. As a first step, desk studies are performed to define the demands and requirements for each section of the road consequent upon the legislation. Similar sections are then grouped together and field investigations undertaken to gather representative information on each group. Results of the field investigations are used to determine the specific level of water protection that is required. If the particular requirements defined in any existing protection ordinances are not as strict as those obtained as a consequence of the investigations performed during the scheme development, technical protection measurements must be adopted to ensure that the water body is protected according to the natural conditions that have now been revealed. If prescribed demands in the ordinances are stricter than established by field investigations, the protection requirements are retained and usually the designer does not oppose them. The designer would need to be certain that every section within the group of sections being analysed was less susceptible to degradation. Even then, the time taken to get the water protection zones re-classified to a lower risk category will often be longer than the time taken to commence construction, negating the benefit that might otherwise accrue to the road scheme.

Among locally- and state-valid legislation, constraints to road construction and operation are usually defined in ordinances defining drinking water source protection zones. They are defined as requirements, prohibitions and restrictions and they can be grouped in the following way:

- restrictions on areas for construction of roads and manoeuvring areas;
- technical engineering requirements for the protection of groundwater (e.g. runoff collection / safety bunds / soakaways / pipe integrity);
- traffic speed restrictions, restrictions of certain traffic types;
- management control measures (e.g. cleanliness of drainage facilities, controlled disposal of plant cuttings, periodic water sampling, etc.);
- restrictions of transport of certain dangerous and harmful substances; and
- demands regarding the setting up of road signs in protected areas.

1.13 Terminology

Pavement sub-surface drainage lies at the boundary of several disciplines each having their own special terms and notations. The book does not avoid these but, rather, seeks to define them when they are used. To help readers, a "Glossary" is included

(Annex C) as well as a list of terms in several languages (Annex B) and a list of symbols (Annex D).

1.14 Conclusion

Water and road construction do not make for a harmonious couple! While water is needed to allow efficient compaction of most of the earthworks and pavement layers and some moisture held in pores can act to develop strengthening suction due to capillarity effects, the overall picture is that water in the road and road sub-structure is undesirable. Water should, if possible, be kept out. If that is impossible (and it usually is impossible to achieve this) then efficient drains must be provided to convey the water away from the loaded areas. To bring about this happy condition, the road engineer has to understand about the response of pavement and geotechnical materials in the presence of water, about flow routes, about the drivers of water movement – climatic and hydrogeological – about the contaminants that can be moved in the water and about the regulatory framework in which he or she is obliged to operate. To successfully and economically deliver a well-behaved road is not easy to do. Therefore the following chapters aim to provide basic and more advanced information in all these areas. Not only do they aim to help the hard-pressed road engineer, but also to provide environmental engineers, hydrogeologists and others with a "language" in which to address the topics that have such an impact on every road user – on all of us!

References

Cedergren, H. R., 1974, "*Drainage of Highway and Airfield Pavements*", Wiley, New York, 285pp.

Cedergren, H. R., 1994, "*America's Pavements: World's Longest Bathtubs*", Civil Engineering, Sept., pp. 56–58.

Lay, M. G., 1992, "*Ways of the World*", Rutgers University Press, Piscataway, 401pp.

Ksaibati, K. & Kolkman, L.L., 2006, "Highway Drainage Systems and Design", in *The Handbook of Highway Engineering*, Fwa, T. F. (ed.), CRC Press, Boca Raton, FL. 888pp.

McKnight, T. L. & Hess, D., 2000, "*Physical Geography: A Landscape Appreciation*", Prentice Hall, Upper Saddle River, N.J., 629pp.

Telford, T., 1838, "*Atlas to the Life of Thomas Telford*", Rickman, J. (ed.), Payne and Foss, London.

Chapter 2
Water Flow Theory for Saturated and Unsaturated Pavement Material

Sigurður Erlingsson*, Mihael Brenčič and Andrew Dawson

Abstract This chapter describes the relation between road structures and water giving the general water balance equation for the pavement structure. Aquifers are briefly introduced. The pavement and its associated embankment are divided into the saturated zone and the unsaturated zone. Porous media are also described briefly together with their grain size distributions and fundamental properties related to water movements. A short summary of water flow theory for saturated and unsaturated soils is then presented, including relevant discussion of the soil water characteristic curve and permeability of unsaturated soils.

Keywords Roads · water flow · porous media · saturated · unsaturated · permeability · soil water characteristic curve

2.1 Introduction

During the planning, design, construction, operation and maintenance of roads, water can be an important environmental and constructional constraint that can significantly influence the bearing capacity of the pavement, the safe operation of traffic and have a large influence on the operational costs of roads.

Due to their length, roads interact with various water phenomena. Interaction between roads and water phenomena can be conceptualized on the basis of the recharge area of the water that intercepts the road. These phenomena of water road interaction can be divided into three groups:

(i) the road's own waters;
(ii) hinterland waters; and
(iii) remote waters.

* *Co-ordinating Author:*
✉ S. Erlingsson
Háskóli Íslands/University of Iceland, Iceland & Statens väg-och transportforskningsinstitut/ Swedish National Road and Transport Institute, Sweden
e-mail: sigger@hi.is

A.R. Dawson (ed.), *Water in Road Structures,* DOI 10.1007/978-1-4020-8562-8_2,
© Springer Science+Business Media B.V. 2009

The road's own water is the runoff that arises as a consequence of precipitation falling onto the carriageway and onto the associated embankment. Hinterland waters come from the near-road environment (e.g. slopes from a cutting) and those which are flowing towards the road embankment. Remote waters have recharge areas far away from road but are crossing the line of the road (e.g. rivers, lakes, subterranean groundwater flow, etc.).

2.2 Water Balance

For a better understanding of the qualitative and quantitative relationships between water and roads, the general interaction of road to water should first be established. The approach adopted here for this interaction analysis is by considering the water balance.

The relation between a road and water can be defined in system theory terms where input and output values are observed. Thus, the road and its pavement can be defined as a system into and out of which water flows. In normal conditions the input to the road pavement is represented by precipitation. Rainfall, or water due to thawing, infiltrates the pavement and flows through into the surrounding environment. Reverse flow of surface or groundwater towards the pavement embankment is also possible.

The general water balance equation can be simply defined as

$$P = R - ETR + IR \tag{2.1}$$

or

$$P = R - ETR + G + \Delta S \tag{2.2}$$

each term having units of volume/time/area [L/T] and where P represents the precipitation, R is the surface runoff, ETR represents evapotranspiration, G is the deep percolation or the groundwater recharge, IR is the surface infiltration and ΔS is the water storage change (or the net volume flux, thus $q_{out} - q_{in}$) of the pavement structure or the embankment. When some external inflow, q_{ext}, to the system is present, the water balance equation is defined as

$$P + q_{ext} = R - ETR + G + \Delta S \tag{2.3}$$

The presence of external flow q_{ext} is particularly important when we want to define the water balance of a road that is interacting with its hinterland and, especially, remote water bodies.

The water balance of roads and embankments is complex, depending on the structure of the system and on the goals that have to be achieved with the water balance model. A conceptual water balance model of a pavement-embankment system is represented in Fig. 2.1.

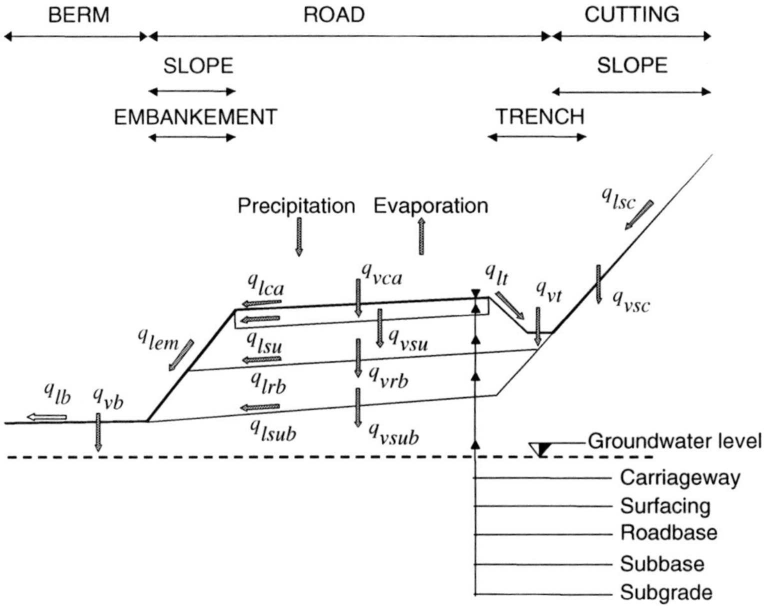

Fig. 2.1 Conceptual model of water balance on pavement – embankment system. v = vertical, l = lateral, ca = carriageway, su = surfacing, rb = roadbase, sub = sub-base, sc = slope/cutting, t = trench, em = embankment and b = berm

The water balance of the road depends on the geometry of the road and on the course of the road through the land. The water balance differs for roads where the pavement is completely above the ground level from those roads where the pavement or the complete road is below surface of the ground or even in tunnels or covered galleries.

A detailed water balance model of the pavement addresses water flow for each pavement component. In Fig. 2.1, water balance components are defined for the carriageway, surfacing, roadbase, sub-base and subgrade. Also reflected in this figure is the important influence on the water balance of the action of water flowing in the adjacent drainage ditches and on the slopes.

In general components of the water balance in the pavement can be divided into two general types: the vertical component (subscript v in Fig. 2.1) and the lateral component (subscript l in Fig. 2.1). Vertical components of the water balance represent the recharge of, or loss from, the groundwater while the lateral components present the contribution to the total surface outflow from, or inflow to, the pavement system.

2.3 Relation Between Road and Groundwater

According to the literature, various definitions of groundwater exist. For some authors, groundwater is defined as water only in those pores that are completely saturated, but for others a more general definition is acceptable with the groundwater being all the water below the ground surface either in the saturated or in the unsaturated part.

A geological medium containing groundwater is defined as an aquifer. Aquifers can be described as rock or soil (sediment) of high permeability that are able to store and transmit significant quantities of groundwater. In hydrogeological terms, the limit between an aquifer and less permeable geological media is frequently defined with a coefficient of permeability, K, of 10^{-6} m/s.

Figure 2.2 shows a schematic view of the regions of subsurface water in a pavement structure. The upper part represents the unsaturated or the vadose zone and the lower part represents the saturated or the phreatic zone. The groundwater table, defined as the surface where the water pressure is equal to the atmospheric pressure, separates them. The water content within the vadose zone is at saturation near its base while at its upper extent it is dependent on the characteristics of the soil. The vadose zone is divided into:

- a **capillary zone**. Above the groundwater table is the capillary zone or the capillary fringe because water is pulled upward from the water table by surface tension. The capability is linked to the pore size distribution of the material, the smaller the pores the greater the extent of the capillary rise. The thickness of the capillary fringe can vary from a few centimetres in coarse grained soils to a few metres in fine grained soils. The pores are saturated but the water pressure is less than atmospheric.
- an **intermediate vadose zone**. The second sub-region of the vadose zone is the intermediate vadose zone where water is held by capillary forces. In sealed pavements in good conditions, the surface layer is relatively impermeable and the water held here should be relatively stable with water content at or near

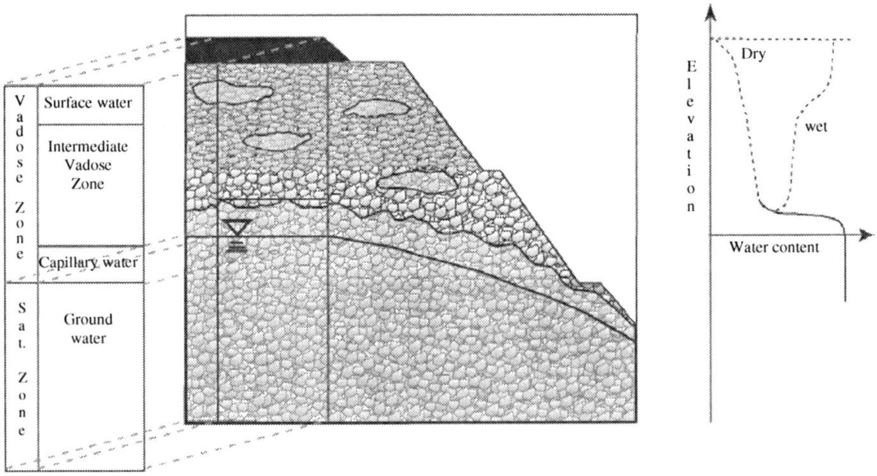

Fig. 2.2 Conceptual model of the relation between road and groundwater

field capacity (the remaining water content held by a soil after it has been allowed to drain freely), but through spring thaw or wet periods water could migrate inwards from the shoulders resulting in temporary higher water contents. In cracked pavements, on the other hand, water can move downward from the surface through the intermediate vadose zone to the capillary zone resulting, spatially, in periods of higher water content than the field capacity stipulates.

- a **surface water zone**. The layer closest to the surface is the surface water zone. Again, as for the intermediate vadose zone for sealed pavements in good conditions the water content should be relatively constant close to the field capacity or lower depending on atmospheric conditions. In cracked pavements on the other hand water can enter to the granular layer through cracks or other openings during periods of rainfalls and part of the layer may, therefore, include high water content or even become fully saturated.

In the unsaturated zone pore spaces are only partially filled with water and the direction of the groundwater flow is predominantly vertical. In the saturated zone, pores are completely filled with water and groundwater flow is nearly horizontal. The saturated aquifer zone is underlain by a low permeability basement stratum, known as the confining bed, which acts as a hydrogeological barrier. This is a necessary condition for the aquifer existence: without the barrier there would be no saturated zone.

2.3.1 Hydrodynamic Types of Aquifers

Based on their ambient material, aquifers can be classified into three types, i.e. open, confined and semi-confined, see Fig. 2.3.

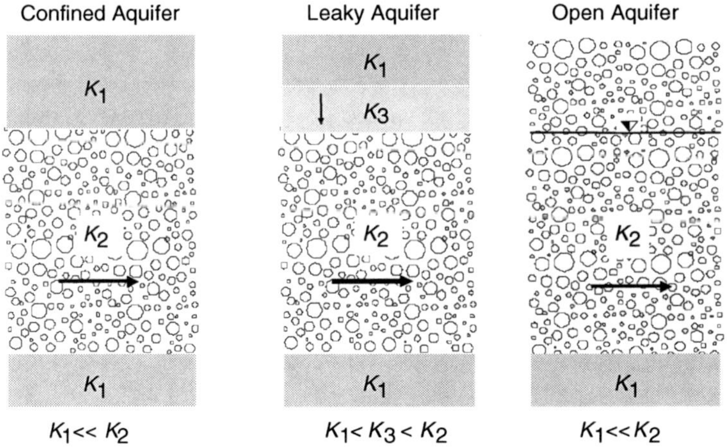

Fig. 2.3 Hydrodynamic classification of aquifers. K is permeability

- A confined aquifer is bounded both above and below by relatively impermeable materials or confining beds (such as clay or unfractured rock.) The confined water is under pressure thus a tube extended from the surface down into the aquifer would allow the water to rise inside the tube to a level above the top of the aquifer.
- An open aquifer is one without a confining bed above so it can be directly recharged by rainfall.
- A semi-confined aquifer is a confined aquifer where one of the confining beds is saturated material with low permeability which, thus, impedes the movements of the water.

Most road structures or road embankments can be considered open aquifers. Only under some special circumstances do they act as a semi-confined or confined aquifers.

For pavements and their relation to groundwater (from the geometrical point of view) perched-type aquifers are of importance. These aquifers are the consequence of some lower permeable lenses (e.g. clays) inside a permeable aquifer. As a consequence of a high coefficient of permeability of the aquifer material and favourable infiltration into the ground, water mounds appear above the lenses. These aquifers have no direct connections with lower lying groundwater. Usually they are very near the surface and may cause several problems for the construction and maintenance of the roads.

2.4 Porous Media

Roads and embankments are made up by a finite number of layers. They can be considered as porous media that consist of aggregates or granular materials and soils through which fluid can flow. The road layer can appear either unbound or stabilized with bitumen or cement to increase their strength. In roads, most surface layers have very low permeability properties and can often be treated as impervious, at least in roads in good conditions. Usually, all others layers are permeable. The fluid flow behaviour of the different layers is strongly dependent on their particle size distribution and pore space openings.

2.4.1 Grain Size Distribution

The grain size distribution of unbound aggregates or soils is determined by either sieving or by the rate of settlement in an aqueous suspension. Table 2.1 shows the classification of soil particles or aggregates by size according to EN ISO 14688-1 (CEN, 2002). The distribution of coarse and very coarse soil fractions can be estimated through sieving but the size distribution of the fine soil particles needs to be estimated in a settlement rate test (hydrometer test).

Table 2.1 Classification of soil particles according to the size (CEN, 2002)

Soil fractions	Sub-fractions	Symbols	Particle sizes (mm)
Very coarse soil	Large boulders	LBo	> 630
	Boulders	Bo	> 200–630
	Cobble	Co	> 63–200
Coarse soil	Gravel	Gr	> 2.0–63
	Coarse	CGr	> 20–63
	Medium	MGr	> 6.3–20
	Fine	FGr	> 2.0–6.3
	Sand	Sa	> 0.063–2.0
	Coarse	CSa	> 0.63–2.0
	Medium	MSa	> 0.2–0.63
	Fine	FSa	> 0.063–0.2
Fine soil	Silt	Si	> 0.002–0.063
	Coarse	CSi	> 0.02–0.063
	Medium	MSi	> 0.006–0.02
	Fine	FSi	> 0.002–0.006
	Clay	Cl	≤ 0.002

Figure 2.4 shows then the grain size distribution curve for three soils where the particle size is plotted on the x-axis and the percent mass retained (percent larger than the given size) is plotted on the y-axis.

The uppermost curve in Fig. 2.4 includes the highest proportion of fines (typically defined as silt particles and smaller, i.e. those less than 0.06 mm (60 μm) in size), about 60%, and the rest are sand sized. The fines content of the middle curve is about 4% and about 54% is sand and what remains is gravel. Finally the lowest curve only has about 1% fines, sand forms about 23% of the material and the rest is gravel.

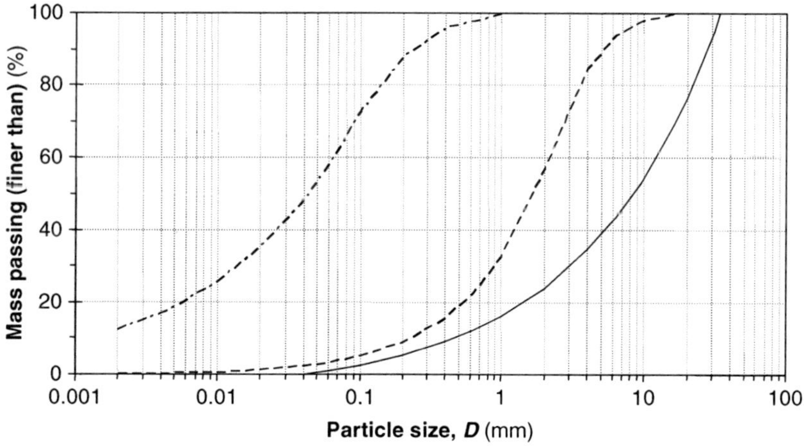

Fig. 2.4 Three different grain size distribution curves

Based on the grain size distribution curve three parameters are frequently determined, that is the effective size D_{10}, the uniformity coefficient C_u and coefficient of gradation C_c. The parameter D_{10} is the diameter of the largest particle that can be found in the smallest 10% of the particle-size distribution curve. The uniformity coefficient C_u and coefficient of gradation C_c are given as:

$$C_u = \frac{D_{60}}{D_{10}} \tag{2.4}$$

$$C_c = \frac{D_{30}^2}{D_{60} \cdot D_{10}} \tag{2.5}$$

where D_{30} and D_{60} are the diameter on the particle-size distribution curve, similar to D_{10} but corresponding to 30% and 60% finest fractions respectively. These three parameters are sometimes used to estimate the saturated coefficient of permeability of some types of soils.

2.4.2 Porosity

Porosity is defined as the space inside a rock or sediment (soil), consisting of pores. The total volume of pores is defined as the total porosity. For water, only those pores that are interconnected are important. The interconnected part of the pore system is defined as the effective porosity. The porosity can be described as a three phase system comprising solids, water (liquid) and air (gas), see Fig. 2.5.

The definition of porosity, n, of an aggregate skeleton is the ratio of volume of voids and its bulk volume or:

$$n = \frac{V_v}{V} \ (\mathrm{m}^3/\mathrm{m}^3) \tag{2.6}$$

Void ratio, e, is, on the other hand, defined as the ratio of the same volume of voids but now over its aggregate volume or:

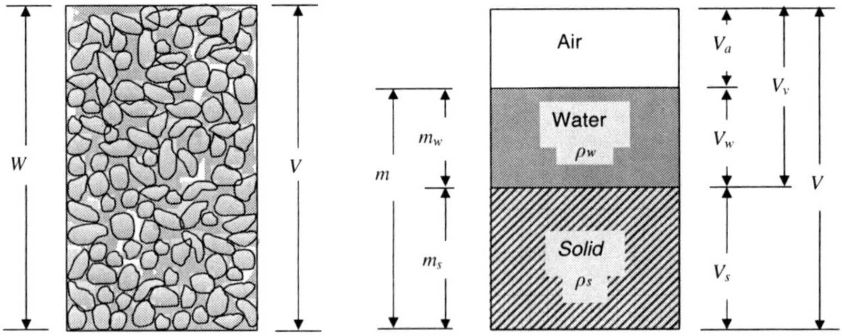

Fig. 2.5 A three phase diagram of an unsaturated element of aggregates

$$e = \frac{V_v}{V_s} \text{ (m}^3\text{/m}^3\text{)} \tag{2.7}$$

A simple relationship exists between the porosity and the void ratio, that is:

$$e = \frac{V_v}{V_s} = \frac{V_v}{V - V_v} = \frac{\frac{V_v}{V}}{1 - \frac{V_v}{V}} = \frac{n}{1-n} \Rightarrow n = \frac{e}{1+e} \tag{2.8}$$

The porosity represents the total amount in a unit volume that can be filled with water. The volumetric water content, θ, is defined as the volume of water of a specific element to its total volume or:

$$\theta = \frac{V_w}{V} \text{ (m}^3\text{/m}^3\text{)} \tag{2.9}$$

and gives therefore the actual fraction of water in the pores. It is therefore obvious that the volumetric water content must lie within the range $0 \leq \theta \leq n$.

The gravimetric water content, w (also referred to as the gravimetric moisture content), is on the other hand defined as the ratio of mass of water of an element to its mass of the solids or

$$w = \frac{m_w}{m_s} \text{ (kg/kg)} \tag{2.10}$$

Gravimetric water content is a much more commonly used measure than the volumetric water content. Therefore, unless stated otherwise or as required by the context, the gravimetric water content is indicated by the short form 'water content'.

There is a simple relationship between the gravimetric water content, w, and the volumetric water content, θ:

$$w = \frac{m_w}{m_s} = \frac{m_w \cdot g}{m_s \cdot g} = \frac{\rho_w \cdot V_w}{\rho_d \cdot V} = \frac{\rho_w}{\rho_d} \cdot \theta \tag{2.11}$$

where ρ_w is the water density, ρ_d is the dry density of the material and g is the acceleration due to gravity. The degree of water saturation is the volume of water per volume of void space, or:

$$S_r = \frac{V_w}{V_v} = \frac{\theta_w}{n} \tag{2.12}$$

The saturation range becomes, therefore $0 \leq S_r \leq 1$ (100%).

Porosity of a geological material can change with time. According to their origins two types of porosity can be recognized, that is a primary and secondary porosity. The primary porosity refers to the original porosity of the material where the secondary porosity refers to the portion of the total porosity resulting from diagenetic processes such as dissolution, post compaction, cementation or grain breakage.

2.5 Darcy's Law

Water flows though porous media from a point to which a given amount of energy can be associated to another point at which the energy will be lower (Cedergren, 1974, 1977). The energy involved is the kinetic energy plus the potential energy. The kinetic energy depends on the fluid velocity but the potential energy is linked to the datum as well as the fluid pressure. As the water flows between the two points a certain head loss takes place.

From an experimental setup as shown in Fig. 2.6, the total energy of the system between points A and B is given from Bernoulli equation as

$$\frac{u_A}{\rho_w g} + \frac{v_A^2}{2g} + z_A = \frac{u_B}{\rho_w g} + \frac{v_B^2}{2g} + z_B + \Delta h \tag{2.13}$$

where u and v are the fluid pressure and velocity respectively, z is elevation above the datum line and h is head loss between point A and B that is generating the flow. As velocities are very small in porous media, velocity heads may be neglected, allowing head loss to be expressed as:

$$h = \left(\frac{u_A}{\rho_w g} + z_A\right) - \left(\frac{u_B}{\rho_w g} + z_B\right) \tag{2.14}$$

Darcy related flow rate to head loss per unit length through a proportion constant referred to as K, the coefficient of permeability (also known as the coefficient of hydraulic conductivity) as:

$$v = -K\frac{h}{L} \tag{2.15}$$

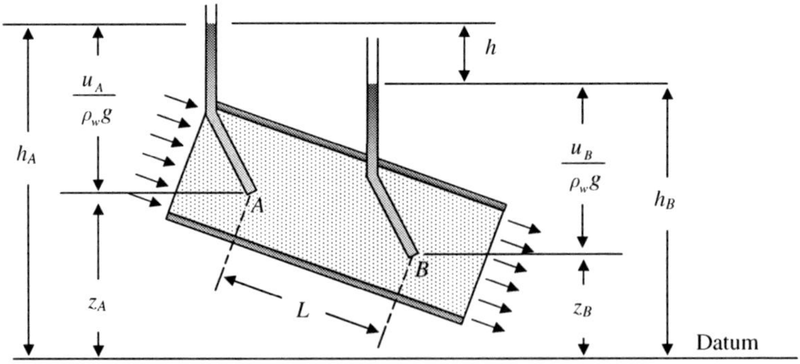

Fig. 2.6 Head loss as water flows through a porous media. Where u = pore water pressures, h = heads, z & L = distances

or in more general terms, at an infinitesimal scale:

$$v = -K\frac{dh}{dl} = -Ki \tag{2.16}$$

where dh is the infinitesimal change in head over an infinitesimal distance, dl, and i is the hydraulic gradient of the flow in the flow direction. The above equation is known as Darcy's law and governs the flow of water through soils (see Eq. 1.2).

It should be pointed out that Darcy's law applies to laminar, irrotational flow of water in porous media. For saturated flow the coefficient of permeability may be treated as constant provided eddy losses are not significant (see below). Above the groundwater table, in the unsaturated zone, Darcy's law is still valid but the permeability will be a function of the water content, thus $K = K(\theta_w)$, as described in Section 2.8.

Darcy's law can easily be extended to two or three dimensions. For three dimensions using a Cartesian coordinate system Darcy's law is given for a homogeneous isotropic medium as

$$v_x = -K\frac{\partial h}{\partial x}, \qquad v_y = -K\frac{\partial h}{\partial y}, \qquad v_z = -K\frac{\partial h}{\partial z} \tag{2.17}$$

which, in matrix notation, can be written as:

$$\begin{bmatrix} v_x \\ v_y \\ v_z \end{bmatrix} = -\begin{bmatrix} K & 0 & 0 \\ 0 & K & 0 \\ 0 & 0 & K \end{bmatrix} \begin{bmatrix} \frac{\partial h}{\partial x} \\ \frac{\partial h}{\partial y} \\ \frac{\partial h}{\partial z} \end{bmatrix} \tag{2.18}$$

which can be, in turn, be written as:

$$\mathbf{v} = -\mathbf{K} \cdot \nabla \mathbf{h} \tag{2.19}$$

In the case that the porous media shows anisotropy, the permeability matrix becomes in the general case:

$$\mathbf{K} = \begin{bmatrix} K_{xx} & K_{xy} & K_{xz} \\ & K_{yy} & K_{yz} \\ sym & & K_{zz} \end{bmatrix} \tag{2.20}$$[1]

If, further, the medium is inhomogeneous each of the above given coefficients in the permeability matrix may vary in space (Bear and Verruijt, 1987). For most

[1] sym = symmetric.

highway engineering problems the porous media, that is each layer, can be assumed homogenous and isotropic.

For very coarse grained soils or aggregates, some of the voids in the material become quite large and the assumption of a laminar flow of water is no longer valid. Instead of irrotational flow, eddy currents develop in the larger voids and/or the flow may become turbulent involving more energy loss than in a laminar flow. For these circumstances the hydraulic gradient in Darcy's law can be replaced with Forcheimer's law:

$$-i = \frac{\nu}{K_1} + \frac{\nu^2}{K_2} \tag{2.21}$$

where two coefficient of permeability, K_1 [LT^{-1}] and K_2 [L^2T^{-2}], are now required to describe the behaviour.

In highway practice this means that coarse aggregates with large pores – such as those which comprise typical granular base courses – must be tested at low hydraulic gradients to ensure laminar flow is maintained and that appropriate values of K are obtained. This aspect is covered further in Chapter 3, Section 3.3.1 (see Fig. 3.7 in particular).

2.5.1 Factors Affecting Permeability

Predicting the saturated permeability of soils or aggregates based on theoretical considerations has turned out to be difficult as permeability is dependent on a number of factors such as grading, void ratio, soil texture and structure, density and water temperature (Cedergren, 1977). Therefore, several empirical equations for estimating the permeability have been proposed in the past. These equations frequently include parameters linked to the grading curve of the material or their void ratio.

Hazen (1911) proposed an equation of the permeability of loose clean filter sand as:

$$K = cD_{10}^2 \tag{2.22}$$

where c is a constant that varies from 1.0 to 1.5 when the permeability K is in cm/s and the effective size D_{10} is in mm. Hazen's equation is only valid for limited grain size distributions of sandy soils. A small quantity of silt or clay particles may change the permeability substantially. It is seldom a good means of estimating K as illustrated by Fig. 2.7.

Another form of equation that has been frequently used in estimating the permeability of sandy soils is based on the Kozeny-Carman equation (Das, 1997; Carrier, 2003):

$$K = c\frac{e^3}{1+e} \tag{2.23}$$

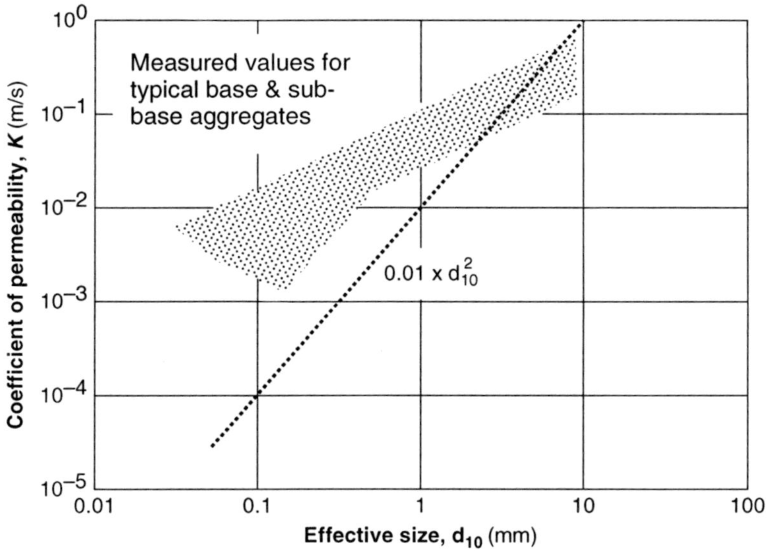

Fig. 2.7 Illustration of mis-match between Hazen's estimation and measured values from road aggregates (adapted from Jones and Jones, 1989)

where c is a constant. Samarasinghe et al. (1982) proposed a similar equation for normally consolidated clays:

$$K = c\frac{e^n}{1+e} \tag{2.24}$$

where n and c are experimentally determined parameters. These equations are almost certainly improvements over Eq. 2.22.

2.6 Filter Design

Pavement structures consist of material layers with different grain size gradations and different mechanical as well as permeability properties. As water will flow through the structure it is important that migration of a portion of the fines from one layer to the next will not take place. To achieve this, the principles of filtration/separation must be applied at each interface (Cedergren, 1977). This is of special interest where water flows from fine grained material into a coarser grained material on its way out of the structure.

In applying the filter criteria the material to be drained is frequently referred to as the base material and the new material to be placed against it the filter material. The filter criteria stipulate that the filter needs to fulfil two functions:

- water needs to drain freely through it (filtration function or permeability criteria); and
- only a limited quantity of solid particles are allowed to move from the base layer into or through the filter layer (separation function or piping requirement). This criteria is set as otherwise the coarse grained material could be filled or clogged with time.

These two functions are in conflict with each other as filtration requires a high discharge through the filter while separation requires this to be small. The conditions of these two requirements can be expresses as:

$$\frac{D_{15}^f}{D_{15}^b} > 4 \text{ to } 5 \quad \& \quad \frac{D_{15}^f}{D_{85}^b} < 4 \text{ to } 5 \tag{2.25}$$

where D_{15}^f is the diameter in the particle-size distribution curve for the filter material corresponding to 15% finer and D_{15}^b and D_{85}^b are the diameter in the particle-size distribution curve for the base material corresponding to 15% and 85% finer respectively. The application of these issues is presented in Chapter 13, Section 13.3.9.

2.7 Water in the Vadose Zone

The groundwater table is defined as the locus of points at atmospheric pressure. Below the water table the pore water pressure is positive and in a hydrostatic state, while the pore water pressure increases linearly with depth. Above the groundwater table, in the vadose zone, water only remains in the pores due to capillary action. The water pressure is then negative or less than the atmospheric pressure and capillary pressures, known as matric suctions, exist. Large matric suctions correspond to large negative water phase pressures and soils under such conditions usually have low water saturations. Furthermore, both the water content and therefore the permeability are nonlinear functions of these capillary conditions.

In unsaturated soils, pore spaces are filled with both water and air. The water is held in the soil pores by surface tension forces. These lead to a pressure difference between the air and water in the porous medium, as long as the interface is curved. As the curvature of the interface between the two phases – the air and the water – changes with the amount of water (or moisture content) in the soil, the matric suction is dependent on the water content. This relationship is non-linear, greatly complicating analyses of water flow through unsaturated soils.

The capillary conditions are described by the matric suction, s (expressed in pressure terms) or can be described in terms of the matric potential, Ψ, which represents the height (head) of a column of water that could be induced by such a suction. Although negative with respect to zero (i.e. atmospheric pressure), Ψ and s are usually expressed as positive quantities and are simply related as follows:

$$\Psi = \frac{s}{\rho_w g} \tag{2.26}$$

where ρ_w is the density of water and g is the acceleration due to gravity. Matric suction is related to the phase pressures and interface curvature by the relationship:

$$s = u_a - u = \frac{2\sigma_i}{r_c} \tag{2.27}$$

where u_a and u are the air and water pressures, respectively, σ_i is the interfacial energy or the surface tension and r_c is the average radii of curvature as illustrated in Fig. 2.8.

In the unsaturated zone, because part of the soil pores are filled with water while the rest is filled with air, therefore the sum of the volumetric water, θ, and air, θ_a, contents must be equal to the total porosity, n:

$$\theta + \theta_a = n \tag{2.28}$$

This equation may also be written in terms of water saturation, S_r, that is

$$S_r + S_a = 1 \tag{2.29}$$

where S_a is the air saturation content.

To understand the distribution of water in the unsaturated soils we may consider a soil mass that is initially dry. Upon addition of water, the water is first adsorbed as film on the surface of the soils grains. This thin skin of adsorbed water is usually called pellicular water (Bear, 1972) and is held strongly by van der Waal's forces and can hold very high matric suctions. Thus even if a matric suction of tens of atmospheres were applied this water would not be removed from the soil.

If water is further added, water starts to accumulate at the contact point between the grains that represent the smallest pore space openings. This water is referred to as the pendular water. The pendular water is held at the contact points by capillary forces. Capillary forces are caused by the presence of surface tension between the air and the water phases within the soil voids and causes water to move into the

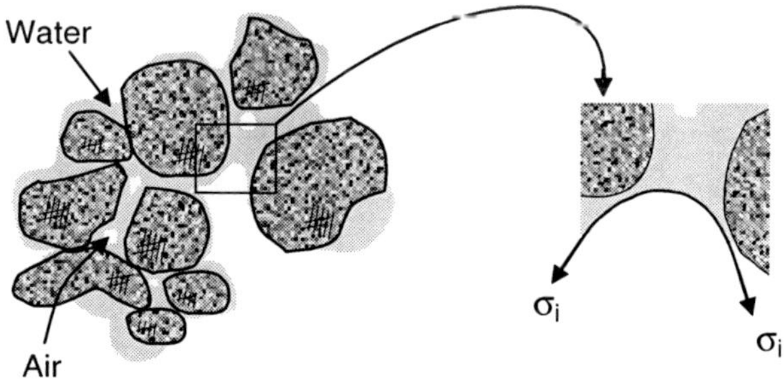

Fig. 2.8 (**a**) Distribution of water in a porous media. (**b**) Curved interface separating water and air phases

smallest pores. At this stage water movement through the soil skeleton is slow even under large hydraulic gradients because the water is forced to move along the thin film of adsorbed water. As the water content increases the film get thicker and water moves more easily.

As the water content continues to increase the water saturation reaches the stage where air becomes isolated in individual pockets in the larger pores and flow of air is no longer possible. This saturation is called insular saturation. These air pockets may dissolve leading to full water saturation of the soil.

2.7.1 The Soil Water Characteristic Curve (SWCC)

The soil water characteristic curve (SWCC) provides the relationship between the matric suction and water content for a given soil. In fact, calling the curve "characteristic" is something of a misnomer as the relationship is not solely a function of the soil type, but varies with (for example) temperature, pressure and pore water chemistry. A typical soil water characteristic curve for sand and clay can be seen in Fig. 2.9a.

Figure 2.9 shows clearly that, even at very high matric suctions (capillary pressures), all the water cannot be removed from the soil. The residual (or the irreducible) water content, usually denoted θ_r (and in a similar way the irreducible water saturation, S_{rr}) is the water content that is not removed in the soil even when a large amount of suction is applied.

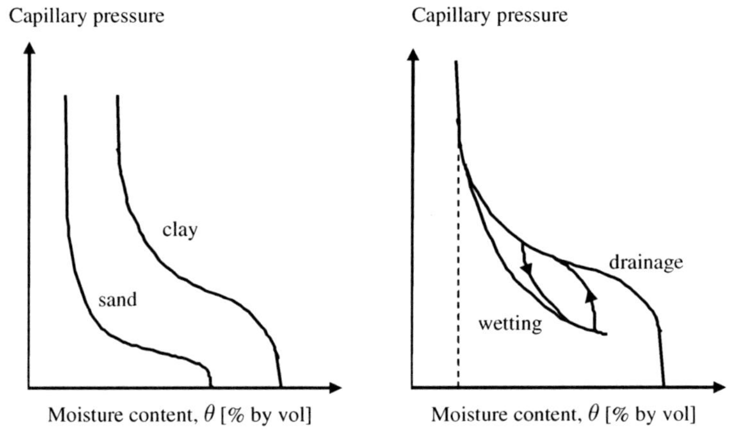

Fig. 2.9 (**a**) Typical characteristic curves for coarse grained (gravel, sand) and fine grained soils (clay, silt). (**b**) Soil water characteristic curve showing drainage, wetting and scanning (intermediate) curves. The dotted line represents the irreducible (residual) water content

2.7.2 Hysteresis Behaviour

For most soils the soil water characteristic curve (SWCC) shows hysteresis. This means that the $s - \theta$ (or the $\Psi - \theta$) relationship depends on the saturation history as well as on the existing water content. Figure 2.9b shows the SWCC for drainage and wetting conditions. The upper curve corresponds to a soil sample that is initially saturated and is drained by increasing the matric suction (capillary pressure), hence the drainage curve. The lower curve is called the wetting curve and gives the re-wetting of the soils with corresponding decrease in capillary pressure. If a wetting or a drainage process is stopped between the two endpoints and a reverse process is starting the scanning curves (indicated by arrows) are followed.

2.7.3 Analytical Models of the SWCC

There is a number of parametric models that have been suggested in the literature for describing the matric potential's dependency on water content (matric potential being defined in Eq. 2.26). The models are all empirical and two frequently used are the power law model suggested by Brooks and Corey (1964) and the model suggested by van Genuchten (1980). For further details, see Fredlund and Rahardjo (1993), Fredlund and Xing (1994) and Apul et al. (2002).

The Brooks and Corey (1964) model is given as:

$$\Theta = 1 \quad \text{for} \quad \Psi \leq \Psi_b \quad \Theta = \left(\frac{\Psi_b}{\Psi}\right)^{\lambda} \quad \text{for} \quad \Psi > \Psi_b \tag{2.30}$$

where the normalized water content, Θ, is defined as:

$$\Theta = \frac{\theta - \theta_r}{n - \theta_r} = \frac{S_r - S_{rr}}{1 - S_{rr}} \tag{2.31}$$

The parameters θ_r and S_{rr} are the irreducible (residual) volumetric water content and irreducible (residual) saturation respectively. The parameter Ψ_b is the air entry value and the parameter λ is called the pore size distribution index. The air entry value of the soil, Ψ_b, is the matric suction at which air starts entering the largest pores in the soil (Fredlund et al., 1994). The parameter λ characterizes the range of pores sizes within the soil, with larger values corresponding to a narrow size range and small values corresponding to a wide distribution of pore sizes.

The van Genuchten (1980) model relates the water content to the suction characteristics and is given as:

$$\Theta = \left[\frac{1}{1 + (\alpha\Psi)^N}\right]^M \tag{2.32}$$

where α, N and M are experimentally determined parameters. Based on Mualem's (1976) relative permeability model the parameters N and M are related as follows:

$$M = 1 - \frac{1}{N} \quad \text{or} \quad N = \frac{1}{1 - M} \tag{2.33}$$

According to Fredlund and Xing (1994) this constraint between the parameters M and N reduces the flexibility of the van Genuchten model. They further claim that more accurate results can be achieved leaving the two parameters without any fixed relationship. Their model is a somewhat more general than the others and is based on the pore size distribution of the medium. It is given as:

$$\theta = n\left[\frac{1}{\ln\left(e+\left(\frac{\Psi}{\alpha}\right)^{N}\right)}\right]^{M} \tag{2.34}$$

where α, N and M are now different parameters to those of the van Genuchten model and also need to be estimated experimentally, but where n is the porosity and Ψ is the matric potential in metres, as before (see Eq. 2.32). Figure 2.10 shows two typical soil water characteristic curves (SWCC) according to the van Genuchten model for coarse grained (unbroken line) and fine grained soils (dotted line) respectively. The parameters used for the two curves are given in Table 2.2.

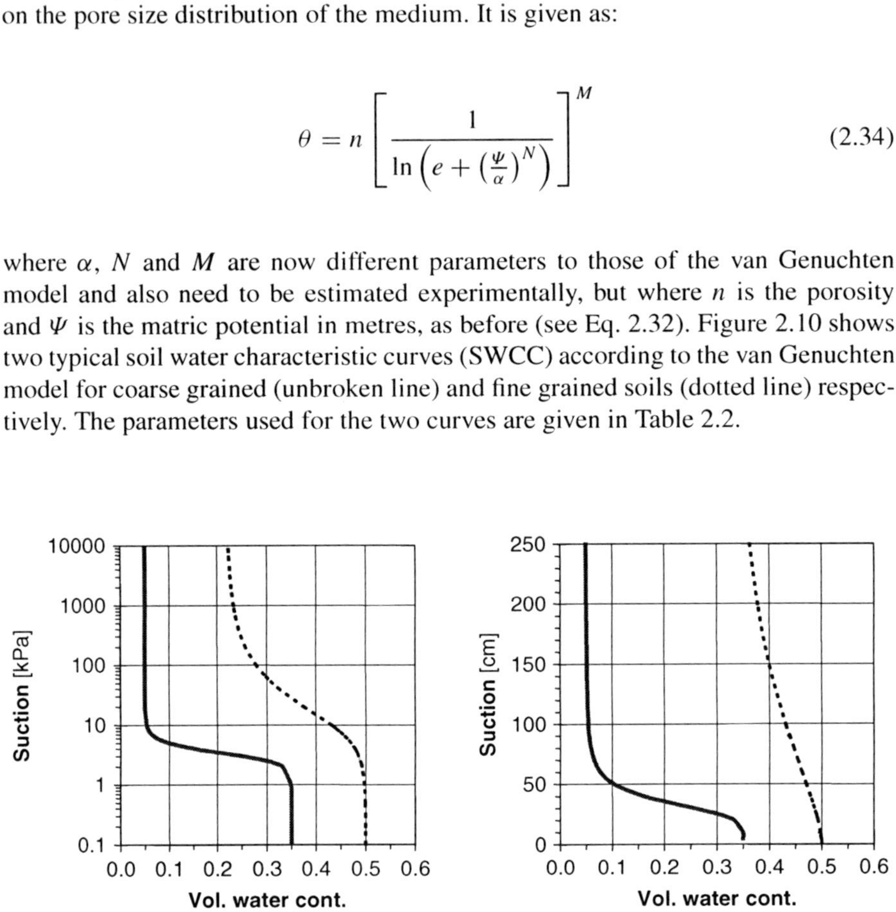

Fig. 2.10 Typical SWCC for a coarse (*unbroken line*) and fine grained (*dotted line*) soil using the van Genuchten's model. The parameter used to plot the curves are given in Table 2.2

Table 2.2 Typical SWCC parameter for a coarse and fine grained soils according to the van Genuchten model

	Porosity	Residual volumetric water content	Residual Saturation	van Genuchten model parameters		
	n (-)	θ_r (-)	S_{rr} (-)	M (-)	N (-)	α (cm^{-1})
Sand	0.35	0.05	0.143	0.8	5.00	0.03
Clay	0.50	0.22	0.440	0.4	1.67	0.01

2.8 Permeability in Unsaturated Soil

Water flow in unsaturated soils is primarily dependent on the volumetric water content, matric suction and on the gravitational potential. Due to the presence of air within part of the pores, water movements are obstructed and flow is only achieved through the finer pores or in films around the soil particles. The permeability (or "hydraulic conductivity") of unsaturated soils is, therefore, reduced compared with fully saturated soils due to the presence of air in the porous media. Usually the permeability of unsaturated soils is given as the water-relative permeability defined as the ratio of the permeability at a specific water content to its permeability under fully saturated conditions, thus:

$$k_{rw}(\theta) = \frac{K_w(\theta)}{K} \tag{2.35}$$

where $k_{rw}(\theta)$ is the water-relative permeability, $K_w(\theta)$ is the water permeability – both being functions of the volumetric water content, θ – while K is the saturated coefficient of permeability. Brooks and Corey (1964, 1966) suggested that the water-relative permeability could be estimated as

$$k_{rw}(\theta) = \Theta^{\left(3+\frac{2}{\lambda}\right)} \tag{2.36}$$

where λ is a the pore size distribution index and Θ is the normalized water content. Based on Mualem's model (Mualem, 1976) van Genuchten (1980) expressed the water-relative permeability as:

$$k_{rw}(\theta) = \sqrt{\Theta} \cdot \left(1 - \left(1 - \Theta^{\frac{1}{M}}\right)^{M}\right)^2 \tag{2.37}$$

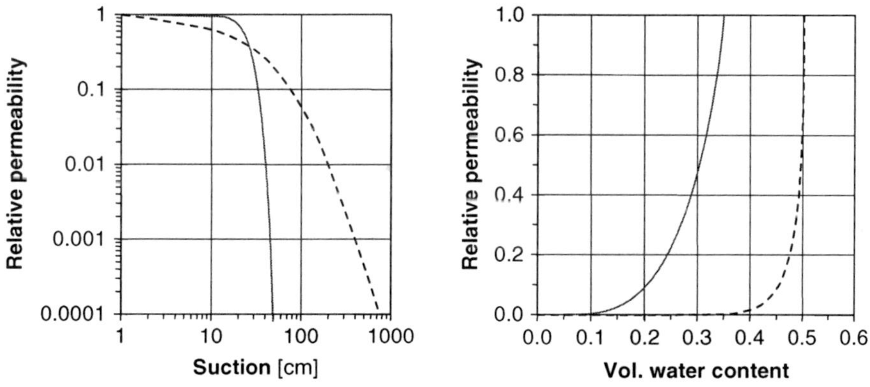

Fig. 2.11 Water-relative permeability for a coarse (*unbroken line*) and fine grained (*dotted line*, mostly to the right in each figure) soil using van Genuchten's equation. The parameters used to plot the curves are given in Table 2.2

where M is the same experimental parameter as given in the van Genuchten's SWCC.

Figure 2.11 shows clearly that the permeability of a porous medium varies significantly with the suction or the volumetric water content (or degree of saturation as $S_r = \theta/n$). A reduction in the degree of saturation from fully saturated soil to 80% results in a relative permeability of only 36% of the saturated permeability for the sand and 18% for the clay respectively in the above figure. Hence, modelling water movements in pavement structures needs to address this to obtain a realistic movement of the water flow.

2.9 Drainability

Broadly graded aggregates, as typically used in granular base and sub-base layers of the road construction have relatively small pores as the large pores between the coarser particles are mostly filled with smaller particles. This means that coefficient of permeability values, as characterised by the Darcy coefficient, K, are relatively low. Laboratory testing of typical sub-base aggregates has revealed hydraulic conductivity values that are usually less than 10^{-3} m/s in the unlikely even of saturation (Jones and Jones, 1989) and as low as an effective value of 10^{-6} m/s when, more normally, in a partially saturated state.

A second implication is that the pores will have a measurable suction ability. Hence dryer aggregates will usually show an ability to pull water towards them. There is a draft CEN standard (1999) for the evaluation of the suction height (i.e. of matric potential which is an indirect measure of the suction capacity of the aggregate). Figure 2.12 shows a result obtained by a similar method in which a column of

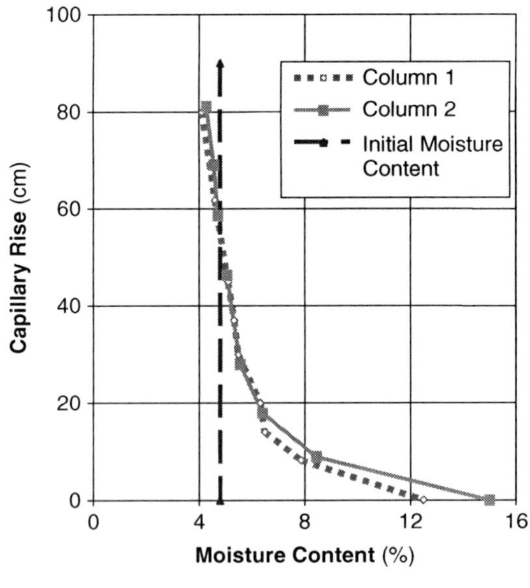

Fig. 2.12 Typical moisture profile with height for a compacted column specimen with water provision at the base (adapted from Jessep, 1998)

aggregate, compacted near its optimum water condition into a pipe, was stood with its lower end in a free supply of water (Jessep, 1998). It will be seen that almost no water drains from the material as originally compacted, even 0.5 m above the water supply level, but there is evidence of a large suction capability pulling in significant volumes of water near the base such that the lower aggregate is close to being saturated. This observation leads to the questioning of drainage design practice that is based on the assumption of the free drainage of saturated aggregate. Such an assumption often ignores an aggregate's matric (suction) potential.

The type of result in Fig. 2.12 confirms that the theoretical approach prediction of McEnroe (1994) is broadly correct. McEnroe applied the van Genuchten (1980) characterisation of the moisture – suction relationship, Eq. 2.32, to obtain a saturation ratio-suction relationship. A parallel saturation ratio-permeability relationship allows a relationship between permeability and suction to be established. On this basis McEnroe showed that more than 50% of a typical granular material's void space will be undrainable if the hydraulic conductivity is less than 7.5×10^{-4} m/s. Taking McEnroe's and Jesseps's (and similar observations) together, it follows that most graded aggregate bases and sub-bases are difficult or impossible to drain by gravity. With highway aggregate layers often less than 1m above the water table this would mean that they could more easily act to attract water than to drain it.

2.10 Conclusions

In the highway and near-highway environment, much of the road construction and the naturally occurring or imported soils are in a partially saturated condition. As the chapter has shown, the effect of this partial saturation is to cause suctions to exist within the pores of the component soils and aggregates. Water flow is also made more complex to predict as the basic theory of permeability has to be adapted to model the real movements of water through the pores. The degree of saturation that exists is very dependent on the boundary conditions and small changes, e.g. by rainfall infiltrating through cracks in an asphalt road's surface, may significantly effect the saturation levels and the retained soil suction. As will be shown in Chapters 8 and 10, these changes can have major effects on the mechanical behaviour of the soil or aggregate concerned and, thus, the behaviour of the road and earthworks can be very sensitive to such changes. Accordingly, it is important that we are able to predict the suction and the ease of drainage. These two abilities are very important tools with which drainage schemes may be assessed and the consequent performance of the highway asset evaluated. They are also required if we are to be able to compare alternative remedial strategies or to make long-term predictions of material and soil behaviour.

With increasingly sophisticated water regime modelling programs available for personal computer use, the parameters discussed in this chapter are becoming the basic inputs for ever-more routine analyses. Furthermore, the water flows that such computer codes predict now allow users to undertake environmental analyses on contaminant movement. Once again, to achieve, this the basic parameters outlined in this chapter will be needed.

References

Apul, D. S., Gardner, K., Eighmy, T., Benoit, J. & Brannaka, L., 2002, "*A Review of water movement in the highway environment: Implications for recycled materials use*", Recycled Materials Resource Center, University of New Hampshire, Durham.

Bear, J., 1972, "*Dynamics of fluids in porous media*", Elsevier, New York.

Bear, J. & Verruijt, A., 1987, "*Modeling groundwater flow and pollution*", D. Reidel Publishing Company, Dordrecht, Holland.

Brooks, R. H. & Corey, A. T., 1964, "Hydraulic properties of porous media", *Hydrology Paper* no. 3, Civil Engineering Dept., Colorado State Univ., Fort Collins, Ohio.

Brooks, R. H. & Corey, A. T., 1966, "Properties of porous media affecting fluid flow", *J. Irrig. Drain. Div., Am. Soc. Civil Eng.*, **92**(IR2), pp. 61–88.

Carrier, W. D., 2003, "Goodbye, Hazen; Hello, Kozeny-Carman", *Am. Soc. Civil Eng., J. Geotech. Geoenv. Eng.*, **129**(11), Nov., pp. 1054–1056.

Cedergren, H. R., 1974, "*Drainage of highway and airfield pavements*", Wiley, New York.

Cedergren, H. R., 1977, "*Seepage, drainage and flow nets*", 2nd ed., John Wiley & Sons, New York.

CEN, 1999, "*Determination of water suction height*", prEN 1097-10.

CEN, 2002, "*Geotechnical investigation and testing – identification and classification – Part 1: Identification and description*", EN ISO 14688-1.

Das, B. M., 1997, "*Advanced soil mechanics*", 2nd Edn., Taylor & Francis, Washington, D.C.

Fredlund, D. G. & Rahardjo, H., 1993, "*Soil mechanics for unsaturated soils*", John Wiley and Sons, New York.

Fredlund, D. G. & Xing, A., 1994, "Equations for the soil-water characteristic curve", *Can. Geotech. J.* **31**, pp. 521–532.

Fredlund, D. G., Xing, A. & Huang, S., 1994, "Predicting the hydraulic conductivity function for unsaturated soils using the soil-water characteristic curve", *Can. Geotech. J.* **31**, pp. 533–546.

Hazen, A., 1911, "Discussion of 'Dams on Sand Foundations', by A. C. Koenig," *Trans. Am. Soc. Civil Eng.*, **73**, p. 199.

Jessep, R., 1998, "*Suction Characteristics of Compacted Unbound Aggregate*", Masters Thesis, Univ. Nottingham.

Jones, H. A. & Jones, H. A., 1989, "Horizontal permeability of compacted aggregates", in *Unbound Aggregates in Roads*, ed. Jones, R.H. & Dawson, A.R., Butterworths, pp. 70–77.

McEnroe, B. M., 1994, "Drainability of granular bases for highway pavements", *Transportation Research Record*, Transportation Research Board, National Academy Press, Washington, DC., **1434**, pp. 23–28.

Mualem, Y., 1976, "A new model for predicting the hydraulic conductivity of unsaturated porous media", *Water Resources Research*, **12**, pp. 513–522.

Samarasinghe, A. M., Huang, Y. H. & Drnevich, V. P., 1982, "Permeability and consolidation of normally consolidated soils," *J. Geotech. Eng. Div., Am. Soc. Civil Eng.*, **108**(GT6), pp. 835–850.

van Genuchten, M. Th., 1980, "A closed form equation for predicting the hydraulic conductivity of unsaturated soils," *Soil. Sci. Soc. Am. J.*, **44**, pp. 892–898.

Chapter 3
Measurement Techniques for Water Flow

Sigurður Erlingsson*, Susanne Baltzer, José Baena and Gunnar Bjarnason

Abstract The chapter describes different measurement techniques for water-flow-related phenomena in pavements and embankments, i.e. water content, permeability and suction. For estimating the water content the gravimetric method is described as well as non-destructive methods such as neutron scattering, time domain reflectometry and ground penetrating radar. Then methods for estimating both the saturated and the unsaturated permeability of soils and granular materials are described. Both steady-state and unsteady methods are mentioned. Finally, common methods for measuring soil suction are briefly introduced.

Keywords Water flow · measurements · water content · permeability · suction

3.1 Introduction

Water entering a pavement structure migrates as moisture through the structure. The amount of water penetrating the pavement is dependent on precipitation, drainage, design of the road structure, type and condition of the surface layer (cracks, joints) and shoulders and the materials in the pavement, subgrade and subsoil. Seasonal fluctuations in temperature, i.e. freezing and thawing, can also provoke moisture flow inside the pavement structure. An excess of water can cause a lower bearing capacity of the pavement structure and reduces pavement life (see Chapters 8 and 10). Also of concern is transport of contaminants with the liquid fluxes from the road and into the environment (see Chapter 6). As pavement deterioration can be reduced by proper drainage it is important to be able to measure the water content in order to understand how water moves within the road structure.

Predictions of water flow and contaminant transport in unsaturated soils should be based on an accurate description of the subsurface conditions including the aggregates

* *Co-ordinating Author:*
✉ S. Erlingsson
Háskóli Íslands/University of Iceland, Iceland & Statens väg-och transportforskningsinstitut/ Swedish National Road and Transport Institute, Sweden
e-mail: sigger@hi.is

A.R. Dawson (ed.), *Water in Road Structures,* DOI 10.1007/978-1-4020-8562-8_3,
© Springer Science+Business Media B.V. 2009

and hydrological properties of the materials involved. Therefore measurements of the fundamental parameters regarding flux of water need to be carried out. This also helps to set the initial conditions for analysis in water flow modelling.

The most important parameters related to the movement of water through pavement structures are the quantity and spatial distributions of moisture inside the pavement, the coefficient of permeability of individual layers and their matric suctions. These parameters can be determined experimentally, either in the field or in the laboratory, by various techniques. The most common of these methods used in road engineering will be briefly described in this chapter.

3.2 Water Content

A fundamental parameter that characterises the water movement in pavements is the water content. This provides information on the condition of the road layers regarding the moisture saturation stage, which controls the main parameters in the governing equations for water flow (see Chapter 2, Section 2.8). A number of methods are available for measuring water content. They can be divided into destructive methods (gravimetric methods) and non-destructive methods that provide indirect measurements of the water content.

3.2.1 Gravimetric Method

The simplest and most widely used method to measure the soil water content is the gravimetric method where a soil sample is taken and weighed, dried in an oven at 105°C for 24 h and then reweighed. To shorten the drying period a microwave oven can be used as an alternative, although this method introduces the possibility of removing chemically bonded water which would lead to an over-estimation of water content. The gravimetric water content, w, is the mass of water per mass of dry soil (see Eq. 2.10) or

$$w = \frac{W_b - W_d}{W_d} \tag{3.1}$$

where W_b is the total (or "bulk") weight of the soil and W_d is the dry weight of the soil.

This method has two major drawbacks: the sampling is destructive for the road and the method can not be used to make in-situ measurements in real time. However it is an accurate method and is often used to calibrate other measurement techniques.

3.2.2 Non-destructive Methods

A number of methods exist for estimating the soil water content of road materials in a non-destructive way assuming that the instruments are placed in the road during the construction phase. They are all indirect methods as they involve measurements of some property of the material affected by the water content or they measure a property of some object placed in the material. Some of the more common indirect methods used in the highway environment are briefly described here.

3.2.2.1 Neutron Scattering Method

In the neutron scattering method a tube acts as a radioactive source and a detector. The radioactive source is placed at the end of a rod inserted into a pre-made hole to a depth of, typically, 150 and 300 mm. High energy neutrons are emitted from the source and the neutrons collide with nuclei of atoms in the surrounding soil, thus reducing the energy level of the neutrons. They are slowed substantially by collision with nuclei of similar mass, usually hydrogen atoms, making this technique sensitive to water content. Therefore the proportion of neutrons returning to the tube's detector is proportional to the water content of the soil the neutrons have travelled through (Hignet & Evett, 2002; Veenstra et al., 2005). As the neutron scattering method is based on radioactive decay, any other radioactive elements that are present, for example inside the pavement structure, may affect the results.

The neutron scattering method is frequently used along with gamma ray attenuation in nuclear moisture-density devices where both water content and density measurements are provided (see Fig. 3.1). The gamma ray attenuation method uses a beam of gamma rays emitted from a radioactive isotope of caesium that is sent through a soil sample of known volume and measured by a detector. The hydrogen atoms in the water scatter neutrons and the amount of scatter is proportional to the total unit weight of the material.

Neutron scattering is an accurate and precise method for soil water content measurements. However it cannot be left unattended due to its radioactive source and can therefore not be used in an automated monitoring programme.

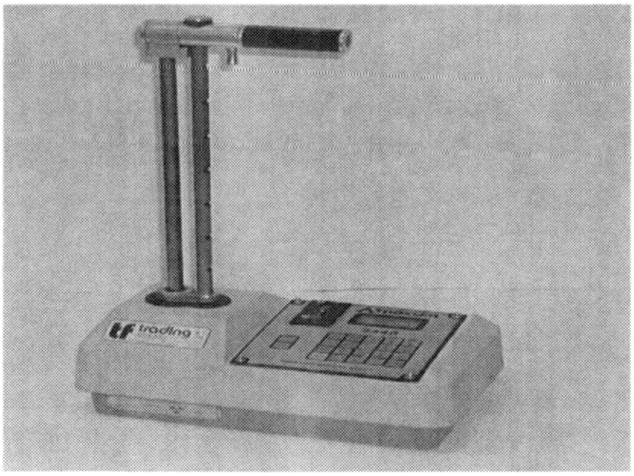

Fig. 3.1 An example of equipment where the neutron scattering method is combined with the gamma ray attenuation method is this Troxler instrument. Courtesy of the Danish Road Institute

3.2.2.2 Time Domain Reflectometry Techniques

Time Domain Reflectometry (TDR) is a non-destructive electromagnetic technology that utilises the relationship between the relative permittivity (usually known as the dielectric constant, κ_r) of porous materials and their water content. Dry soils have values of dielectric constant of around 2–6 but water about 79–82 depending on wave frequency and water temperature. As the water content of the soil increases the dielectric constant also increases and is therefore an indirect indicator of the soil water content (Topp et al., 1980; Topp & Davis 1985; Svensson, 1997; Hillel, 1998; O'Connor & Dowding, 1999).

Time Domain Reflectometry is nowadays the most common technique for monitoring water content in pavement structures and subgrades. Topp et al. (1980) used TDR to measure permittivity of a wide range of agricultural soils and developed an empirical relationship between the permittivity and the volumetric water content which made it possible to use TDR in monitoring the water content of a given soil. A schema of a TDR probe is shown in Fig. 3.2 (Ekblad, 2004). Usually the probes have 2 or 3 rods of length, L (normally approximately 300 mm in length). They act as a wave guide while a transmitter inside the probe generates a pulse of frequencies up to 1 GHz which propagates along the metal conductors of the sensors. An electromagnetic field is therefore established around the probe. The pulse is reflected back to the source at the end of the conductors. The transit time of the pulse is therefore estimated according to

$$c_e = \frac{2L}{t} \tag{3.2}$$

where c_e is the velocity of the electromagnetic pulse, L is the probe length and t is the transit time.

The determination of the dielectric constant is thereafter achieved from the basic equation

$$c_e = \frac{c_0}{\sqrt{\kappa_r \cdot \mu_r}} \tag{3.3}$$

in which c_0 is velocity of light, κ_r is the dielectric constant and μ_r is the relative magnetic permeability. Most soils are practically nonmagnetic, thus their relative magnetic permeability is close to unity. Roth et al. (1992) investigated ferric soils

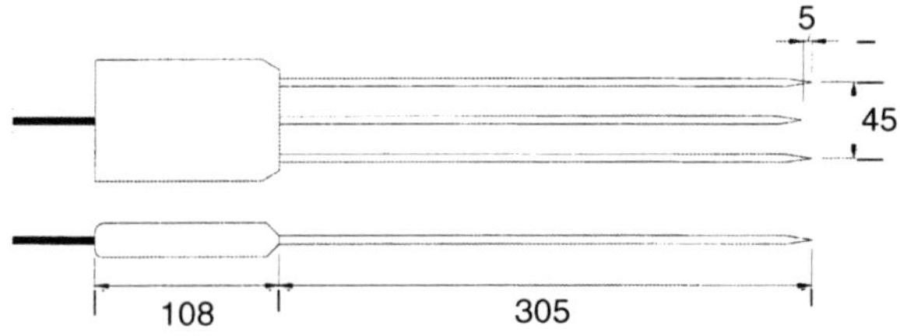

Fig. 3.2 Schema of a TDR-probe. Dimensions are in mm (Ekblad, 2004). Reproduced by permission of Jonas Ekblad

and could not find any influence of magnetic properties on the volumetric water content. By inserting c_e from Eq. 3.2 and rearranging, the dielectric constant can be estimated as

$$\kappa_r = \frac{(c_0 t)^2}{4L^2} \tag{3.4}$$

where the relative magnetic permeability has been set to unity.

TDR moisture probes express their readings as a volumetric water content θ, defined as in Eq. 2.9 which can be related to gravimetric water content, w, as in Eq. 2.11. The volumetric water content is obtained via a relationship with the relative dielectric constant which is based on an empirical approach. A number of relationships exists which, frequently, have been derived using regressions analyses. Topp et al. (1980) gave the relationship as:

$$\theta = -5.3 \times 10^{-2} + 2.92 \times 10^{-2}\kappa_r - 5.5 \times 10^{-4}\kappa_r^2 + 4.3 \times 10^{-4}\kappa_r^3 \tag{3.5}$$

Their analysis was based on a variety of soil types although many were low density agricultural soils. Jiang and Tayabji (1999) derived a similar third-order polynomial relationship for coarse grained soils as:

$$\theta = -5.7875 \times 10^{-2} + 3.41763 \times 10^{-2}\kappa_r - 1.3117 \times 10^{-3}\kappa_r^2 + 2.31 \times 10^{-5}\kappa_r^3 \tag{3.6}$$

Thus, by combining either Eq. 3.5 or Eq. 3.6 with Eq. 3.4, it is possible to relate volumetric water content to transit time.

It is not uncommon that TDR sensors can measure transit time with a resolution of 10 picoseconds, which corresponds to a water content of approximately 0.1% by volume. Figure 3.3 shows readings from a TDR probe over a seven month period,

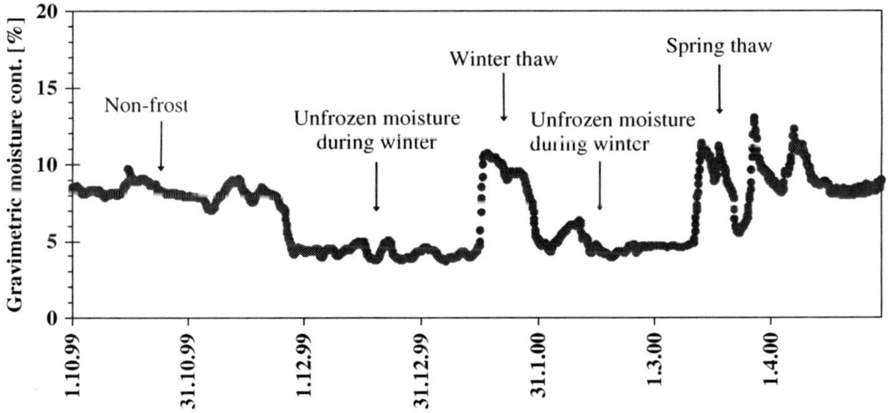

Fig. 3.3 Gravimetric water content in a sub-base layer at 25 cm depth in a low volume road. The readings were carried out at three hour intervals

starting in the autumn of 1999, at a depth of z = 25 cm in a sub-base layer in SW Iceland (Erlingsson et al., 2000; 2002). Gravimetric water content is plotted after Eq. 2.11 has been applied to the TDR readings. In the period October to late November the actual water content decreased at a slow rate, with irregularities due to rainfalls. The freezing period started in late November which can be seen as a rapid drop in the water content. As the soil was frozen the TDR registration represents the unfrozen water content but not the actual water content in the layer. A two week long thawing period can be seen in January 2000. The spring thaw period started in early March and influenced the water content in the sub-base for more than a month.

3.2.2.3 Ground Penetrating Radar

In Ground Penetrating Radar (GPR), electromagnetic waves are sent out from a transmitter on or above the ground surface and picked up by a receiver after penetrating and returning from the soil. The velocity of the electromagnetic wave propagation in soils is dependent on the soil bulk permittivity modulus (Grote et al., 2003). Thus the underlying principles of the GPR soil moisture measurements are the same as those of Time Domain Reflectometry except that in TDR the electromagnetic waves travel along a waveguide whereas with GPR the propagated electromagnetic waves are unconstrained. GPR therefore has the potential to cover a much larger soil volume than does TDR. GPR can be air launched or surface launched or used in boreholes and is completely non-invasive, whereas TDR requires the penetration of rods (waveguides) into the pavement structure.

GPR is primarily used to estimate material thickness but can also detect cables, culverts, steel wire net, the water table and frost depth, but as mentioned it can also be used to measure the dielectric value and therefore water content.

The resolution and depth range of the electromagnetic wave depends on the frequency used and the properties of the medium. Low frequency antennas 100... 500 MHz have good penetration and depth range but lower resolution, whereas high frequency antennas 500... 1000 MHz have a lower depth range, but they give better resolution. With a 400 MHz antenna the depth range is typically 3.5–5 m and the resolution 8–10 cm. With a 1000 MHz antenna the depth range is below 1 m, but the minimum resolution is 3–4 cm.

Figure 3.4 shows results from a GPR survey from Finland for a low-volume road structure. In the figure only the overlays can be clearly identified as the underlying layers are an old "unconstructed" local road. The figure allows comparison of GPR data with the results of a Falling Weight Deflectometer (FWD) assessment[1]. FWD testing can give an indication of weak subsoil in the form of a high BCI-index value as seen in Fig. 3.4 at station 5050 m (middle histogram). The figure also shows a high Surface Curvature Index (SCI) and a low stiffness (E_2). At the same location, the 400 MHz ground radar image shows a strong reflection on the (wet?) boundary

[1] Reference should be made to Chapter 10, Section 10.3, and to Eqs. 10.1–10.4 for an explanation of FWD assessment and of BCI and SCI.

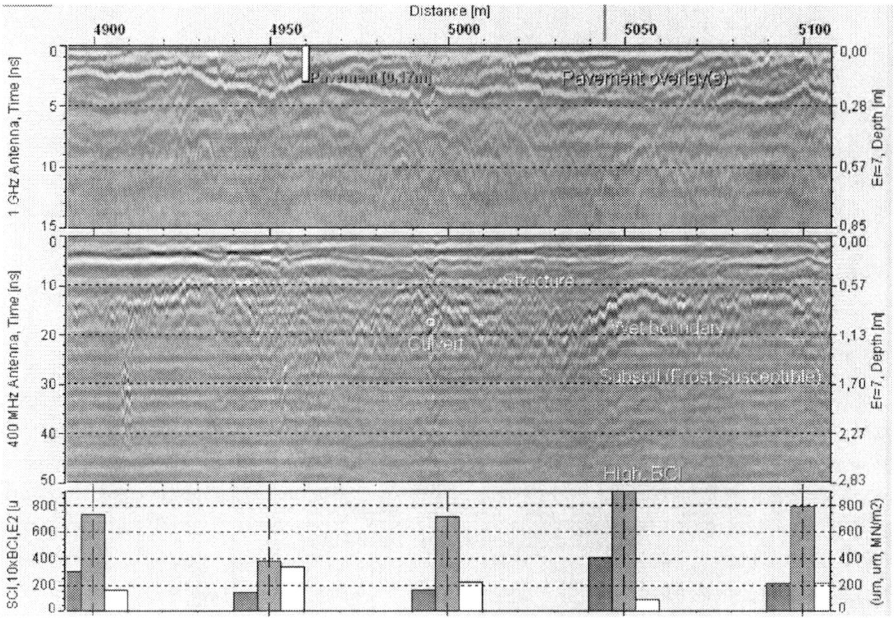

Fig. 3.4 Ground penetrating radar images using a 1 GHz antenna (*top*) and 400 MHz antenna (*middle*) along with FWD parameters (SCI, 10×BCI, E_2) (*bottom*) from a 200 m long section on a low volume road in Finland. SCI and BCI are the Surface and Base Curvature Indices, respectively, and E_2 is the stiffness of the pavement structure. See text for explanation

of the subsoil at a depth of 0.6–1 m. Elsewhere the boundary is not clear, which may then indicate mixing of subsoil and structure material by frost. The 1 GHz GPR image (top) shows that attempts to compensate for the weak road conditions have been made by adding new surface pavement layers, one after another, up to a thickness of 13–25 cm. Perhaps a better drainage system is the proper solution. The 400 MHz image also shows a culvert at 1m depth close to station 5000 m and that the very wet zone due to a frost susceptible layer continues between stations 5030 and 5100 m.

Radar images can give information on the appropriate depth and location for such a solution. The information is even more clear after the radar image is transferred onto a true longitudinal profile. What can be seen from ground radar images partly depends on the conditions (e.g. wet season and more clear wet boundaries; ground water level) or time of the year (e.g frost boundary or its absence at rock).

3.2.2.4 Capacitance Measurements

Capacitive sensors measure the resonant frequency of an inductance-capacitance (LC) tuned circuit where the soil located in between two flat waveguides is the dielectric material. The inductance is kept constant and the resonant frequency f measured and therefore the capacitance can be calculated from

$$f = \frac{1}{2\pi\sqrt{L_e C_e}} \tag{3.7}$$

where L_e is the inductance and C_e is the capacitance. The capacitance C_e is a measure of the relative bulk dielectric constant of the soil and is a function of the water content of the soil (Veenstra et al., 2005). As with all dielectric moisture-based sensors, calibration is necessary for an accurate determination of the water content. Starr and Paltineanu (2002) give an overview of the current capacitance methods, their instrumentation and procedures.

3.2.2.5 Other Methods

A number of other methods exist for estimating soil water content such as nuclear magnetic resonance (NMR), which can detect nuclear species that have a magnetic moment or spin. As hydrogen has a nuclear spin of 1/2 the NMR technique can be used to estimate water content in soils. This is a fast and non-destructive method with high accuracy in uniform samples. However the method is costly, not suitable for field use and highly dependent upon sample calibration and is therefore not used in soil studies or in applications related to roads (Veenstra et al., 2005).

Near infrared reflectance spectroscopy (NIRS), seismic methods and thermal properties are all methods that can be used for estimation of soil water content. Although they are in many respects good and accurate methods, they all have some drawbacks making them non-suitable as routine methods to be used in the pavement environment. In the first two methods the calibration process is complex or difficult to perform due to the influence of other factors, and assessing the thermal properties of the soil is costly and needs a long measurement time relative to other methods (Veenstra et al., 2005).

3.3 Permeability Testing

The permeability of soils is a material parameter that relates the rate of water flow to the hydraulic gradient in the soil and, therefore, determines the material's suitability for drainage layers. An embankment usually consists of compacted materials. The compaction often results in anisotropy such that the vertical and the horizontal permeability properties are not equal. For road construction layers, water movements below the ground water table are almost entirely horizontal and thus it is the horizontal permeability that should be measured. Above the groundwater table in the unsaturated zone the movement of water is much more complex, involving vertical as well as horizontal components depending on material parameters such as temperature, water content and matric suction.

Some typical values of the coefficient of permeability for saturated soils are shown in Table 3.1.

The permeability of soils can either be estimated in saturated conditions or for partially saturated conditions. If the permeability of soils is estimated from saturated

Table 3.1 Typical values of the coefficient of permeability of saturated soils

Soil	Coefficient of perm. K (m/s)	Degree of permeability
Gravel	$> 10^{-3}$	Very high
Sandy gravel, clean sand, fine sand	$10^{-3} > K > 10^{-5}$	High to medium
Sand, dirty sand, silty sand	$10^{-5} > K > 10^{-7}$	Low
Silt, silty clay	$10^{-7} > K > 10^{-9}$	Very low
Clay	$< 10^{-9}$	Virtually impermeable

samples and the unsaturated permeability is sought, the Soil Water Characteristic Curve (SWCC – see Chapter 2) can be used to predict the permeability for a specified volumetric water content (Fredlund & Rahardjo, 1993).

3.3.1 Permeability Tests of Saturated Soils and Aggregates

Traditionally in geotechnical engineering, the saturated permeability is estimated in the laboratory in a constant head test for coarse grained soils whereas a falling head test is used for fine grained soils. An oedometer test can also provide a measure of the saturated permeability for fine grained soils in the laboratory. Field tests which provide a measure of the saturated permeability are usually a kind of pumping well test, or injection test.

3.3.1.1 Constant Head Permeability Test

A constant head permeability test is usually used for coarse grained soils. The sample is placed in the permeameter where a constant head drop is applied to the sample and the resulting seepage quantity is measured (see Fig. 3.5).

By rearranging and substituting into Eq. 2.15, the permeability K is given as:

$$K = \frac{qL}{A(h_2 - h_1)} = \frac{qL}{Ah} \tag{3.8}$$

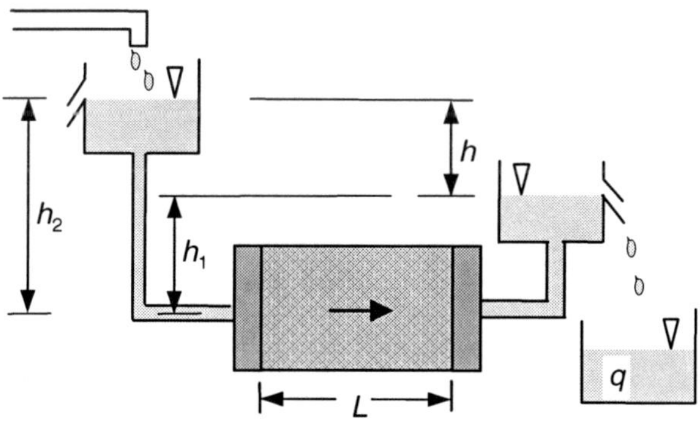

Fig. 3.5 Constant head permeability test

where q is the discharge (L^3/T), L is the specimen length (L), A is the cross section area of the specimen (L^2) and h is the constant head difference (L).

There are limitations to the use of a permeameter test for pavement materials. Sub-bases or drainage layers normally contain particles with a maximum nominal size between about 20 and 80 mm. It has been suggested that to obtain reliable permeability measurements that the value of the ratio of the permeameter diameter to the maximum particle diameter should be between 8 and 12. However, standard permeameters are generally too small and too fragile to allow the largest particles to be included and to achieve correct compaction. Head (1982) describes a 406 mm diameter permeability cell suitable for gravel containing particles up to 75 mm.

To help overcome this limitation, the UK Department of Transport (1990) introduced a large, purpose-designed, permeameter for testing road construction aggregates (Fig. 3.6). It measures horizontal permeability at low hydraulic gradients as these are the hydraulic conditions that might be anticipated in granular pavement layers.

Normally, Darcy flow is assumed to be the regime of permeating water in the soil or aggregate layers under a road, i.e. the water percolates at sub-critical velocities and without eddy-flow when moving from small to large pore spaces. This means that energy losses are only due to friction between the water and the surrounding solids and that a constant value of coefficient of permeability, K, can be defined. When coarse materials, with large pores, are tested for permeability in equipment such as that illustrated in Fig. 3.6, care must be taken to ensure that Darcy conditions are maintained throughout the test. Under many conventional test conditions, high hydraulic gradients are applied (much larger than in-situ) in order to obtain results in a convenient time scale. If such hydraulic gradients are applied to materials with large pores, eddy flows may develop in the large pores and more energy will be lost than Darcy conditions would predict. If the user is unaware of these conditions, the value of coefficient of permeability, K, will be under-estimated (see Fig. 3.7).

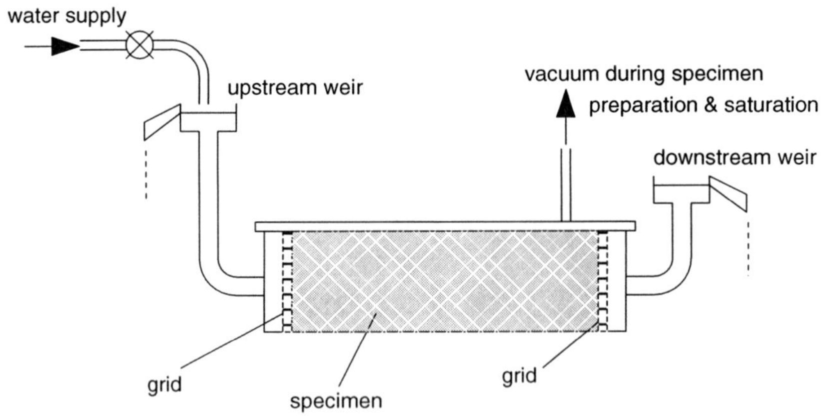

Fig. 3.6 Department of Transport (1990) horizontal aggregate permeameter

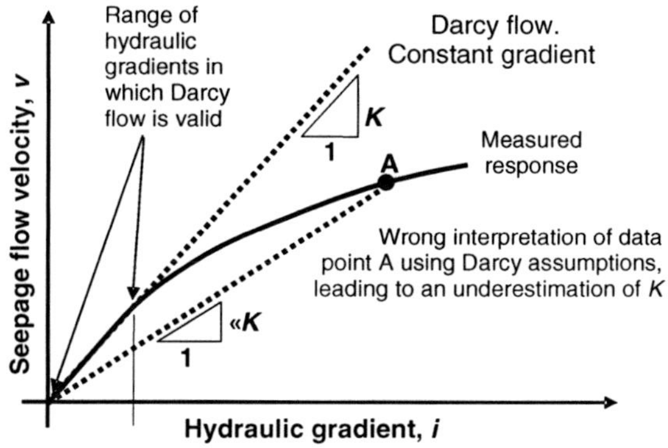

Fig. 3.7 Limits of applicability of Darcy assumptions when testing at variable hydraulic gradients

For this reason, tests should be performed at variable hydraulic gradients on coarse materials. Hydraulic gradients less than 0.1 may be required to achieve Darcy conditions. Alternatively, more advanced permeability formulations may be used such as those given in Eq. 2.21.

3.3.1.2 Falling Head Test

In the falling head test the head is a function of time during testing while water from a standpipe flows through the soil. The falling head test is preferably used for soils with low permeability, i.e. silty or clayey soils (see Fig. 3.8). For such soils the problem of excessively high hydraulic gradients (Fig. 3.7) is avoided as well as the practical issue that, otherwise, the head would fall too rapidly.

Equating the instantaneous flow due flow of water through the specimen according to Darcy's law with the flow necessary for continuity gives:

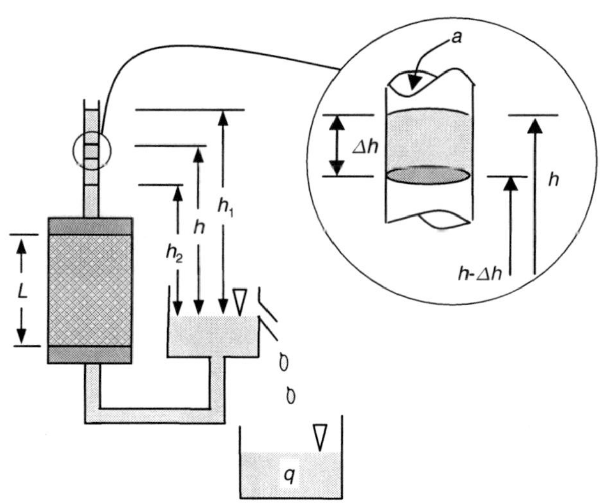

Fig. 3.8 Falling head permeability test

$$q = K\frac{h}{L}A = -a\frac{dh}{dt} \tag{3.9}$$

where q is the flow rate (L^3/T), $h = h(t)$ is the head difference (L) at time t, L is the length of the specimen (L), and a and A are the cross sections (L^2) of the standpipe and the soil specimen, respectively. Integration of the left side with limits of time from t_1 to t_2 and on the right side for the corresponding limits of head in the standpipe, h_1 and h_2, and rearranging results in

$$K = \frac{aL}{A(t_2 - t_1)} \ln \frac{h_1}{h_2}. \tag{3.10}$$

3.3.1.3 Oedometer Test

The coefficient of permeability can be determined as a function of vertical strain in an oedometer test (see Fig. 3.9). The method is normally used for fine-grained (frictionless) soils to estimate their consolidation characteristics. The oedometer test is usually performed either as a constant stress or constant-rate-of-strain test. In a one-dimensional constant stress test a cylindrical specimen of soil enclosed in a stiff metal ring is subjected to a series of increasing static loads, while changes in thickness are recorded against time. The coefficient of permeability is given as:

$$K = c_v \cdot m_v \cdot \gamma_w = c_v \cdot m_v \cdot \rho_w g \tag{3.11}$$

where c_v is the coefficient of consolidation, m_v is the coefficient of volume compressibility, γ_w is the unit weight of water and g is the acceleration due to gravity.

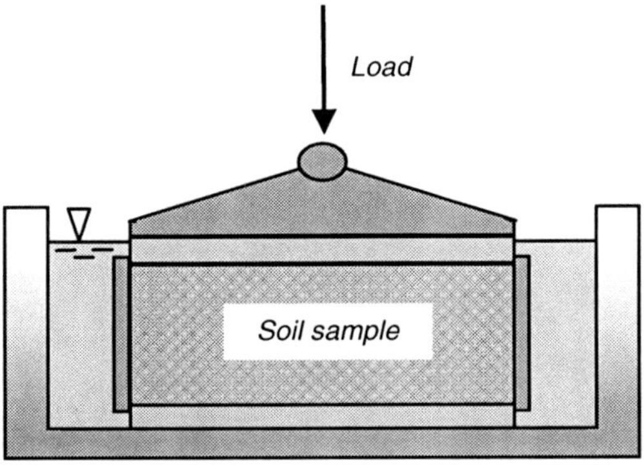

Fig. 3.9 Constant stress oedometer test

3.3.1.4 Pumping Tests

In the field, the average coefficient of permeability of a soil deposit can be determined by performing a pumping test where water is pumped from a well at a constant rate over a certain period of time. With time the drawdown of the water table or the piezometric head will reach a steady state. The rate of discharge, q, from pumping is equal to the rate of flow of groundwater into the well, or

$$q = 2\pi r K h \frac{dh}{dr} \tag{3.12}$$

where h is the piezometric head at the radial distance, r, from the centre of the well.

By separating the two variables r and h and integrating between the two observation wells at distances r_1 and r_2 respectively from the test well, the permeability coefficient can easily be determined (see Fig. 3.10) as:

$$K = \frac{q}{\pi} \cdot \frac{\ln(r_2/r_1)}{(h_2^2 - h_1^2)} \tag{3.13}$$

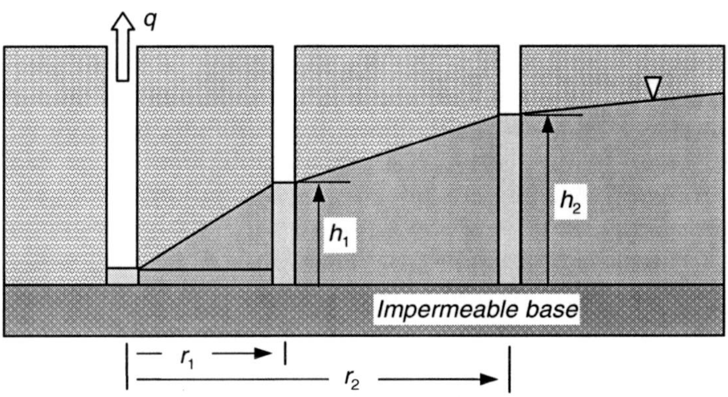

Fig. 3.10 Pumping test from a well with two observation wells in an unconfined permeable soil layer with an impermeable stratum underneath

3.3.1.5 Injection Tests

Injection tests, which are the reciprocal of pumping tests, are normally conducted by pumping water or air into a test section of a borehole and measuring the flow rate. By performing tests at intervals along the entire length of a borehole, a permeability profile can be obtained. Air (or another gas) is used as the "fluid" when testing areas of high permeability, as water cannot be provided at a high enough flow rate for such conditions.

3.3.1.6 Tracer Test

Tracer tests involve the injection of an inert solution, or tracer, into an existing flow field via a borehole or a well. Tracer tests are often desirable because they are passive-type tests and do not place unnatural stress conditions on the flow system.

The dilution rate of the tracer at the injection well or its time of travel to another well can be used to calculate the water velocity and ultimately the permeability. Detection of the tracer, or concentration measurements, can be made by either manual or probe sampling. Commonly used tracers are radioisotopes, salt solutions and fluorescent dyes.

3.3.2 Permeability Tests of Unsaturated Soils

As introduced in Chapter 2, the flow of water in saturated soils is commonly described using Darcy's law which relates the rate of water flow to the hydraulic gradient (Eq. 2.16). Furthermore the coefficient of permeability is relatively constant for a specific soil. Darcy's law applies also to the flow of water through unsaturated soils. However the permeability of unsaturated soils can not be assumed generally to be constant (Richards, 1931; Fredlund & Rahardjo, 1993; Fredlund 1997). Permeability now becomes predominantly a function of either the water content or the matric suction (see Chapter 2, Section 2.8). The main reason for this is linked to the fact that the pores in the material are the channels through which the water flows. In saturated soils all pores are filled with water, allowing the water to move. In unsaturated soils however not all the pores are filled with water. The air-filled pores are therefore not active in transporting water through the material. They can therefore be assumed to behave in a similar way as the solid phase. The permeability of unsaturated soils is therefore lower than in the same soil in a saturated state and decreases as the water content decreases or matric suction increases.

A number of methods exist to measure the unsaturated permeability of soils, both in the laboratory and in the field. As for saturated methods, they can be classified into steady or unsteady methods. In the laboratory the steady state method is recommended as it is relatively simple and has few ambiguities. However the method can be quite time consuming as the flow rate can be very low, especially under conditions of high matric suction. Further it can be difficult to measure the low flow rate accurately due to air diffusion. More recently a faster steady state method has been introduced where a centrifuge is introduced to drive the fluid flow (Nimmo et al., 1987; 1992). The unsteady laboratory methods, such as the thermal method, instantaneous profile method and the multi-step outflow method are usually much quicker than the traditional steady state method but are usually not as accurate. In the field the tension infiltrometer, instantaneous profile method and the cone penetrometer methods can be used. Benson & Gribb (1997) give a comprehensive overview of methods to measure the permeability of unsaturated soils.

The coefficient of permeability of unsaturated soils is not routinely measured in the laboratory as the process is cumbersome and quite time consuming (Fredlund 2006). The permeability of unsaturated soils can also be indirectly estimated from the SWCC. This is attractive as the SWCC can be determined in a much shorter time than the permeability's dependency on matric suction and with greater reliability (Rahardjo & Leong, 1997).

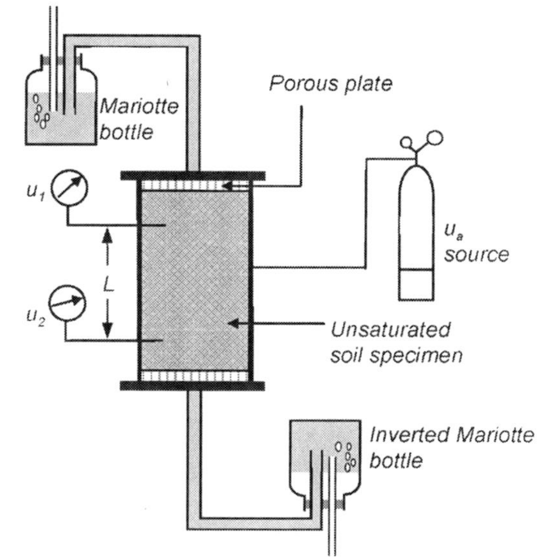

Fig. 3.11 Schematic set-up of steady state measurement of the permeability of an unsaturated soil specimen

In the steady state method, the unsaturated permeability is measured under conditions of a constant matric suction. A constant hydraulic head gradient is applied over an unsaturated soil sample with a constant matric suction to produce a steady state water flow through the specimen (see Fig. 3.11). A Mariotte bottle can be used to provide a constant pore water pressure to deliver a constant rate of flow. When the rates of water flow entering and leaving the sample are equal, the steady state has been reached and the coefficient of the permeability can be calculated according to Darcy's law (Eq. 2.15) as

$$v = -K\frac{h}{L} = K\frac{u_1 - u_2}{\rho_w g L} \tag{3.14}$$

where v is the flow rate of water through the sample, K is the coefficient of permeability and h/L is the hydraulic head gradient across the sample (with h the head difference and L the length of the sample). The head difference can be estimated from u_1 and u_2, the readings from the two pore water pressure sensors, converting pressure to head by dividing by the density of water, ρ_w and the acceleration due to gravity, g.

Now the test is repeated for different suctions in order to establish the relationship between the permeability and the suction. A typical measurement of permeability as a function of the matric suction is given in Fig. 3.12. As matric suction is related to water content through the SWCC, and if that relationship is known, the variation of permeability with water content is also known. Notice that, for the suction range illustrated, the coefficient of permeability changes by 6 orders of magnitude.

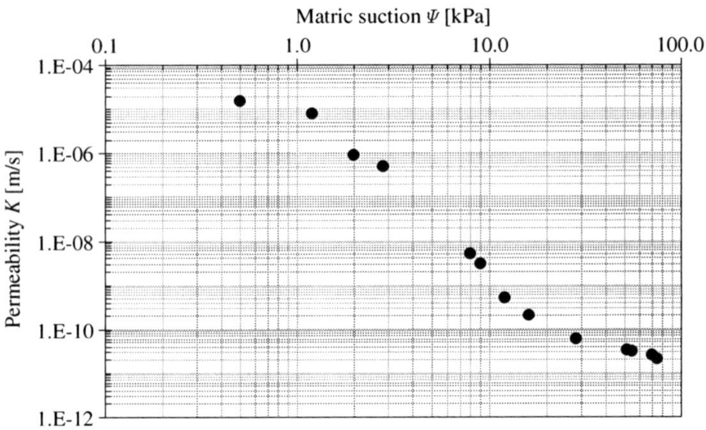

Fig. 3.12 Permeability as a function of matric suction in an unsaturated soil specimen

3.4 Suction

Soil suction or capillary pressure head can be measured either in the laboratory in an undisturbed sample of soil or directly in the field. Soil suction or total suction consists of the matric suction and the osmotic suction. Their magnitudes can range from 0 to 1 GPa (Rahardjo & Leong, 2006). Today no single instrument or technique exists that can measure the entire range with reasonable accuracy. Suction measurement instruments can only measure suction up to about 10 MPa. In the highway environment soil suction in the low range (0–100 kPa) or the mid range (100 kPa–1 MPa) is of most concern. There are different measurement techniques depending on which component of suction one wants to measure, matric or total. Usually in geotechnical engineering it is the matric suction that is measured. Table 3.2 summarises techniques for measuring suction in terms of approximate measuring range and applicability in the laboratory or field (Lu & Likos, 2004; Rahardjo & Leong, 2006). As it is generally not necessary to take osmotic suction into account in routine geotechnical engineering practice, only the main in-situ methods for measuring matric suction are referred to. They include tensiometers, thermal conductivity sensors and contact filter paper techniques.

Direct measurement of suction in aggregates is more difficult than in finer grained soils as it is difficult to establish an effective contact between the measuring device and the pore space in the aggregates. In aggregates with a high proportion of fines, this may be achievable. Alternatively, indirect measures such as discussed in Chapter 2, Section 2.9, can be employed.

3.4.1 Tensiometers

One of the most common devices for measuring suction is a tensiometer. A tensiometer consists of a fine porous ceramic cup connected by a tube to a vacuum

Table 3.2 Summary of common laboratory and field techniques for measuring soil suction. Based on Lu & Likos (2004) and Rahardjo & Leong (2006)

Technique/Sensor	Suction component measured	Measurement range (kPa)	Equilibrium time	Laboratory/Field
Tensiometers	Matric	0–100	Several minutes	Laboratory and field
Axis translation techniques	Matric	0–1500	Several hours–days	Laboratory
Electrical/ thermal conductivity sensors	Matric	0–1500	Several hours–days	Laboratory and field
Contact filter paper method	Matric	0–10000	2–5 days	Laboratory and field
Non contact filter paper method	Total	1000–10000	2–14 days	Laboratory and field

gauge (see Fig. 3.13). The entire device is filled with de-aired water. The porous tip is placed in intimate contact with the soil and the water flows through the porous cup (in or out) until the pressure inside the ceramic cup is in equilibrium with the pore water in the soil. The reading on the pressure measuring device, once corrected for the water column in the device, is the matric suction (Apul et al., 2002). The water pressure that can be measured by this method is limited to approximately −90 kPa, otherwise water will begin to boil inside the tensiometer ("cavitation"). Tensiometers have been found to provide the best measuring technique for low-range suction as they measure the pore pressures directly and respond promptly to pore water pressure changes (Rahardjo & Leong, 2006).

Lately, "high-capacity" tensiometers have been developed (e.g. Ridley & Burland, 1993; Guan & Fredlund, 1997; Tarantino and Mongiovi, 2001). When coupled with specialised operating procedures, for example, cyclic prepressurization techniques, they have been shown to be applicable for matric suction up to 1500 kPa. Comparisons with established measurement systems have shown high-capacity tensiometers to be relatively reliable and quite rapid in terms of response time.

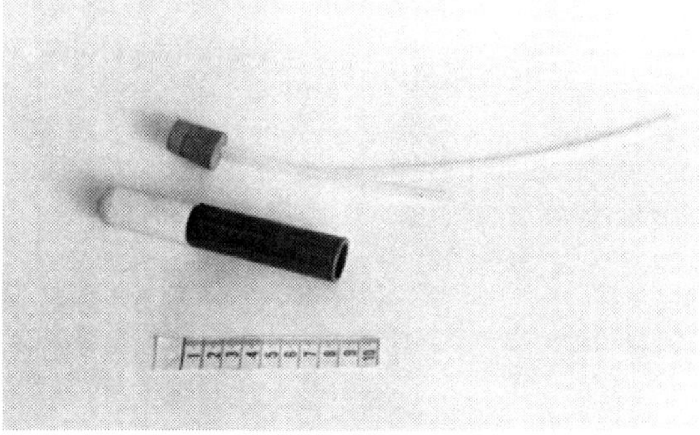

Fig. 3.13 Example of Tensiometer, consisting of ceramic cup and plug connected to tubes (Krarup, 1992). Reproduced with permission of the Road Directorate / Danish Road Institute

3.4.2 Thermal Conductivity Sensors

Thermal conductivity sensors (TCS) are used to indirectly relate matric suction to the thermal conductivity of a porous medium embedded in a mass of unsaturated soil. Any change in the soil suction results in a corresponding change in the water content of the porous medium (governed by its characteristic curve). The thermal conductivity of a rigid porous medium is a direct function of the water content. Therefore, if the thermal conductivity of the porous medium is measured, the matric suction of the soil may be indirectly determined by correlation with a predetermined calibration curve. Figure 3.14 shows the main components of a modern commercial TCS.

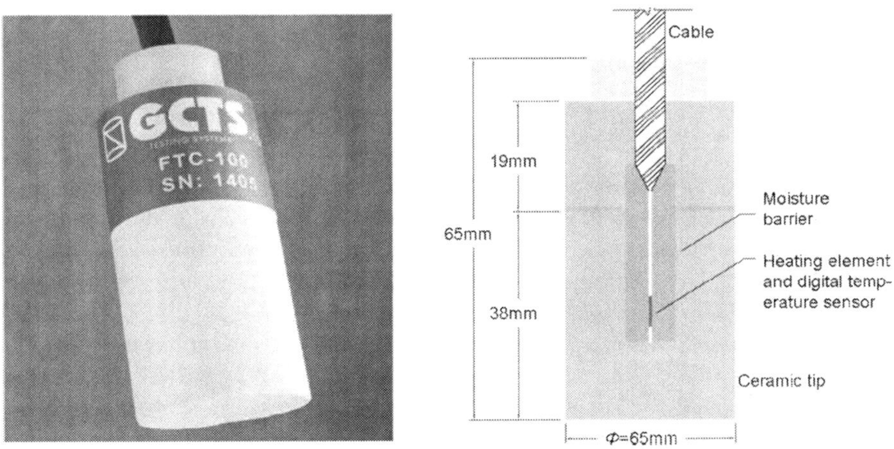

Fig. 3.14 Picture and schema of a thermal conductivity matric suction sensor (TCS). Reproduced with permission from GCTS Testing Services

Some disadvantages usually imputed to the old-fashioned TCS included the problems associated with drift, and, for many sensors, deterioration in the sensor body over time, as well as uncertainties concerning the drying and rewetting processes due to hysteretic effects in the sensor calibration. But most of these problems have been resolved. The major advantages of the more recent TCS include the relative ease with which the sensors may be set up for automated data acquisition, their relatively low cost and their present capability to measure matric suction over a wide range (0–1200 kPa). Figure 3.15 shows some laboratory soil suction measurements carried out with both tensiometers and TCS, where tensiometer failures due to cavitation were noticeable. For further details regarding TCS see Rahardjo & Leong (2006).

3.4.3 Suction Plate

A simple laboratory variant of the tensiometer method for measuring matric suction of fine-grained soils uses a semi-pervious sintered glass plate. A small soil sample

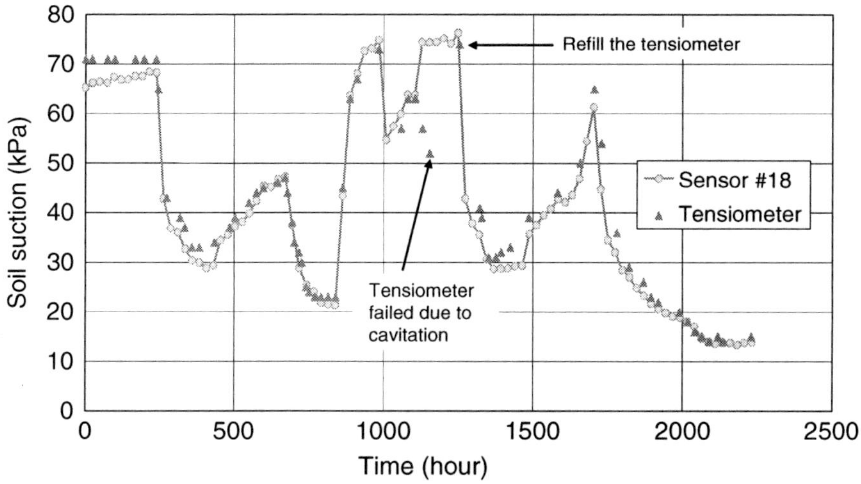

Fig. 3.15 Comparison between laboratory soil suction measurements with tensiometers and TCS. Reproduced with permission of D. Fredlund

is placed on the glass plate and covered immediately with a cap so that the vapour pressure around the soil comes to equilibrium with the suction in the soil, preventing drying. On the other side of the plate de-aired water is provided in a small chamber. The soil attempts to suck water across the glass plate from the chamber but the only way this can happen is by water being drawn into the chamber via a narrow-bore tube. In the tube, beyond the water, there is a small length of mercury and beyond that, air and a small hand-operated suction pump and vacuum gauge (Fig. 3.16). As the water (and, hence, the mercury) is pulled towards the soil, the operator applies a partial vacuum to oppose this and to keep the mercury in the same position. Once the operator no longer needs to apply additional suction, the suction in the soil specimen and applied by the operator are in equilibrium and the suction value may be read from the gauge. Typically, readings may be obtained in around 15 min (though this depends a lot on the soil type). The device is limited to measuring in the range from atmospheric pressure to approximately 70 kPa of suction.

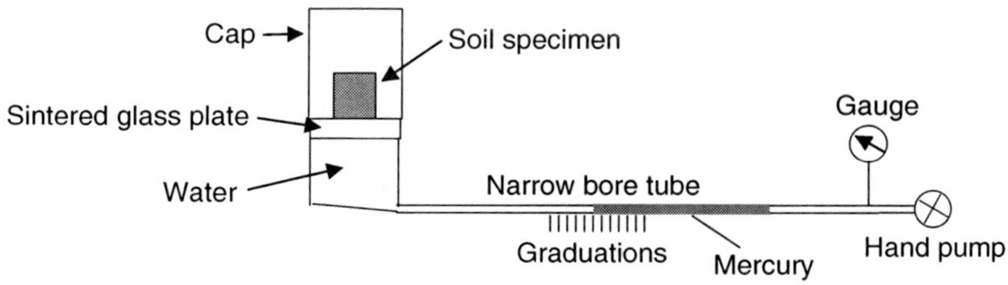

Fig. 3.16 Suction plate device

3.4.4 Contact Filter Paper Techniques

Although far less common than in the laboratory, techniques for in-situ matric suction measurements using the contact filter paper (CFP) method and in-situ total suction measurements using the non-contact filter paper method have also been described (Greacen et al., 1989). The filter paper technique is, in theory, applicable over the entire range of total suction, but the method tends to be impractical for both extremely high and extremely low suction values. Reliable measurements tend to be limited to a range spanning about 0–10 MPa matric suction for the contact filter paper techniques and between 1 to 10 MPa of total suction for the non-contact technique.

The filter paper method is used as an indirect means of measuring soil suction. The advantages of the method include its simplicity, its low cost, and its ability to measure a wide range of suction. Although this technique is more often used in the laboratory, the filter paper method has also been used in the field to measure soil suction. The CFP technique relies on measuring the equilibrium water content of small filter papers in direct contact with unsaturated soil specimens. Figure 3.17 shows the filter paper setup and installation to put it in direct contact with the soil specimen. In the laboratory the filter paper is placed in contact with the soil specimen in an airtight container for seven days and thereafter the water content of the filter paper is determined and the matric suction of the soil specimen is inferred from a calibration curve. Filter paper not in contact with the specimen permits water exchange only in the vapour phase and therefore measures the total suction (Rahardjo & Leong, 2006).

The water content of the filter paper at equilibrium is measured gravimetrically and related to matric soil suction through a predetermined calibration curve for the particular type of paper used. Commonly used types of papers include Whatman No. 42 and Schleicher and Schuell No. 589. Calibration and test procedures for the measurement of matric suction using the contact filter paper technique are described in the ASTM Standard D5298-94 (1997). However, different researchers have suggested different calibration curves for the same filter paper (Leong et al., 2002; Rahardjo and Leong, 2006).

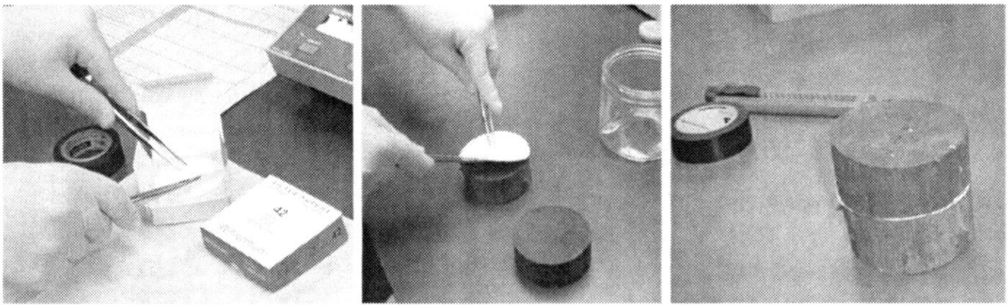

Fig. 3.17 Contact filter paper setup procedure (in laboratory). Reproduced with permission of R. Bulut

3.5 Conclusions

A considerable variety of test and assessment procedures are available for measuring the volumetric and gravimetric water contents of both laboratory and in-situ road construction and geotechnical materials. The simplest tests to perform are usually destructive, but sophisticated geo-physical techniques are becoming increasingly common and usable, not only as identification tools, but also as quantitative measurement techniques.

Suction, which has such a large effect on the mechanical properties of soils and aggregates, is probably the quantity most difficult to measure successfully and must usually be monitored indirectly by the response of, e.g., water content and vapour monitoring. As the relationships between these secondary responses and the primary cause, suction, may both be imprecisely described and hysteretic there is usually some uncertainty in value of suction determined.

Permeability, another major quantity that needs evaluating, is more readily measured using flow tests, but difficulties arise when measuring coarse-grained materials, such as road aggregates. Producing samples that are representative with respect to density and grading can be a challenge and the devices available for testing can allow water to preferentially flow along the edges, introducing further uncertainties into the assessment.

Nevertheless, a knowledge of permeability, suction and water content is indispensable for effective design and assessment of the movement of water in the highway and its adjacent environment.

References

Apul, D. S., Gardner, K., Eighmy, T., Benoit, J. & Brannaka, L. 2002, *A Review of Water Movement in the Highway Environment: Implications for Recycled Materials Use*", Recycled Materials Resource Center, Environmental Technology Building, University of New Hampshire, Durham, 104 pp.

ASTM, 1997, "Standard test method for measurement of soil potential (suction) using filter paper", Standard D5298-94, *Annual Book of ASTM Standards*", Vol. 04.09, Soil and Rock (II): D4943-latest, pp. 157–162.

Benson, C. H. & Gribb, M. M., 1997, "Measuring unsaturated hydraulic conductivity in the laboratory and field, *Proceedings of the Unsaturated Soil Engineering Practice*", Logan, Utah, pp. 113–168.

Department of Transport, 1990, "A permeameter for road drainage layers", HA 41/90, *Design Manual for Roads And Bridges*, **4**(2), April, London.

Ekblad, J., 2004, "*Influence of Water on Resilient Properties of Course Granular Materials*", Licentiate Thesis, Division of Highway Engineering, Department of Civil and Architectural Engineering, Royal Institute of Technology, Stockholm, 107 pp.

Erlingsson, S., Bjarnason, G. & Thorisson, V., 2000, "Seasonal variation of moisture and bearing capacity on Icelandic test road sections", *Fifth International Symposium on Unbound Aggregates in Roads* (UNBAR 5), Nottingham, UK, A.A. Balkema, Rotterdam.

Erlingsson, S., Bjarnason, G. & Thorisson, V., 2002, "Seasonal variation of moisture and bearing capacity in roads with a thin surface dressing wearing course", *Proceedings from the 9th International Conference on Asphalt Pavements*, Copenhagen CD-ROM.

Fredlund, D. G., 2006, "Unsaturated soil mechanics in engineering practice", *J. Geotechn. Geoenviron. Eng.* **132**(3), pp. 286–321.

Fredlund, D. G., 1997, "An introduction to unsaturated soil mechanics", *Proceedings of the Unsaturated Soil Engineering Practice*, Logan, Utah, pp. 1–37.

Fredlund, D. G. & Rahardjo, H., 1993, "*Soil Mechanics for Unsaturated Soils*", John Wiley & sons, New York.

Greacen, E. L., Walker, G. R. & Cook, P. G., 1989, "Procedure for the filter paper method of measuring soil water suction", CSIRO Division of Soils, Div'l. Rep. No. 108, 7pp.

Grote, K., Hubbard, S. & Rubin, Y., 2003, "Field scale estimation of volumetric water content using ground penetrating radar ground wave techniques", *Water Resources Research*, **39**(11) pp. SBH5.1–SBH5.13.

Guan, Y & Fredlund, D. G., 1997, "Direct measurements of high soil suction", *Proceedings 3rd Brazilian Symp. on Unsaturated Soils*, NSAT'97, **2**, Rio de Janeiro, Brazil, pp. 543–550.

Head, K. H., 1982, "*Manual of Soil Laboratory Testing Volume 2: Permeability, Shear Strength and Compressibility Tests*", Pentech Press, London.

Hignet, C. & Evett, S. R., 2002, "Neutron Thermalization", In: Dane, J. H. and Clark, G. C. (eds.), *Methods of Soil Analysis: Part 4 – Physical Methods*, Soil Sci. Soc. Amer., Madison, Wisconsin, pp. 501–521.

Hillel, D., 1998, "*Environmetal Soil Physics*", Academic Press, San Diego, CA.

Jiang, Y. J. & Tayabji, S. D., 1999, "Evaluation of in situ moisture content at long-term pavement performance seasonal monitoring program sites", *Transportation Research Record*, **1655**, pp. 118–126.

Krarup, J., 1992, "*Bearing Capacity and Water, Part 1: Materials, Construction and Instrumentation*", IVTB, Institute of Roads, Transport and Town Planning, Technical University of Denmark and Danish Road Institute.

Leong, E. C., He, L., & Rahardjo, H., 2002, "Factors affecting the filter paper method for total and matric suction measurements", *Can. Geotech. J.*, **25**(3), pp. 321–332.

Lu, N. & Likos, W. J., 2004, *Unsaturated Soil Mechanics*", Wiley, New York.

Nimmo, J., Rubin, J. & Hammermeister, D., 1987, "Unsaturated flow in a centrifugal field: measurement of hydraulic conductivity and testing of Darcy's Law," *Wat. Res. Res.*, **23**(1), pp. 124–134.

Nimmo, J., Akstin, K. & Mello, K., 1992, "Improved apparatus for measuring hydraulic conductivity at low water content, *Soil Sci. Amer. J.*, **56**, pp. 1758–1761.

O'Connor, K. M. & Dowding, C. H., 1999, "*GeoMeasurements by Pulsing TDR Cables and Probes*", CRC Press, Boca Raton, Florida.

Rahardjo, H. & Leong, E. C., 1997, "Soil-water characteristic curves and flux boundary problems, *Proc. of the Unsaturated Soil Engineering Practice*", Logan, Utah, pp. 88–112.

Rahardjo, H. & Leong, E. C., 2006, "Suction Measurements, *Proceedings of the 4th International Conf. on Unsaturated Soils*", Carefree, Arizona, pp. 81–104.

Richards, L. A., 1931, "Capillary conduction of liquids through porous medium", *J. Phys.*, **1**, pp. 318–333.

Ridley, A. M. & Burland, J. B., 1993, "A new instrument for the measurement of soil moisture suction", *Geotechnique*, **43**(2), pp. 321–324

Roth, C. H., Malicki, M. A. & Plagge, R., 1992, "Empirical evaluation of the relationship between soil dielectric constant and volumetric water content as the basis for calibrating soil moisture measurements by TDR", *J. Soil Science*, **43**, pp. 1–13.

Svensson, J., 1997, "*Moisture Content in Road Pavements. State of the art and development of a simple moisture measurement equipment*", VTI report 809A, Linköping, Sweden.

Starr, J.L. & Paltineanu, I.C., 2002, "Capacitance Devices", in: Dane, J.H. and Clark, G.C. (ed.), *Methods of Soil Analysis: Part 4 – Physical Methods*", Soil Science Society of America, Madison, Wisconsin, pp. 463–474.

Tarantino, A. & Mongiovi, L., 2001, "Experimental procedures and cavitation mechanisms in tensiometer measurements, unsaturated concept and their application in geotechnical practice", *Geotech. Geological Eng.* **19**(3), pp. 189–210.

Topp, G. C. & Davis, J. L., 1985, "Measurement of soil water using time-domain reflectometry (TDR): Field evaluation", *Soil Sci. Soc. Am. J.*, **49**, pp. 19–24.

Topp, G. C., Davis, J. L. & Annan, A. P., 1980, "Electromagnetic Determination of Soil Water Content: Measurements in Coaxial Transmission Lines", *Water Resources Research*, **16**, pp. 574–582.

Veenstra, M., White, D. J. & Schaefer,V. R., 2005, "*Synthesis of Nondestructive Testing Technologies for Geomaterial Applications*", Report no. CTRE Project 03-146, Iowa State University.

Chapter 4
Heat Transfer in Soils

Åke Hermansson*, Robert Charlier, Frédéric Collin, Sigurður Erlingsson, Lyesse Laloui and Mate Sršen

Abstract Temperature highly affects pavement performance. High and low temperatures not only affects the viscosity of asphalt concrete but also has an impact on the moisture flow within pavements. At temperatures below 0°C the freezing of pavements dramatically changes the permeability and frost action might occur forcing water to flow upwards to the freezing front resulting in frost heave and other pavement distress.

Keywords Heat transfer · conduction · temperature · frost

4.1 Introduction

The thermal state may have a major influence on the moisture condition of a pavement or foundation. Thermal gradients due to temperature changes on the surface will induce not only heat flow in the pavement but also moisture flow.

Freezing and thawing are definitely the most important aspects linking temperature to water flow. Furthermore, a freezing temperature significantly reduces the permeability of soils but also increases the moisture flow caused by hydraulic gradients due to ice lens formation in the frozen soil.

Moreover, the water viscosity depends on the temperature; at higher temperature, some water will flow in the vapour phase, and this depends on the temperature gradient. Heat flow and moisture flow are, therefore, linked processes with complex interaction between them. This chapter will describe the basis of heat transfer laws and models.

4.2 Basic Principles of Heat Transfer

The basic principles to model the complete time-dependent heat transfer in soils are described in this section. More details of these and the associated effects can be

---

* *Co-ordinating Author:*
✉ Å. Hermansson
Väg-och Transportforskningsinstitut / Swedish Road and Traffic Institute (VTI)
e-mail: ake.hermansson@vti.se

A.R. Dawson (ed.), *Water in Road Structures*, DOI 10.1007/978-1-4020-8562-8_4,
© Springer Science+Business Media B.V. 2009

found in some of the better or more specialist geotechnical textbooks, for example Mitchell and Soga (2005) or Fredlund and Rahardjo (1993).

The water content in various road structures and the underlying soil are subject to climatic (temperature) effects. An example of the thermal variation in a road structure is given in Fig. 4.1. It can be clearly seen that the thermal state is definitely changing, which means that it is characterised by the heat transfer properties of the material. Temperatures may be positive or negative (inducing freezing). Heat transfer in soils is due to conduction, radiation, convection and vapour diffusion. A general overview of thermal transfer may be found in many textbooks, e.g. Selvadurai (2000) or Lewis and Schrefler (1998).

As pavement surface temperature depends greatly on the weather, typically changing hourly and daily, the physical process never reaches a steady state (i.e. there would be only a long-term equilibrium).

Geostructures, like road pavements and embankments, are made up of porous materials with solid and fluid phases. At micro-scale, i.e. the grains and pores scale, the heat transfer is highly complex, involving convection and radiation in the pores

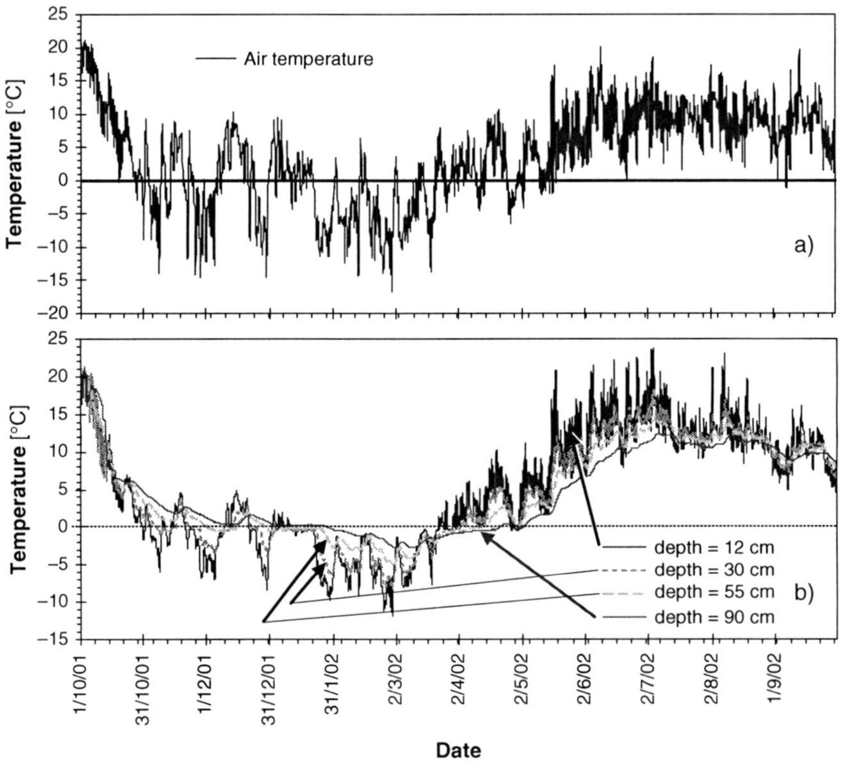

Fig. 4.1 Hourly measurement during one year in SW Iceland (**a**) of air temperature and (**b**) at four depths in a pavement structure

and conduction in the grains. At the engineering scale (i.e. at macro-scale, at the layers thickness scale), the solid, liquid and gas phases may be considered separately as a continuum.

As discussed by Selvadurai (2000), it is known by experience that heat flows from points of higher temperature to points of lower temperature. Heat is, therefore, a form of energy, which is released by warmer regions of a body to the cooler regions.

Although heat flow cannot be measured directly, its presence manifests itself in terms of a measurable scalar quantity, which is the temperature. From the knowledge of the temperature distribution $T(x,t)$ within a region, it is possible to calculate the heat flow within the region. As a consequence, the major part of the study of heat transfer in geostructures focuses on the determination of the temperature distribution within the medium which is subjected to appropriate boundary conditions and initial conditions applicable to $T(x,t)$.

4.2.1 Conduction

Conduction refers to the mode of heat transfer where this energy transport takes place in the solid phase of the porous media and fluids, which are at rest. This is in contrast to convective heat transfer, which involves motion of a conducting fluid.

The grains of a soil, typically, are in contact with each other at distinct contact points and pores between grains are filled with a mixture of air and water. When completely dry the heat flow passes mainly through the grains, but has to bridge the air-filled gaps around the contact points. At very low water contents, thin adsorbed water layers cover the grains. The thickness of these layers increases with increasing water content as introduced in Chapter 2. At higher water contents, rings of liquid water form around the contact points between the grains. From this point on, the thermal conductivity increases rapidly with increasing water content. At even higher water contents, complete pores are filled with water resulting in a further but slower increase in thermal conductivity.

For a conducting medium which possesses homogeneity and isotropy with respect to its heat conduction characteristics (all parameters which govern the heat conduction process are assumed to be independent of direction and position), it is assumed that the amount of heat crossing an element of material in a given time, Δt, is proportional to the difference in temperature, T, and related to the material properties. Generalizing this concept leads to Fourier's equation that defines a vector of heat flow $\underline{Q}$ (W/m^2):

$$\underline{Q} = -\lambda \nabla T = -\lambda \mathbf{grad} T \tag{4.1}$$

where λ(W/m°C) is defined as the thermal conductivity of the medium (see Section 4.3). Normally, the thermal conductivity of grains is in the range of 1–3 W/m°C. Water has a thermal conductivity of about 0.6 W/m°C at room temperature. In comparison with solids and liquids, the thermal conductivity of air is very small, being about 0.025 W/m°C.

For anisotropic materials like soils, Eq. 4.1 is generalized in matrix form:

$$\mathbf{Q} = -\boldsymbol{\lambda}\nabla T = -\boldsymbol{\lambda}\,\mathbf{grad}T \tag{4.2}$$

where $\boldsymbol{\lambda}$ is the matrix of thermal conductivities.

4.2.2 Radiation

Radiation is a process by which energy moves through a medium or vacuum without the movement of any molecules and without heating any medium through which it passes. The quantity of energy radiated from a grain surface increases with increased surface temperature and neighbouring grains increase their temperature by absorbing the radiation emitted. Because the higher-temperature grains radiate more energy, radiation results in a net transfer of energy to the lower-temperature grains. For coarse dry soils heat conduction is low and 10–20% of heat transfer can be due to radiation. Generally though, radiation plays a negligible role for heat transfer in soils.

4.2.3 Vapour Diffusion

Vapour moves towards a lower vapour pressure by a molecular process known as diffusion (fundamentally the same process that redistributes contaminants in static water as described in Chapter 6, Section 6.3.1.). There is no need for there to be a conventional pressure differential, only a difference in the concentration of the vapour. Vapour pressure decreases with decreasing temperature and with decreasing relative humidity. That is, cold and dry areas attract vapour. Typically, at micro-scale, water evaporates at the warm end of a pore and condenses at the cold end, thereby transferring latent heat from the warm to the cold end of the pore.

4.2.4 Convection

Convection is energy transfer by macroscopic motion of fluid (liquid and gas) particles. The fluid motion is the result of a force. The force may be due to a density gradient or due to a pressure difference generated by, for example, a pump, by gravity or induced differences of density. The moving particles bring their high or low energy with them. In soils, typically, the moving particles which transport heat are water molecules. Density gradients may be due to a gradient of water or of salt content. Because the density of water varies with temperature, the density gradient may be due to temperature gradient only.

4.2.5 Relative Importance of the Different Mechanisms of Heat Transfer in Soils

4.2.5.1 Temperatures Below 0°C

The transfer of heat by conduction is the dominating factor at temperatures below 0°C (Sundberg, 1988). In the small pores of frost susceptible soils though, due to freezing point depression, some water remains unfrozen at temperatures below 0°C. This allows convection caused by so-called cryo-suction effects (see Section 4.6.2 below) and a small amount of heat transfer at temperatures below 0°C.

4.2.5.2 Temperatures Above 0°C and Below Approximately 25°C

At this temperature range, conduction of heat is still the dominating factor (Sundberg, 1988). In highly permeable soils there may be more forced convection – like groundwater flow that is natural or caused by water abstraction. High temperature gradients in permeable soils may also cause significant natural convection.

4.2.5.3 Temperatures Above Approximately 25°C

For the lower temperatures in this range, conduction is still the dominating factor (Sundberg, 1988). At higher temperatures and relatively low water content, vapour transport gets successively more important. At saturation, heat conduction is always the dominating factor. Again, high temperature gradients in permeable soils may cause more natural convection. In coarse soils at high temperature, there will be more radiation but it will still play a minor role.

4.2.6 Conclusions Concerning Heat Transfer

For pavements, conduction of heat is the most important factor for heat transfer. During warm and sunny summer days though, the temperature of a pavement base layer under a thin asphalt concrete, may reach high values and natural convection in a fairly permeable base layer should not, then, be neglected.

4.3 Thermal Conductivity, λ

Mineral content, porosity, degree of water saturation and temperature affect the thermal conductivity of soils. The total conductivity is a function of the conductivity of each soil phase, solid grains, water and gas. Various equations for these mixtures have been proposed by Keey (1992) and Krischer (1963). Thermal conductivity values range between 1 and 4 W/m°C for saturated soils, and from 0.2 to 0.4 W/m°C for dry soils.

4.3.1 Mineral Content

Because thermal conductivity of quartz is 3–4 times higher than that of other minerals the quartz content of a soil greatly affects the thermal conductivity. Typically, cohesive soils have a low quartz content while the quartz content of a fine sand is normally high.

4.3.2 Porosity, *n*

Because the thermal conductivity of minerals is much higher than that of water and air, thermal conductivity of soil decreases with increasing porosity.

4.3.3 Degree of Water Saturation, S_r

The thermal conductivity of air in a soil or aggregate's pores is negligible but the conductivity increases with increasing degree of water saturation.

Fine soils generally have a high porosity and a low quartz content and, consequently, the thermal conductivity of dry clay and silt is low. However, the fine pores of these soils more easily hold a higher amount of water and fine soils typically deliver thermal conductivities in the same range as other soils.

4.3.4 Temperature, *T*

The thermal conductivity of ice is four times higher than that of water and consequently the thermal conductivity of soils with a high degree of water saturation increases dramatically at or below freezing. Here it should be kept in mind, that fine soils at temperatures below 0°C can still hold a large amount of unfrozen water and that thermal conductivity increases, therefore, progressively with decreasing temperature. Coarse soils typically have low degrees of water saturation and thermal conductivity does not increase significantly at freezing.

A typical relationship between thermal conductivity and water and ice content is shown in Fig. 4.2. The relationships shown assume only ice or only water, respectively.

4.4 Thermal Capacity, *c*

Thermal capacity characterises the ability of a material to store or release heat. It is the important property that relates to the delay in heat transfer. The thermal capacity of water is approximately twice as high as that for most minerals and for ice, while the thermal capacity of air is negligible.

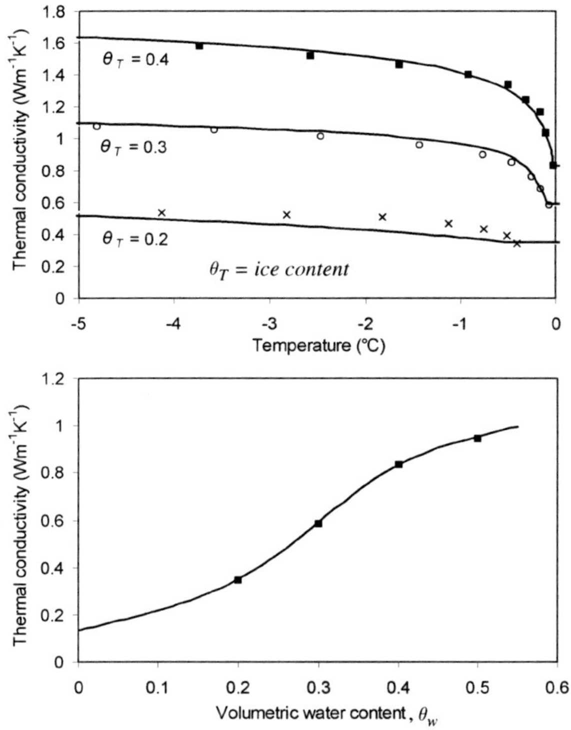

Fig. 4.2 Typical relationships between thermal conductivity and ice content (*top*) and between thermal conductivity and water content (*bottom*) (Hansson et al., 2004). Credit: the Vadose Zone Journal, published by the Soil Science Society of America

The thermal capacity of saturated soils ranges between 800 and 1000 J/(kg°C) – that is between 2000 and 2400 J/(m^3°C) – while dry soils exhibit values of between 300 and 1600 J/(m^3°C).

4.5 Thermal Diffusivity, α

The thermal diffusivity, α (m^2/s), is the ratio between thermal conductivity (λ) and thermal capacity (c):

$$\alpha = \lambda / c \tag{4.3}$$

It, thus, measures the ability of a material to conduct thermal energy relative to its ability to store thermal energy. Soils of large α will respond quickly to changes in their thermal environment, while materials of small α will respond more sluggishly. From a physical point of view the thermal diffusion of a medium is indicative of the speed of propagation of the heat into the body during temperature changes. The higher the value of α, the faster propagation of heat within the medium.

For example, during sunny days the pavement surface temperature will show strong daily oscillation and in soils and pavement materials with a high thermal diffusivity this oscillation penetrates to a greater depth.

4.6 Physics of Frost Heave

4.6.1 Frost Heave and Spring Thaw

Frost heave occurs in roads having fine graded, so-called frost-susceptible, material, at a depth to which the freezing front reaches during the winter. A well-built road of consistent materials and cross-section can be expected to heave relatively evenly. When inconsistencies or inhomogeneities are found in the construction of the affected subgrade, then frost heave is likely to be uneven and may well cause an uneven road surface that results in reduced travelling speed and comfort.

Although such heave can be problematic, a much greater problem usually arises in spring-time when the ice that has formed in the road construction, which was instrumental in causing the frost heave, melts and results in a very high water content in the pavement and subgrade. The increased water content often means reduced bearing capacity. For this reason many countries impose spring-thaw load restrictions on low volume roads to avoid severe pavement deterioration.

4.6.2 Ice Lenses

Frost heaving of soil is caused by crystallization of ice within the larger soil voids. Ice lenses attract water to themselves by the, so-called, cryo-suction process, and grow in thickness in the direction of heat transfer until the water supply is depleted or until freezing conditions at the freezing interface no longer support further crystallization, see Fig. 4.3. As the freezing front penetrates deeper into the pavement, the growth of ice lenses ceases at the previous level and commences at the new level of the freezing front. At some point the heat flow will be reduced so that further freezing slows or the weight of overlying construction will impose sufficient stress to prevent further ice lens growth.

Fundamentally, the so-called cryo-suction process results from the consequences of the phase change of pore water into pore ice and the associated energy changes in the remaining pore water. As water arrives at the point of freezing, the soil skeleton expands to accommodate it while consolidation of the adjacent, un-frozen, soil may occur as water is pulled from that. The particular characteristics of the process are strongly affected by soil porosity, soil-water chemistry, stress conditions at the point of ice formation, temperature, temperature gradients in the adjacent soil(s), water availability, etc. Coussy (2005) gives a more detailed outline of the mass flow and heat flow elements that combine in the processes that are active at the freezing front.

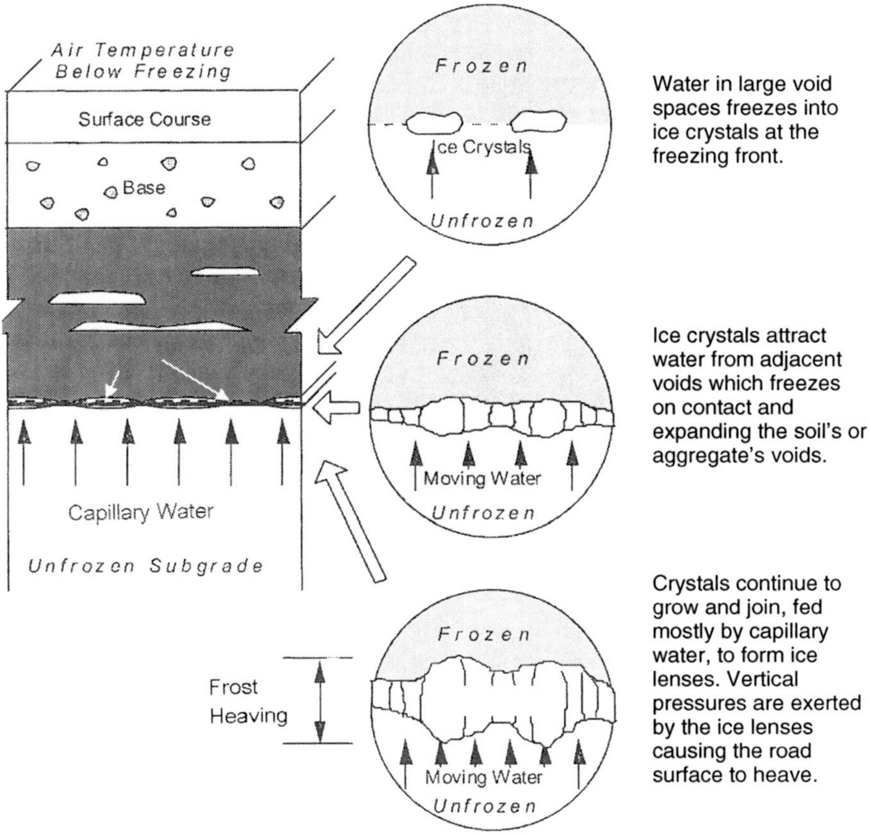

Fig. 4.3 Formation of ice lenses in a pavement structure (WSDOT, 1998)

Capillary water is crucial for frost heaving. The pavement damage range depends on the rapidity of freezing, i.e. if freezing occurs rapidly, ice lenses are distributed over a greater mass of soil, which is somewhat more favourable compared to slow freezing where the capillary inflow of water will cause a high concentration of ice lenses. Frost heave primarily occurs in soils containing fine particles (i.e. frost-susceptible soils). Clean sands and gravel are non-frost susceptible because they cannot hold significant pore suctions, so water cannot be drawn to the freezing front through them. Silty soils represent the greatest problem. For even finer soils, like clays, the pores are very small. These small pores lead to a low coefficient of permeability which does not allow water to travel through the pores at a speed that would allow a fast capillary rise, hence ice lens formation is less.

4.6.3 Recent Research

Researchers interested in frost action in soils agree on the description illustrated in Fig. 4.3 on how ice lenses grow and cause frost heave. Nevertheless, when it comes to the degree of water saturation of the unfrozen soil below the freezing front there are two different conceptions. Some researchers believe that the unfrozen soil is fully saturated while others believe it is unsaturated. Of course, these discrepancies in understanding lead to different explanations of the driving force of the capillary rise of water as well as different opinions on how to run laboratory experiments. Andersland and Ladanyi (2004), for example, Konrad and Morgenstern (1980) and Nixon (1991) give equations where it is obvious that full saturation is assumed for the unfrozen soil. Accordingly, Konrad (1990) refers to experiments where the specimen freezes from below and free access to water is permitted at the top. This, of course, gives full saturation of the unfrozen soil. In contrast to this view, Miller (1980) discusses frost heave as freezing of unsaturated soil and references experiments where water is fed in at the bottom of the specimen and ice lenses are fed through capillary rise in unsaturated soil. Accordingly, Penner (1957) freezes unsaturated soil and Hermansson and Guthrie (2005) present laboratory experiments where freezing and frost heave takes place at a height more than 0.5 m above the level of the water supply. It should also be noted that Hermansson and Guthrie (2005) describe testing where the specimen heaves significantly without addition of any external water at all. This, of course, contradicts the assumption that the soil below the freezing front is fully saturated. The expansion without addition of water is suggested to be an effect of air entering the soil.

In agreement with the laboratory experiments Hermansson (2004) described a field study where the depth to the groundwater table is 6 m. Under such a thoroughly drained condition, it is reasonable to assume that the soil is far from saturated. Despite this Hermansson reported 80 mm of frost heaving over a period of 2 months.

The conclusions from these studies are twofold,

- Firstly, frost heave does not require full saturation; and
- Secondly, even a well drained soil might experience a significant frost heave.

In addition to the different understandings about the importance of saturation, there are also two different schools when proposing equations to describe frost heave (Hansson, 2005). One school neglects the liquid water pressure and the other one neglects ice pressure. The first school, characterised by "Miller-type" models, develops models describing the frost heave on a microscopic scale while the second school, characterised by "hydrodynamic" models, handles equations for the redistribution of water up to the freezing front, supplying the frost heave. No computer code is known that handles both processes realistically.

4.7 Conclusions

Heat transfer in soils involves convection, radiation, vapour diffusion and conduction. For pavements, conduction is the most important factor. During warm and sunny summer days though, natural convection should not be neglected. The heat transfer is closely associated with water movements – evaporation pulls water through the soil to the evaporation surface. Freezing also drives water movement as water is drawn to the freezing front in soils which have moderate pore sizes and moderate permeability.

Frost susceptible soils always experience frost heave at freezing even if there is no saturation. Drains will lower the heave by reducing the water content but a frost susceptible soil will always hold enough water for a significant heave. Chapter 13 describes some drainage techniques that can help to address these problems. Interested readers are also directed to the reports on frost and drainage, mostly in the context of seasonally frost affected roads, available from the ROADEX project (Berntsen & Saarenketo, 2005; Saarenketo & Aho, 2005).

References

Andersland, O. B. & Ladanyi, B., 2004, "*An Introduction to Frozen Ground Engineering*", ASCE & John Wiley & Sons, 363 pp.

Berntsen, G. & Saarenketo, T., 2005, "*Drainage on Low Traffic Volume Roads*", Roadex II Project Report, Highland Regional Council, available from http://www.roadex.org/Publications/pub-RII.htm.

Coussy, O., 2005, "Poromechanics of freezing materials", *J. Mech. Phys. Solids*, **53**(8), 1689–1718.

Fredlund, D. G. and Rahardjo, H., 1993, "*Soil Mechanics for Unsaturated Soils*", Wiley, 544pp.

Hansson, K., 2005, "*Water and Heat Transport in Road Structures, Development of Mechanistic Models*", Ph. D. Thesis, Uppsala Universitet. Geotryckeriet, Uppsala, Sweden.

Hansson, K., Šimůnek, J., Mizoguchi, M., Lundin, L. C. & van Genuchten, M. Th., 2004, "Water Flow and Heat Transport in Frozen Soil – Numerical Solution and Freeze-Thaw Applications", *Vadose Zone J'nl.*, **2**(3), 693–704.

Hermansson, Å., 2004, "Laboratory and Field Testing on Rate of Frost Heave versus Heat Extraction". *Cold Reg. Sci. Technol.*, **38**, 137–151.

Hermansson, Å. & Guthrie, S., 2005, "Frost Heave and Water Uptake Rates in Silty Soil Subject to Variable Water Table Height During Freezing", *Cold Reg. Sci. Technol.*, **43**, 128–139.

Keey, B. B., 1992, "*Drying of loose and particulate materials*", Hemisphere Publ. Corp., 504 pp.

Konrad, J.-M., 1990, "Segregation potential – pressure – salinity relationships near thermal steady state for a clayey silt", *Can. Geotech. J.*, **27**, 203–215.

Konrad, J.-M. & Morgenstern N. R., 1980, "A mechanistic theory of ice lens formation in fine-graded soil". *Can. Geotech. J.*, **18**, 482–91.

Krischer, O., 1963, "*Die wissenschaftlichen Grundlagen der Trocknungstechnik*". 2nd ed'n, Springer-Verlag, pp. 103–105.

Lewis R. & Schrefler B., 1998, "The finite element method in the static and dynamic deformation and consolidation of porous media", 2nd ed'n. Wiley, 508 pp.

Miller, R. D., 1980, "*Freezing phenomena in soils. Applications of soil physics*", Hillel, Daniel, New York, pp. 254–299.

Mitchell, J. K. & Soga, K., 2005, "*Fundamentals of Soil Behavior*", 3rd ed'n., Wiley, 577pp.

Nixon, J. F., 1991, "Discrete ice lens theory for ice lens in soils", *Can. Geotech. J.*, **28**(6), 843–859.

Penner, E., 1957, "*Soil Moisture Tension and Ice Segregation*", Bulletin No.168. Highway Research Board, pp. 50–64.

Saarenketo, T. & Aho, S., 2005, "*Managing Spring Thaw Weakening on Low Volume Roads*", Roadex II Project Report, Highland Regional Council, available from http://www.roadex.org/Publications/pub-RII.htm.

Selvadurai, A.P.S., 2000, "*Partial Differential Equations in Mechanics*", Springer, 698 pp.

Sundberg, J., 1988, "*Thermal properties of soils and rocks*", Chalmers Tekniska Högskola, Geologiska Institutionen, Thesis, Publ. A57, Gothenburg, Sweden.

WSDOT, 1998, "*Pavement Guide*", **2**, Washington State Dept. of Transportation. (Section 2.0, Fundamental Design Parameters, Part 3.1).

Chapter 5
Water in the Pavement Surfacing

Andrew Dawson*, **Niki Kringos, Tom Scarpas and Primož Pavšič**

Abstract Pavement surfaces provide a key route of ingress of rain water into the pavement construction. Thus, permeability of asphaltic materials and the water ingress capacity of cracks in the pavement are very important. A range of equipment exists to determine the permeability of asphaltic mixtures both by *in-situ* and laboratory testing. Sometimes porous asphalt surfacing is provided to deliberately allow water into the pavement to limit spray from vehicles and to limit tyre-pavement noise generation. These porous surfaces can become clogged with fines, but rehabilitating without causing premature damage is a challenge. Except for this planned acceptance of water into the pavement, water is generally undesirable as it often causes ravelling (stripping) of the asphalt whereby aggregate and binder separate. The mechanisms behind this separation are becoming better understood due to advances in computational engineering and mechanical and physio-chemical testing.

Keywords Asphalt · cracking · infiltration · stripping · ravelling · porous asphalt · permeability

5.1 Introduction

The topmost layer of most pavements is comprised of a bound layer. The vast majority of pavements have an asphaltic surface. A far lower proportion have a Portland cement concrete (PCC) surfacing. Whilst it is usually a design aim of these surfaces that they provide an impermeable covering to all the lower pavement layers, water does penetrate such surfacings. It may do so either through intact, but not impermeable, bound material or through cracks and joints in the surfacing. Although the emphasis of this book is on water movement and its impact in the unbound material and subgrade layers of the pavement, some information on the movement and response in the upper, bound layers is indispensable. Apart from any other consideration, any

* *Co-ordinating Author:*
✉ A.R. Dawson
University of Nottingham, Nottingham, UK
e-mail: andrew.dawson@nottingham.ac.uk

A.R. Dawson (ed.), *Water in Road Structures,* DOI 10.1007/978-1-4020-8562-8_5,
© Springer Science+Business Media B.V. 2009

complete analysis of water in the road and foundation structure must consider the input conditions – which are significantly affected, perhaps even controlled, by the surface layer. In this chapter the emphasis is on asphaltic mixtures, their permeability and the damage that they suffer from water. Consideration is also given to the ingress of water through joints and cracks. Some of this may also be applicable to the ingress through construction joints and cracks in Portland concrete surfacings. In addition, some consideration is given to porous asphalt mixtures – surfacing that is designed to prevent runoff flowing over the surface.

5.2 Permeability of Intact Asphaltic Mixtures

The permeability of asphaltic mixtures is controlled by the size and interconnection of the void space. To illustrate this, some recent data is given in Fig. 5.1. This presents results for various hot-mix asphaltic specimens as described in Table 5.1. Figure 5.1 shows that permeability values of intact asphaltic materials are typically in the 0 to 40×10^{-6} m/s range. It is apparent that permeability is insignificant at less than approximately 7% air voids but can then rapidly increase. Probably this is because interconnection of voids becomes possible at these high air void ratios and because mixtures that exhibit such air void proportions may be inadequately prepared leaving permeable fissures in the material's structure. Other authors (Zube, 1962; Brown et al., 1989 and Santucci et al., 1985) conclude that a limit of 8% air voids should be adopted to avoid rapid oxidation and subsequent cracking and/or ravelling and to keep permeability low.

Furthermore, higher permeability values are associated with asphaltic mixtures having larger voids. Larger voids are found both in fine-grained mixtures having high *in-situ* air void contents or in coarser grained mixtures at lower void contents. For example, Table 5.2 summarises the air void content at which a threshold between essentially non-permeable and permeable behaviour was observed *in-situ*,

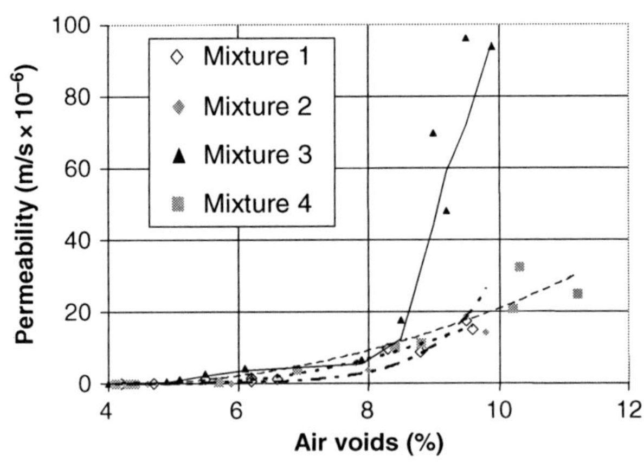

Fig. 5.1 Laboratory determination of permeability of laboratory moulded cylinders of asphaltic mixtures (see Table 5.1 [Vivar & Haddock, 2007]). Reproduced with permission of J. Haddock

Table 5.1 Summary of hot-mix asphaltic specimens for which results are plotted in Fig. 5.1 (after Vivar & Haddock, 2007)

Gradation	Density* (%)	NMAS (mm)	
		9.5	19
Coarse	90 92 94 96	Mixture 1	Mixture 3
Fine	90 92 94 96	Mixture 2	Mixture 4

* expressed as % of maximum theoretical mixture specific gravity
NMAS = nominal maximum aggregate size.

Table 5.2 Relationship between grading, air voids and permeability (after Cooley et al., 2001)

Nominal max. aggregates size (mm)	*In-situ* air void content when permeability increases (%)	Permeability (m/s $\times 10^{-6}$)
9.5	7.7	10
12.5	7.7	10
19	5.5	12
25	4.4	15

together with the permeability coefficient at that point (Cooley et al., 2001). The laboratory derived data of Fig. 5.1 tells a similar story although with a threshold void content of 8–9% for a permeability of 10×10^{-6} m/s. Both data sets reveal that the coarse, low-fines mixtures are least well-performing. In Fig. 5.1, Mixture 3, the poor performance is seen in the rapid increase in post-threshold permeability, while in Table 5.2 the threshold air-void content at which permeability increases is much lower.

5.3 Permeability of Cracked Pavements

In distressed pavements a large proportion of ingress may be through cracks, even if the intact material is relatively impermeable. It has been suggested that there are four factors which influence infiltration rates in cracked asphaltic pavements (Ridgeway, 1976):

- the water-carrying capacity of the crack or joint,
- the amount of cracking present,
- the area that drains to each crack, and
- the intensity and duration of the rainfall.

The first of these is of particular concern in this chapter and is addressed in Section 5.4.2.

5.4 Measuring Permeability

5.4.1 Laboratory Permeability Determination

Both constant head and falling head laboratory methods are available to determine the permeability of asphaltic cores, often with sides sealed using a membrane and/or a confining pressure to prevent edge-leakage (Cooley, 1999). There is some evidence (Maupin, 2000) that the falling-head device is the better device for testing both cores and moulded cylindrical specimens. There are some standardized test procedures of which the recent standard published by FDOT (2006), is an example. A schematic of their laboratory permeameter with flexible walls is given in Fig. 5.2. There is also a European standard available (CEN, 2004).

5.4.2 In-Situ Infiltration Measurement

Because cracks play such an important part in allowing water to enter a pavement through the surfacing, laboratory assessments of the permeability of intact asphaltic mixtures are not overly useful. Therefore, a range of techniques have been developed to assess permeability by infiltrating water into the pavement surface from a device which acts over a limited area of the surface (e.g. Ridgeway, 1976; Cooley, 1999; Fwa et al., 2001; Taylor, 2004; Mallick & Daniel, 2006). Used randomly on the surface, the infiltration observed will be likely to relate to the mean value – being a combination of water entering through intact material and via degradation cracks induced by compaction, by the environment or by traffic. Alternatively, the devices may also be used over specific cracks or joints to assess the water that can enter through that crack or joint (e.g. Mallick & Daniel, 2006). Essentially, two approaches have been adopted – to keep the surface of a specific piece of pavement wet and monitor what water supply rate is required to do this, or to provide a falling head arrangement and note the rate of head drop.

Some of the earliest work on infiltration, in this case through specific joints, involved the fixing of a bottomless wooden box, sealed with clay around its edges, onto an area of pavement containing a measured length of crack crossing the box. Sufficient water is added to the box to maintain a thin layer over the enclosed pavement, the rate at which water is added to maintain this condition can be monitored and the quantity of water infiltrating the pavement structure per unit length of crack calculated. The mean infiltration rate generated by Ridgeway (1976) using this approach was approximately 100 cm^3/h/cm of crack. Site crack lengths and infiltration rates are given in Table 5.3.

An overall infiltration rate for a large area of pavement can be deduced from that measured as infiltrating through particular cracks. To achieve this, Baldwin, et al. (1997) suggested that maintenance intervention occurs when 10% of the surface is cracked which, they said, was equivalent to 0.002 cm of crack/cm^2 of pavement surface This represents a worst-case value for infiltration through cracks if

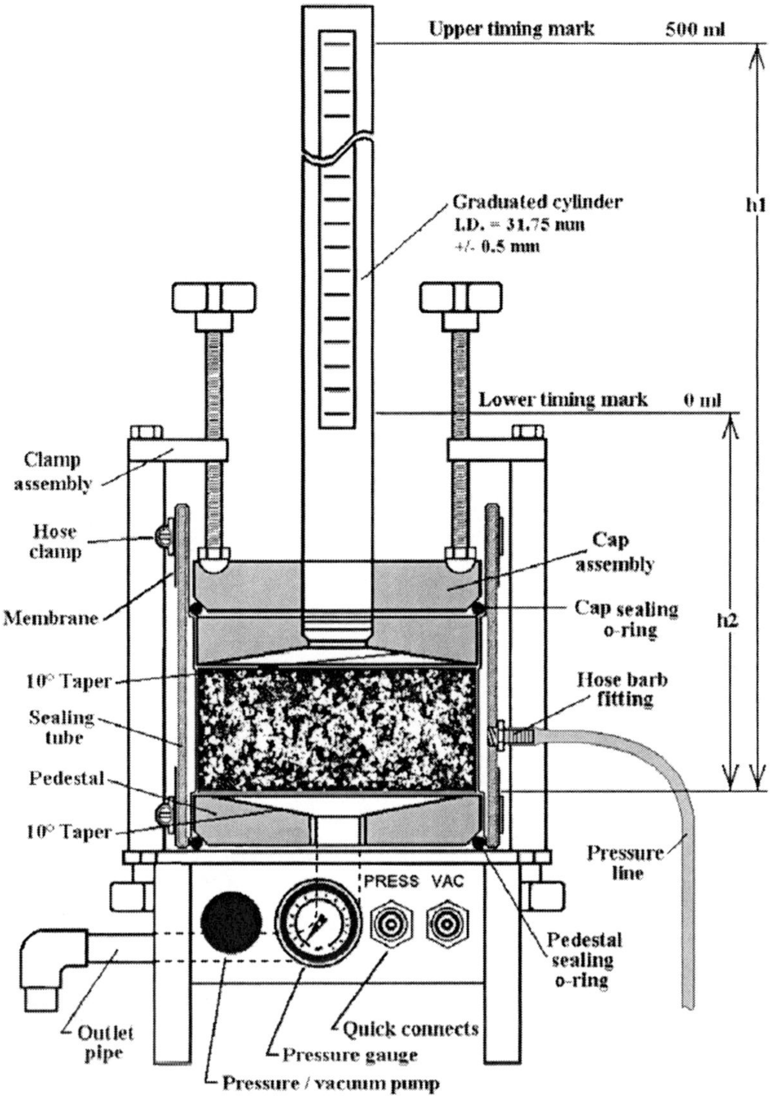

Fig. 5.2 Laboratory permeameter for cores of asphaltic mixtures (FDOT, 2006). Reproduced courtesy of the Florida Department of Transportation

they exist in the same magnitude across the whole of a pavement's surface. Thus the infiltration over such a well cracked pavement would be given by:

$$IR_{max} = 0.02 \times i_c \tag{5.1}$$

where IR_{max} is the maximum anticipated infiltration in units of litres/hour/m^2 of pavement area and i_c is the infiltration measured through one crack in units of

Table 5.3 Site crack lengths and infiltration rate generated by Ridgeway (1976)

Site	Crack length (cm)	Infiltration rate(cm^3/h/cm of crack)	
		Summer 1974	Autumn 1974
1	160	9	28
2	107	620	230
3	183	100	56
4	241	56	37
5	152	2	2
6	208	37	–
7	147	19	84

Approx. mean infiltration rate = 100 (cm^3/h/cm of crack).

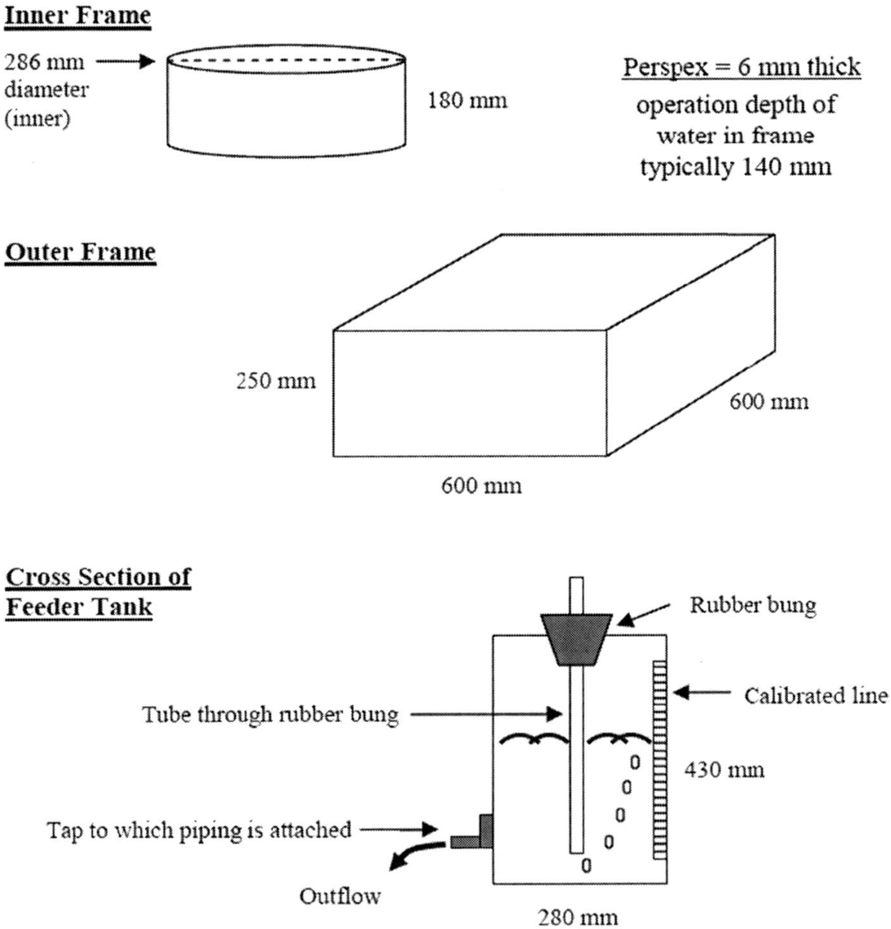

Fig. 5.3 Infiltrometer used by Taylor (2004). Reproduced with permission of J. Taylor

Table 5.4 Infiltrometer results obtained by Taylor (2004)

Site	Distress classification*	Distress severity level*	Area infiltration for well-cracked zones l/hr/m^2	Mean infiltration at maintenance level (single cracks) l/hr/m^2
1a	Fatigue cracking (alligator-type)	Medium	0	
1b	Fatigue cracking (alligator-type)	Medium	0	
2	Longitudinal cracking	Medium		2.70
3	Longitudinal cracking at edge of patch (over service trench)	Good patch but edge seal is lost		0.22
4	Patch (over service trench). Slight ravelling	Low	8.78	
5	Patch (over service trench). No ravelling	None	4.88	
6a	Fatigue Cracking (alligator-type) with some ravelling	Medium	8.70	
6b	Fatigue Cracking (alligator-type) with some ravelling	Medium	3.52	
7a	Fatigue Cracking (alligator-type) with some ravelling	Medium	2.04	
7b	Fatigue Cracking (alligator-type) with some ravelling	Medium	2.50	

n/a = not applicable.
* Types of classification taken from SHRP[1] (1993).

cm^3/hour/cm of crack. It seems likely that direct use of this equation will over-estimate a pavement's potential to accept water. This is because water flowing through one crack may spread laterally below the surface whereas water flowing through one of many adjacent cracks at the same time will be constrained by water flowing through its neighbouring cracks.

To overcome this potential over-reading, a double ring infiltrometer can be used (e.g. Fig. 5.3). The level of water in both a central area and an annular ring are maintained at the same level but only ingress from the central area is monitored as this should flow only vertically through the pavement because of the 'confinement' offered by the ingress from the outer ring. Data obtained in this way by Taylor (2004) is shown in Table 5.4. Eq. 5.1 has been used to bring the data in the last column to the same units as for the data recorded directly in the penultimate column.

The mean infiltration ability of the pavements studied by Taylor (2004) and of the cracks studied by Ridgeway (1976) are similar if Eq. 5.1 is accepted –3.33 and 2 l/h/m^2 respectively, giving confidence of their representativeness.

Another approach, as adopted at the US National Center for Asphalt Technology (NCAT) (Cooley, 1999) employs a device comprising a series of cylindrical standpipes of reducing cross section stacked on top of each other as a form of a

[1] Strategic Highway Research Program (US)

Table 5.5 Summary of infiltrometer data reported by Cooley (1999)

	Calculated permeability, field 10^{-6} m/s				Laboratory measured permeability 10^{-6} m/s			
Site	Mean	St Dev	Max	Min	Mean	St Dev	Max	Min
Mississippi 1	41	26	85	1	94	45	171	58
Mississippi 2	511	385	1526	125	274	214	542	102
Virginia 1	165	98	297	62	134	21	160	107
Virginia 2	35	23	69	5	58	63	166	12
S Carolina 1	518	194	835	271	237	52	284	154
S Carolina 2	169	145	389	20	117	35	169	73

St Dev = Standard Deviation.

falling head permeameter. The lowest cylinder is sealed to the pavement surface using a metal plate over a rubber disc. When the pavement is most permeable only the lowest standpipe is used while when it is least permeable the tallest, narrowest permeameter is employed. Unlike the devices described above, relatively impermeable pavements will, therefore, be subjected to unrealistically high surface water pressures. Whereas even a heavy rainstorm will only impose a few millimetres of water on the surface, the NCAT device can deliver as much as 0.5 m of head. With this device permeability values were obtained as shown in Table 5.5. The comparative laboratory data in the Table were obtained from cores taken from the same pavements and tested using a permeameter as described in Section 5.4.1. The much larger permeability values than given in Table 5.2 is evident, illustrating the important, but often overlooked, need to achieve adequate *in-situ* densities.

For such falling head tests, the coefficient of permeability, K (in m/s), may be estimated using Eq. 3.10. Other procedures for measuring the permeability of porous asphalt have been introduced by Fwa and his co-workers for both laboratory and *in-situ* evaluations (Tan et al., 1999; Fwa et al., 2002).

5.5 Water-Induced Damage in Asphaltic Wearing Surfaces

5.5.1 Introduction: The Problem of Water for Road Surfacings

Practice has shown that asphaltic wearing surfaces which are exposed to water generally start losing aggregates prematurely through a damage phenomenon that has become known as asphaltic ‘stripping’ or ‘ravelling’. Stripping is generally attributed to water infiltration into the asphaltic mixture, causing a weakening of the mastic, and a weakening of the aggregate-mastic bond. Due to the continuing action of water and traffic loading, progressive dislodgement of the aggregates can occur. This initial stripping rapidly progresses into a more severe ravelling of the wearing surface, and ultimately leads to pothole forming, Fig. 5.4.

Sometimes, open-graded mixtures are deliberately designed and laid. As described in Section 5.7, this is to help drain pavement surface water. This tends to

Fig. 5.4 Water-induced damage in asphaltic material (**a**) ravelling (**b**) potholing (de Bondt, 2005). Reproduced by permission of N. Kringos

allow water to reside more or less permanently within the mixture, contributing to the development of water-induced damage.

An additional challenge in the pavement industry is that there is often a big difference between the asphaltic mixture composition and the material characteristics, which are determined in the laboratory, and the asphaltic mixture which is actually constructed on the road. For instance, with regards to water-induced damage, it is not uncommon for the asphaltic mixture components to be exposed to water, even before construction, Fig. 5.5. Since most aggregates and binders do absorb moisture, when exposed to a wet environment, a binder with a significantly reduced stiffness and an initially damaged mixture would end up on the pavement – see Fig. 5.6.

The only real solution to date, for keeping the asphaltic wearing surfaces at an acceptable performance and safety standard, is frequent closure of the major highways for repair and maintenance, implying high costs and frequent road congestion.

Fig. 5.5 Wet asphaltic mixture components before construction (Huber, 2005). Reproduced by permission of N. Kringos

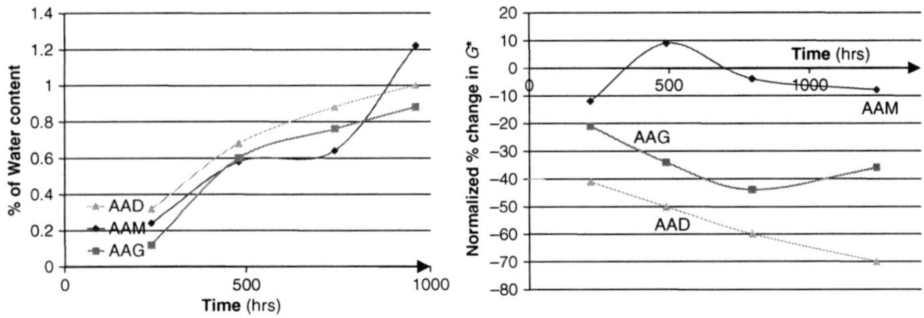

Fig. 5.6 (**a**) water absorption in three SHRP binders (**b**) reduction of binder stiffness, G^*, due to water infiltration (Huber, 2005). Reproduced by permission of N. Kringos

For this reason, it is greatly desired to shift the solution from a repair measure to a preventive measure. This is currently impossible as mixture designers have no prior knowledge of the engineering properties of the mixture at the time of purchase of the bulk materials. Common practice for evaluation of the moisture sensitivity of any particular asphaltic mixture is to perform a set of mechanical tests on dry and moisture-conditioned specimens, giving 'moisture sensitivity ratios' for the engineering mixture properties. Unfortunately, such ratios can only be used to compare case-specific mixtures under a set of pre-determined conditions, but give no insight into the actual water damage phenomena, nor lead to any fundamental remedies.

For this reason, in recent years, these phenomenological studies are giving way to more fundamental studies in which both experimental and analytical investigations on water-induced damage in asphaltic mixtures are combined. Researchers at Delft University of Technology in the Netherlands have focussed on developing a computational tool which allows a study of the interaction between physical and mechanical water damage inducing processes. The tool developed is named RoAM (Kringos & Scarpas, 2004; Kringos, 2007) and operates as a sub-system of the finite element system developed at TU Delft, CAPA-3D (Scarpas, 2005).

5.5.2 Coupled Physical-Mechanical Water-Induced Damage

One of the important realizations is that the problem cannot be solved by mechanical considerations alone. Clearly, water has an effect on the material characteristics of the asphaltic components and their bond, even without mechanical loading. Therefore, both physical and mechanical water damage-inducing processes are included in the model. Another realization is that, in order to acquire a fundamental insight into the processes which cause water damage, the asphaltic mixture needs to be considered at a micro-scale. This implies that the experimental characterization and the computational simulations of the water damage-inducing processes must be dealt with at mixture component level; i.e. the aggregates, the mastic, the bond between the aggregates and the mastic and the (macro) pore space. Each of these

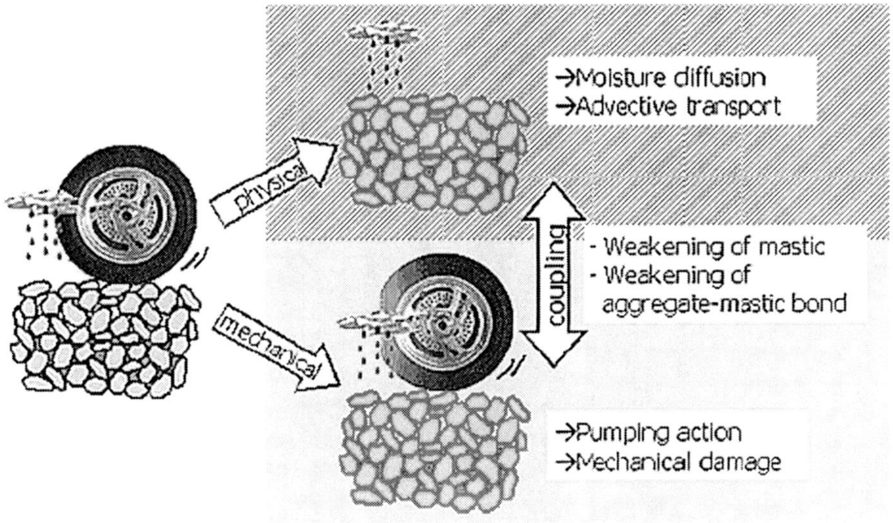

Fig. 5.7 Separation of water damage into physical and mechanical processes (Kringos, 2007)

contributes to the mechanical performance of the mixture as well as to its moisture susceptibility.

The physical processes that have been identified as important contributors to water damage are (c.f. Chapter 6, Section 6.3.1):

- the molecular diffusion of water through the mixture components and
- the advective transport, i.e. 'washing away', of the mastic due to the moving water flow through the connected macro-pores.

A mechanical process that is identified as a contributor to water damage is the occurrence of intense water pressure fields inside the mixture caused by traffic loads and known as the 'pumping action'. In the model, these physical material degradation processes interact with a model for mechanical damage to produce the overall water-mechanical damage in the mixture – see Fig. 5.7.

5.5.2.1 The Physical Processes Contributing to Water-Induced Damage

The existence of a water flow through an asphaltic mixture may cause desorption of parts of the mastic films which are in direct contact with the water flow, Fig. 5.8(a) carrying away elements of the bitumen[2] by advection (see Chapter 6 for a fuller definition of advection). Exposure of an asphaltic mixture to stationary water (i.e. no

[2] In British English, bitumen refers to the binder in an asphaltic material whereas in American English the term asphalt-cement is normally used, or simply 'asphalt'. In this book the term bitumen is used to describe the binder and the word asphalt (which in British English refers to the whole mixture) is not used alone, to avoid confusion.

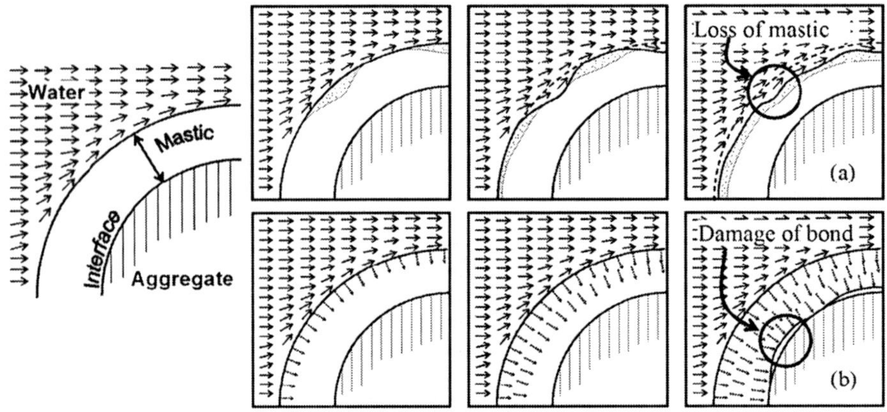

Fig. 5.8 Schematic of physical water damage-inducing processes (Kringos & Scarpas, 2005a; Kringos, 2007) (**a**) Loss of mastic due to advective transport (**b**) Damage of the bond due to water diffusion

water flow) would, therefore, show no advective transport damage. Since practice has shown that exposure of asphaltic mixtures to stationary water does, in time, cause ravelling of the mixture, this process cannot be the only phenomenon causing water damage. In asphaltic mixtures, diffusion of water through the mastic films surrounding the aggregates is a molecular process that may eventually lead to water reaching the interface area between the mastic and the aggregates. Depending on the bond characteristics, the water can then cause an adhesive failure of the mastic-aggregate interface, Fig. 5.8b).

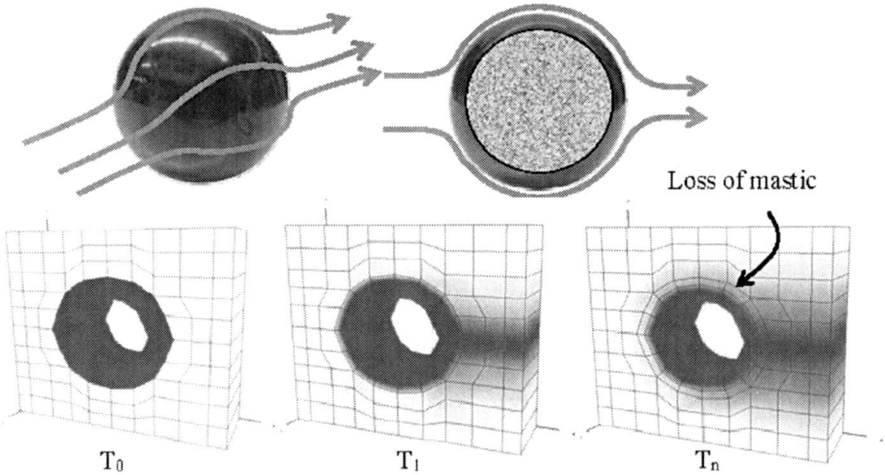

Fig. 5.9 Simulation of loss of mastic from around a coarse aggregate particle due to a fast water flow field
Aggregate is omitted in Finite Element model shown in the lower part of the figure.

As time increases, the diffusion of water through the mastic weakens the cohesive strength and the stiffness of the mastic and may actually aggravate the desorption. These are modelled using the approaches set out in Chapter 11. More details about these mathematical formulations can be found in Kringos & Scarpas (2004; 2005a & 2006), Kringos (2007) and Kringos et al. (2007). Figure 5.9 shows an example of a mastic desorption simulation with RoAM, in which a coated aggregate is exposed to a fast water flow field.

5.5.3 The Mechanical Processes Contributing to Water-Induced Damage

Because water-induced damage influences the dry response of the material, the effects of the physical processes must be coupled with a three dimensional elasto-visco-plastic constitutive model for mastic response (Scarpas et al., 2005). Mastic in asphaltic mixtures is known to be a material whose behaviour, depending on strain rate and/or temperature, exhibits response characteristics varying anywhere between the elasto-plastic and the visco-elastic limits. Constitutive models for such types of materials can be developed by combining the features of purely elasto-plastic and purely visco-elastic materials to create a more general category of constitutive models termed elasto-visco-plastic. This is the approach that was adopted by the Delft researchers, but a detailed description of the formulation is beyond the scope of this book but is available (Kringos & Scarpas, 2006) for interested readers.

5.5.3.1 Aggregate-Mastic Bond Strength as a Function of Water Content

Clearly, such modelling requires a knowledge of the aggregate-mastic bond as a function of water content. The direct tension test provides the means of assessing bond strength and to define a relationship between the mastic-aggregate bond strength and the conditioning time in a water-bath test (see Fig. 5.10). When the purpose of the test is to acquire a comparison between particular mastic-stone combinations, results of the pull-off test may directly provide useful information, provided that similar geometries and water conditioning are used. To determine, however, the fundamental relationship of the influence of water on the bond strength, the quantity of water at the interface is of paramount importance. Since this type of information cannot be determined from the test, an additional procedure was developed (Copeland et al., 2007) to relate the bond strength to the quantity of water in the bond. By simulating the test specimens with the RoAM software (see above), the relationship between the quantity of water at the mastic-aggregate interface and the soaking time can be found, Fig. 5.10 (left). By combining the results of finite element simulations and the pull-off test, a relationship between the bond strength and the water content is determined, Fig. 5.10 (right).

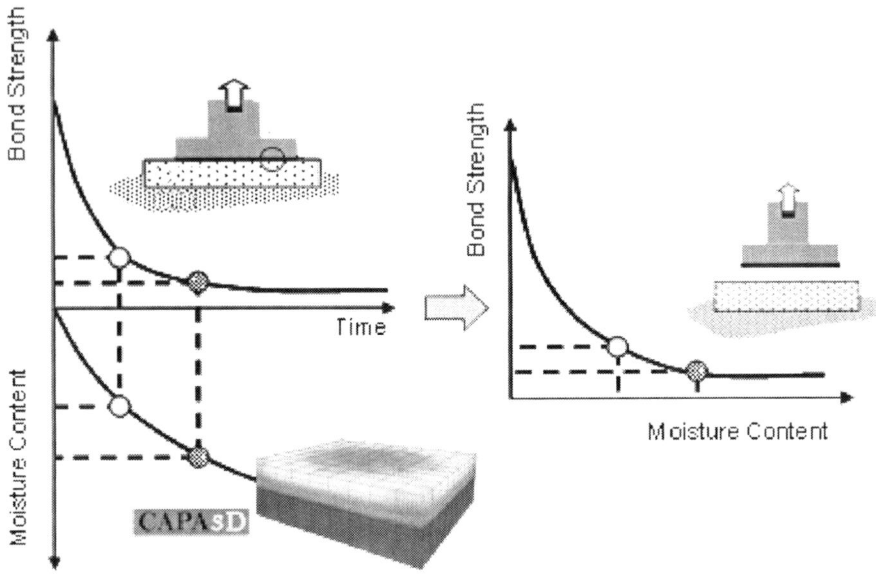

Fig. 5.10 Experimental-computational methodology for determination of the aggregate-mastic bond strength as a function of water content (Kringos & Scarpas, 2005b). Reproduced by permission of N. Kringos

From test results, using for the mastic a SHRP core bitumen AAD[3] (PG[4] 58-28) with a diabase filler material passing the $\leq 75\ \mu$m sieve (#200) and for the aggregate substrate a diabase rock, the tensile bond strength, S_{md}, as a function of volumetric water content, θ, was determined as:

$$S_{md} = e^{\left(\ln S_0 - \alpha\sqrt{\theta}\right)} \tag{5.2}$$

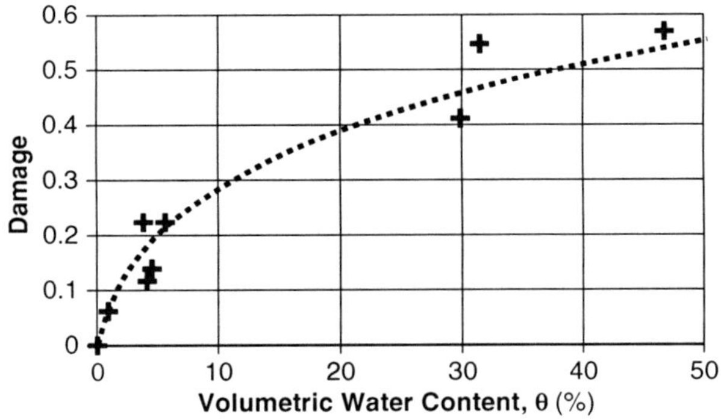

Fig. 5.11 Aggregate-mastic bond strength as a function of water (adapted from Copeland et al., 2007)

[3] AAD is one of the core SHRP bitumen binders

[4] PG = Penetration Grade

where $\ln S_o = 0.30$, S_o being the dry adhesive strength, and $\alpha = 3.76$. This can be reformulated into a water-induced damage parameter, d:

$$d(\theta) = 1 - e^{-\alpha\sqrt{\theta}} \tag{5.3}$$

Figure 5.11 shows the result of the experimental-computational procedure and the regression curve, as expressed by Eq. 5.3.

5.5.3.2 Moisture Diffusion Coefficients

Moisture diffusion measurements are still not performed very commonly for asphaltic materials. There are currently two main test procedures being utilized. The first is an overall measurement of the increase of weight as a sample is exposed to a controlled moisture conditioning (Cheng et al., 2003). The second is a slightly more complicated procedure using Fourier transform infrared spectroscopy (Nguyen et al., 1992). It is quite challenging though, to utilize the available test data, since the values can vary greatly. For instance, for the AAD-1 bitumen, values have been published ranging from 4.79 mm^2/h (Cheng et al., 2003) to 9.0×10^{-5} mm^2/h (Nguyen et al., 1992). Comparing these values with published diffusion coefficients of, rubber, PVC and polyethylene (Abson & Burton, 1979), the lower diffusion value for the AAD-1 asphalt binder (bitumen) seems to be more plausible. Since the mastic matrix in an asphaltic mixture generally consists of asphalt binder as well as sand particles and filler material, a higher diffusion value than for the binder alone can be expected. In the simulations described in this chapter, a diffusion coefficient of 1.0×10^{-3} mm^2/h has been utilized for the mastic film.

5.5.4 Micro Scale Simulation of Combined Mechanical-Water Induced Damage

The motivation for the following micro-scale finite element simulation, is the ongoing discussion about cohesive versus adhesive failure mechanisms in asphaltic mixtures. It is the authors' belief that, depending on the ability of the individual components and the bond between them, either one of these failure mechanisms may be dominant. It is, therefore, of paramount importance to establish the fundamental relations between environmental weathering and material strength and stiffness. These relations can be used to assist the designer to optimize the water damage resistance characteristics of the mixture at purchase time on the basis of the response of the individual components.

In Fig. 5.12 details are shown of a micro-mechanical mesh that has been utilized for simulation of the results of pumping action due to traffic loading in a porous mixture (Kringos & Scarpas, 2005a).

Depending on the specified characteristics of the individual mixture components, cohesive Fig. 5.13(a) or adhesive Fig. 5.13(b) failure can occur.

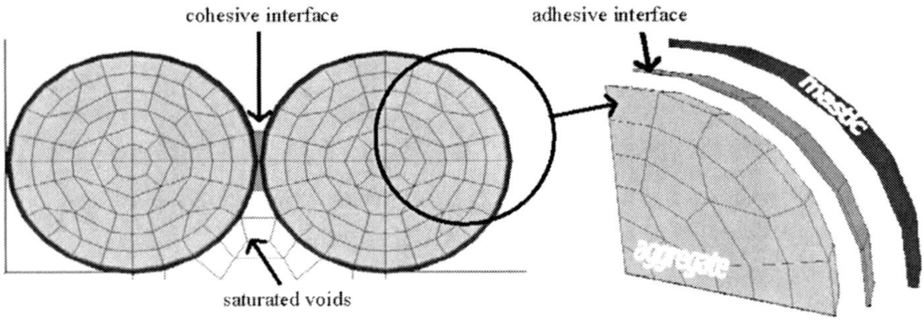

Fig. 5.12 Cohesive versus adhesive interface elements (Kringos & Scarpas, 2005a)

Using the same approach, it is possible to study the combined effect of water-induced damage and traffic loading. Because the time-scale of water-induced damage accumulation compared to the time-scale of the mechanically-induced damage differs by several orders of magnitude, mechanical damage can be combined with water-induced damage, at discrete time-intervals. In terms of the finite element simulation, this implies that diffusion studies can be performed until a desired level of water content is attained and the associated water-induced damage in the specimen can be computed. Subsequently, mechanical loading is imposed on the specimen at discrete time intervals and the total damage at a particular time interval can be computed. As can be seen from Fig. 5.14, for the chosen set of material parameters, stripping of the mastic film from the aggregate can be simulated.

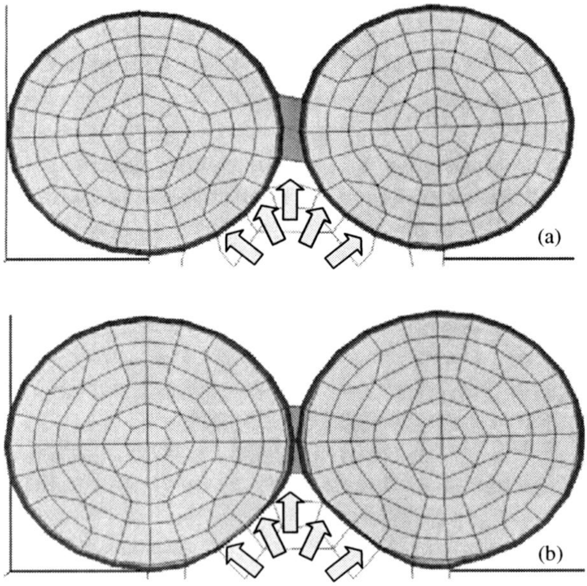

Fig. 5.13 Finite element simulation of (**a**) cohesive versus (**b**) adhesive failure (Kringos & Scarpas, 2005a)

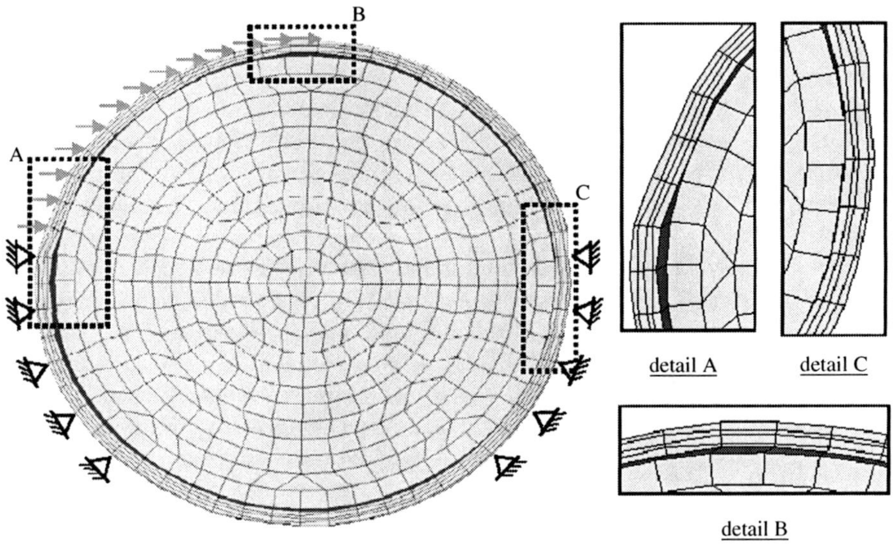

Fig. 5.14 Simulation of ravelling in an aggregate coated in a mastic film, subjected to a mechanical and water-induced damage (Kringos & Scarpas, 2006)
The dark areas in the details are areas of de-bonding.

5.6 Pollution-Induced Degradation of Bound Layers

Performance characteristics of bound pavement layers are known to be influenced by water-borne pollutants that cause changes in mechanical behaviour, ageing and degradation. With the exception of Portland cement concrete (PCC) pavements, this has not yet received much attention. Asphalt pavements are not seriously affected by inorganic pollutants, but most of the organic chemicals, including gasoline and motor oil, soften up or break down the asphalt binder leaving the asphaltic layer vulnerable to further degradation. Damage of the surface layer, due to ageing (stiffening because of ultra violet light), traffic induced cracks and chemical degradation, opens an ingress route in the pavement system for pollutants from the surface.

In hot climates, salt can be moved by evaporating water to near the pavement surface. This may result in the expansive crystallisation of salt in voids in asphaltic mixtures (or other bound layers) just below the road pavement's surface leading to blistering of the running surface, to cracking of the pavement and to overall degradation. Guidelines to the understanding and treatment of this issue are available (Obika, 2001). In temperate climates, salt (NaCl) seems, usually, not to be a significant contributor to damage of asphaltic materials. It may, sometimes, accelerate deterioration of poor quality materials, but it appears that it is water damage itself which is the primary cause (see Section 5.5). However, the chemistry of the water in the pores of asphaltic materials can have an important influence on whether stones and binder adhere efficiently. Greater alkalinity (i.e. higher pH) potentially results in increased rates of moisture damage, although Calcium Hydroxide (slaked lime) dissolved in the water doesn't have this effect even though pH rises (Little & Jones, 2003). In cold climates, salt has been implicated in causing damage. According to Hudec and Anchampong (1994) certain fine grained aggregates degrade rapidly during wetting and drying cycles and during freeze-thaw cycles especially if deicing salts have been used abun-dantly. The extensive use of chlorides has also been reported to cause accelerated pavement deterioration (Dore et al. 1997, Saarenketo, 2006).

In PCC, deterioration is related to complex processes associated with physical and chemical alteration of the cement paste and aggregates. One major chemical degradation mechanism resulting from the long-term application of the popular chemical de-icer sodium chloride (NaCl) is the dissolution of calcium hydroxide ($Ca(OH)_2$). Another common de-icer, $CaCl_2$, is associated with a deleterious chemical reaction with PCC. The chemical attack is accompanied by the formation of hydrated calcium oxy-chloride according to the following reaction:

$$3Ca(OH)_2 + CaCl_2 + 12H_2O \rightarrow 3CaO \cdot CaCl_2 \cdot 15H_2O$$

This reaction is considered to be disruptive to the concrete matrix because of the expansive pressures generated. Another potential detrimental effect of the application of chemical de-icing salts is increased alkali – silica reactivity (ASR), which is a distress caused by undesirable chemical reactions between alkalis in the cement paste (Na_2O and K_2O) and the reactive siliceous components of susceptible aggregates. The product of the reaction is expansive in the presence of moisture, destroying the integrity of the weakened aggregate particle and the surrounding cement paste. When aggregates like dolomitic limestone are used there is a possibility of alkali-carbonate reactivity, where alkalis react with carbonate aggregates. Besides these mentioned processes, there is also the possibility of external and internal sulphate attack, which can cause deterioration. Other de-icing chemicals (magnesium chloride, calcium magnesium acetate, Ca-acetate, Mg-acetate, urea, etc.) may also have damaging effect on PCC pavement layers (MTTI, 2002).

Hydraulically bound mixtures may be considered as low strength PCC. From this point of view, the effects of the pollutants (mainly de-icers and sulphate) are similar to those on PCC pavements, except that chemical degradation, deterioration and loss of strength of the hydraulically bound layer will be quicker than in PCC layers, leading to higher stress on the layers beneath and faster degradation of the pavement.

5.7 Porous Asphalt

In countries that suffer from large amounts of rainfall, the asphaltic wearing surfaces are often constructed of open graded asphaltic mixtures. The high permeability of these wearing surfaces ensures a fast drainage of the water away from

Fig. 5.15 (**a**) bad road visibility conditions (**b**) hydroplaning and 'splash and spray' (Erkens, 2005). Reproduced by permission of N. Kringos

the surface, avoiding hydroplaning and bad visibility conditions due to 'splash and spray', Fig. 5.15, and thus improving the overall road safety.

Porous asphalt uses aggregate with a moderate to coarse median particle size and a very steep grading curve – i.e. the majority of the stones in the mixture are of a similar size. This has the effect of developing a mixture with a very high air void volume (20–30%) with stones only adhering to each other by virtue of the films of bitumen at their point of contact. In this way its porosity is very high compared with conventional asphaltic material and water does not easily collect on the surface during rainstorms.

An added benefit not included in the original concept, but now an important driving force for the wider adoption of porous asphalt surfacings is the reduced traffic noise from pavements with these surfacings. The porous nature reduces tyre-surface interaction sounds and acts as a partial absorbent of other vehicle induced noise. Typically they provides a 3–5 dB(A) noise reduction over conventional pavement surfacings. Even greater benefits can be achieved by using two-layer porous asphalt with a finer, filter, layer over a coarser, drainage layer. Noise reduction may then be 8 or 9 dB(A) quieter than conventional asphaltic mixtures and 4 dB(A) quieter than a single-layer porous asphalt.

As the surfacing is so permeable, rain can drain vertically into the porous asphalt layer before being conveyed laterally within the pavement. Typically a porous asphalt surfacing will have a thickness of between 20 and 100 mm and be placed on top of an impermeable asphaltic base. Hence, water flowing in the surfacing cannot continue to flow vertically but is forced to travel sideways, exiting from the layer at its edge. Unless this edge is free, special attention must be paid as to how the water is to be collected and led away from the pavement. An impermeable edge, such as a conventional kerb, would dam the water within the layer. Consequently special kerbs with inlets and pipe systems have been developed to lead the water into a conventional surface drainage system (Highways Agency, 1997).

Despite the advantages of the material in providing relatively dry surfaces in wet weather, the material and its use pose a number of problems:

1. Lack of durability. Careful mixture design is needed to ensure that there is enough bitumen to coat the stones and ensure longevity of performance – too much and the mixture may rut too readily and the pores become blocked by bitumen (preventing drainage). Too little bitumen and ravelling will be likely (as described in Section 5.5.1), particularly in cold weather when ice could form in the pore space forcing the topmost layers of stone loose. An added issue is that the greater opportunity for bitumen to react with atmospheric oxygen, because of air in the voids, tends to lead to early embrittlement of the bitumen (Herrington et al., 2005). Bitumen film thickness is, thus, of particular importance. Hence, both design and construction practice require careful attention, perhaps more so than for conventional asphaltic mixtures. More detailed coverage of this topic is beyond the scope of this chapter, but interested readers may wish to consult NAPA (2004).
2. Clogging due to ingress of particulates. Small particles and dust, that comes from the environment, blown soil, engine wear, brake wear and from cargoes (see Chapter 6, Section 6.2), tend to get washed into the pore space of the porous asphalt, thereby blocking it. In thin porous surfacings on high-speed roads it appears that reduction in permeability is not of great concern. After some initial deterioration, further clogging is often not significant, probably because high-speed traffic develops high transient water pulses in the pores of the asphaltic mixture during wet weather, causing a self-cleansing action (Bendtsen et al., 2005). In slower speed roads this action is not evident and clogging is, typically, progressive. These authors monitored an urban test road in Denmark comprising 3 porous asphaltic surfacings and a control surface (Table 5.6).
   Using an infiltrometer somewhat like that of Cooley (1999), see Section 5.4.1, Bendtsen et al. (2005) observed, Fig. 5.16, that clogging developed quite rapidly in the finer asphaltic pavements. The two porous pavements with 5 mm aggregate, Sections II and III, were effectively clogged after 15–20 months whereas Pavement I with 8 mm aggregate remained in a much better condition. The reason for this clogging is believed to be the dirt and fine material from the adjacent dense asphaltic concrete pavement being dragged onto the porous pavements by vehicle tyres, since clogging first appeared at the position nearest to the reference section.

Table 5.6 Danish test road construction (after Bendtsen et al., 2005)

No	Type	Top-layer		Bottom-layer	
		Thickness [mm]	Aggregate size [mm]	Thickness [mm]	Aggregate size [mm]
I	Porous asphalt	25	5/8	45	11/16
II	Porous asphalt	20	2/5	35	11/16
III	Porous asphalt	25	2/5	65	16/22
IV	Dense asphaltic mixture	30	0/8	–	–

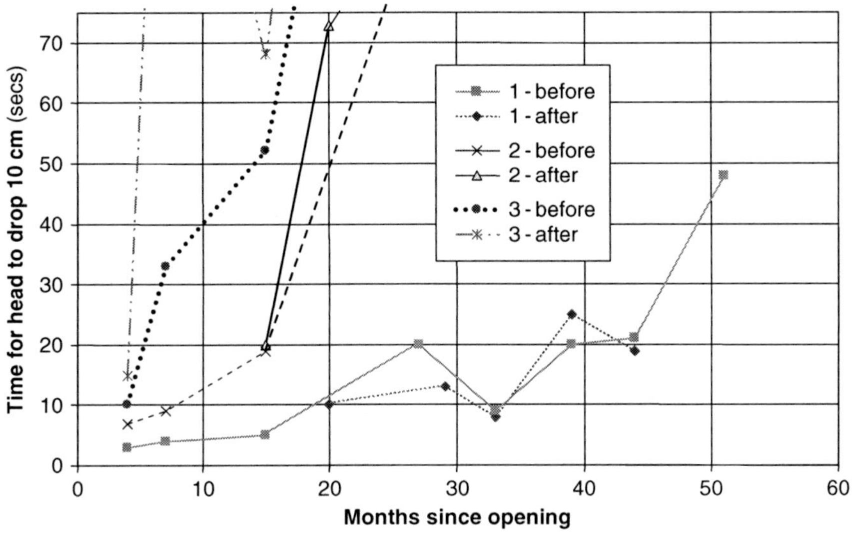

Fig. 5.16 Out-flow time (s/10 cm) in the wheel-track as a function of pavement age in months before and after cleaning (adapted from Bendtsen et al., 2005)

Fig. 5.17 Porous asphalt cleaning machine. Reproduced by permission of the Danish Road Institute

3. Cold weather clogging by ice and snow. Bäckström and Bergström (2000) evaluated the function of porous asphalt in cold climates using a climate room. At the point of freezing point, the infiltration capacity of porous asphalt was approximately 50% of the infiltration capacity at +20°C. They simulated conditions of snowmelt by exposing the porous asphalt to alternating melting and freezing over a period of 2 days and found that the infiltration capacity was reduced by approximately 90%.

To overcome clogging, cleaning devices have been developed. These (Figs. 5.17 and 5.18) typically comprise high pressure water jets that aim to disturb and erode fines resting in the pores of the porous asphalt, washing them to the surface from where they may be vacuum-collected by the cleansing machine. It is interesting to note that the results of Bendtsen et al. (Fig. 5.16) show that there is not a significant reduction in the level of clogging due to cleaning (compare 'before' (b) to 'after' (a) readings in the figure) and that cleaning did not bring back performance close to the original function.

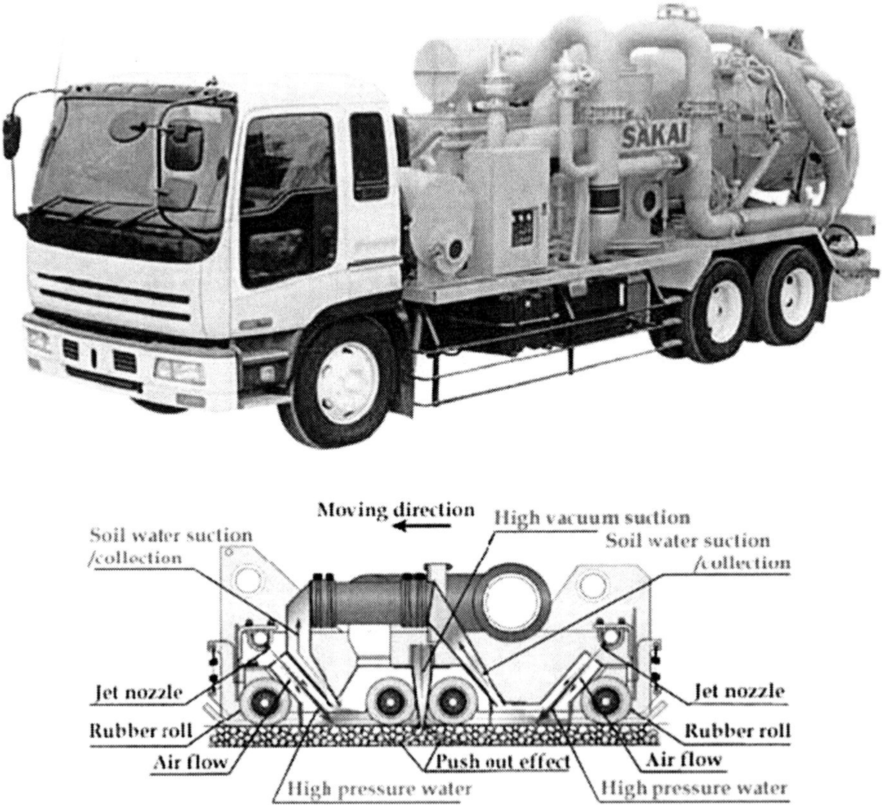

Fig. 5.18 Porous asphalt cleaning machine and diagram of active part. Reproduced by permission of Sakai Heavy Industries

5.8 Conclusions

A range of equipment exits to determine permeability of asphaltic mixtures both by *in-situ* and laboratory testing. At present the values of permeability collected seem, mostly, to be being used for relative performance assessments and they are not much integrated in whole-pavement water regime modelling. Current advances in computational engineering and mechanical and physio-chemical testing enable the identification of the actual physical processes of water-induced damage in asphaltic mixtures and the evaluation of their effects on the total mixture response. It is hoped that, over a period of time, availability of these resources will enable a gradual transition in mixture design from a design-by-testing approach to an approach in which design is by identification of the optimal choice of individual mixture components on the basis of their physio-chemical and mechanical characteristics and interactions. The increasing use of porous asphalt for noise reduction and spray reduction purposes is an important challenge to the pavement engineer. Purposely allowing water into the structure provides the opportunity for much greater and faster ravelling. There are also concerns about clogging due to washed-in fines and due to ice formation. Rehabilitating clogged porous asphalt without causing premature damage is a challenge.

Acknowledgments Some of the work described in this chapter has been performed with financial, computational or experimental support of Ooms Nederland Holding Bv, the Section of Structural Mechanics of TU-Delft (both in the Netherlands) and the Turner Fairbanks Highway Research Center of the US Federal Highway Administration. The authors wish, therefore, to express their thanks to Dr. A. de Bondt, ir. C. Kasbergen and Dr. J. Youtcheff respectively for this.

References

Abson, G. & Burton, C., 1979, "*Physical tests and range properties, Bituminous Materials: Asphalts, Tars and Pitches*", Vol.1, Krieger Publishing Company, N.Y., pp. 213–288.

Bäckström, M. & Bergström, A., 2000, "Draining Function of Porous Asphalt during Snowmelt and Temporary Freezing", *Canadian J. Civ. Eng.*, **27**(3), pp 594–598.

Baldwin G., Addis, R., Clark J. & Rosevear, A., 1997, "*Use of Industrial By-Products in Road Construction – Water Quality Effects*", CIRIA, London.

Bendtsen, H., Andersen, B., Raaberg, J., Larsen, L.E. & Kragh, J., 2005, "*Two-layer porous asphalt – for urban roads*", Danish Road Institute, Report 144, 18pp. Downloadable from http://www.vejdirektoratet.dk/publikationer/VIrap144/index.htm.

Brown, E.R., Collins, R & Brownfield, J.R., 1989, "Investigation of segregation of asphalt mixtures in the State of Georgia", Paper presented to *68th Annual Meeting of Transportation Research Board*, Washington DC.

Cheng, D., Little, D.N., Lytton, R. & Holste, J.C., 2003, "Moisture Damage Evaluation of Asphalt Mixtures by Considering Both Moisture Diffusion and Repeated-Load Conditions", *Transportation Research Record*, N°. 1832, TRB, National Research Council, Washington, D.C., pp. 42–49.

Comité European de Normalisation (CEN), 2004, "*Test methods for hot mix asphalt – Part 19: Permeability of Specimen*", standard EN 12697–19.

Cooley, L. Allen, Jr., Brown, E.R. & Maghsoodloo, S., 2001, "*Development of Critical Field Permeability and Pavement Density Values for Coarse-Graded Superpave Pavements*", National Center for Asphalt Technology, Auburn University, NCAT Report No. 01–03, September.

Cooley, L.A., Jr., 1999, "*Permeability of Superpave Mixtures: Evaluation of Field Permeameters*", National Center for Asphalt Technology, Auburn University, NCAT Report No. 99-1, February.

Copeland, A., Kringos, N, Youtcheff, J. & Scarpas, A., 2007, "Measurement of Aggregate – Mastic Bond Strength in the Presence of Moisture: A Combined Experimental – Computational Study", *Proc. 86th Transportation Research Board.*

Copeland, A., Kringos, N., Scarpas, A. & Youtcheff, J. 2006, "Determination of Bond Strength as a Function of Moisture Content at the Aggregate-Mastic Interface", *Proc. 10th Conf. Int'l Soc. Asphalt Pavements* (ISAP).

de Bondt, A., 2005, Personal Communication.

Dore, G., Konrad, J-M. & Marius, R., 1997, "*The Role of Deicing Salt in Pavement Deterioration by Frost Action*", Paper 97–0598 presented to the Transportation Research Board, 76th Annual Meeting, Washington, DC.

Erkens, S., 2005, "Raveling in the Netherlands". Proc. *Int'l Workshop on Moisture Induced Damage of Asphaltic Mixes.*

FDOT, 2006, "*Measurement of Water Permeability of Compacted Asphalt Paving Mixtures*" Florida Method of Test FM 5-565, Florida Dept. of Transportation Revised: January 26, 10pp.

Tan, S.A., Fwa, T.F. & Chuai, C.T., 1999, "Automatic field permeameter for drainage properties of porous asphalt mixtures", *J. Test. Eval.*, **27**(1), pp. 57–62.

Fwa, T.F., Tan, S.A. & Guwe, Y.K., 2002, "Laboratory evaluation of clogging potential of porous asphalt mixtures", *Transportation Res. Rec.*, **1681**, pp. 43–59.

Fwa, T.F., Tan, S.A. & Guwe, Y.K., 2001, "Expedient Permeability Measurement for Porous Pavement Surface", *Int. J. of Pavement Eng.*, **2**(4), pp. 259–270.

Herrington, P.R., Reilly, S. & Cook S., 2005, "*Porous Asphalt Durability Test*", Transfund New Zealand Research Report 265, 32pp.

Highways Agency, 1997, "Edge of Pavement Details for Porous Asphalt Surface Courses", *Design Manual for Roads & Bridges*, UK Highways Agency, **4**, Section 2, Part 4, Ref. HA 79/97, 20pp.

Huber, G., 2005, "Tenderness Caused by Moisture", Proc. *Int'l Workshop on Moisture Induced Damage of Asphaltic Mixes.*

Hudec, P.P. & Anchampong, F., 1994, "Improving Aggregate Quality by Chemical Treatment", in *Proc. 2nd Int'l Aggregates Symp.*, Erlangen, Germany, October 22–27, 1990, ed. Luttig, G.W., pp. 135–146.

Kringos, N., 2007, "*Modeling of Combined Physical-Mechanical Moisture Induced Damage in Asphaltic Mixes*", PhD dissertation, Delft University Press.

Kringos, N. & Scarpas, A., 2004, "Development of a Finite Element Tool for Simulation of Ravelling of Asphaltic Mixes". *Proc. 2004 Int'l Conf. Computational & Experimental Eng'g & Sciences* (ICCES '04), Madeira, Portugal, July, Tech Science Press, in CD ROM.

Kringos, N. & Scarpas, A., 2005a, "Ravelling of asphaltic mixes: Computational identification of controlling parameters", *Transportation Res. Rec.*, **1929**, pp.79–87.

Kringos, N. & Scarpas, A., 2005b, "Simulation of Combined Mechanical-Moisture Induced Damage in Asphaltic Mixes". *Proc. Int'l Workshop on Moisture Induced Damage of Asphaltic Mixes.*

Kringos, N. & Scarpas, A., 2006, "Numerical Simulation of the Physical Processes Inducing Moisture Damage in Asphaltic Mixes", *Proc. 10th Conf. Int'l Soc. Asphalt Pavements* (ISAP).

Kringos, N., Scarpas, A. & Kasbergen, C., 2007, "Three Dimensional Elasto-Visco-Plastic Finite Element Model for Combined Physical-Mechanical Moisture-Induced Damage in Asphaltic Mixtures", *J. Assoc. Asphalt Paving Technologists*, **76**, pp. 495–517.

Little, D.N. & Jones, D.R., 2003, "Chemical and Mechanical processes and moisture damage in hot-mix asphalt" in "*Moisture sensitivity of asphalt pavements*", Transportation Research Board, Washington, DC., pp37–90.

Mallick, R.B. & Daniel, J.S., 2006, "Development and evaluation of a field permeameter as a longitudinal joint quality indicator", *Int'l J. Pavement Eng'g*, **7**(1), 11–21.

Maupin, G.W., Jr., 2000, "Investigation of Test Methods, Pavements and Laboratory Design Related to Asphalt Permeability", Report, VTRC 00-R2, Charlottesville, Virginia, June.

MTTI, 2002, "*Guidelines for Detection, Analysis, and Treatment of Materials-Related Distress In Concrete Pavements*", Final Report, Michigan Tech Transportation Institute, Michigan Technological University, Civil & Env. Eng'.g Dept., U.S. Dept. of Transportation Federal Highway Administration, **1**.

NAPA, 2004, "*Porous Asphalt Pavements, Design, Construction & Maintenance Guide*", Information Series 121, National Asphalt Pavement Association January, 16pp.

Nguyen, T., Byrd, W., Bentz, D & Seiler, J. Jr., 1992, "*Development of a Technique for In Situ Measurement of Water at the Asphalt/Model Siliceous Interface*", SHRP report NISTIR 4783, March.

Obika, B., 2001, "*The prevention and repair of salt damage to roads and runways*", Roads, Dep't., Ministry of Transport & Communications, Gabarone, Botswana, 48pp. Available on-line at www.vegvesen.no in the International Activity/Downloads sub-section of the web-site.

Ridgeway, H.H., 1976, "Infiltration of Water Through the Pavement Surface", *Transportation Research Record* **616**, Transportation Research Board, Washington, DC., pp.98–100.

Santucci, L.E., Allen, D.D. & Coates, R.L., 1985, "The effects of moisture and compaction on the quality of asphalt pavements", *J. Assoc. Asphalt Paving Technol.*, **54**, pp. 168–208.

Scarpas, A., 2005, "*Mechanics based Computational Platform for Pavement Engineering*", PhD thesis, Delft University of Technology.

Scarpas, A., Kasbergen, C. & Kringos, N., 2005, *International Course on Advanced Constitutive Modeling of Asphaltic Materials*, Course Notes, January, Maryland, USA.

Saarenketo, T., 2006, "*Electrical Properties of Road Materials and Subgrade Soils and the Use of Ground Penetrating Radar in Traffic Infrastructure Surveys*", PhD thesis, Univ. Oulu., 123 pp.

SHRP, 1993, "*Distress Identification Manual for the Long-Term Pavement Performance Project*", SHRP-P-338, Strategic Highway Research Program, Washington, DC. 147pp.

Taylor, J.V., 2004, "*Migration of Contaminants Associated with Pavement Construction*", PhD thesis, School of Civil Engineering, University of Nottingham.

Vivar, E. & Haddock, J.E., 2007, "Hot-Mix Asphalt Permeability and Porosity", paper to *82nd Annual Meeting of Association of Asphalt Paving Technologists*, San Antonio, Texas.

Zube, E., 1962, "*Compaction studies of asphalt concrete pavement as related to water permeability test*", Publ'n 62-22, State of California, Dept. of Public Works, Div. of Highways.

Chapter 6
Sources and Fate of Water Contaminants in Roads

Lennart Folkeson*, Torleif Bækken*, Mihael Brenčič, Andrew Dawson, Denis François, Petra Kuřímská, Teresa Leitão, Roman Ličbinský and Martin Vojtěšek

Abstract This chapter gives an overview of sources, transport pathways and targets of road and traffic contaminants. Pollution sources include traffic and cargo, pavement and embankment materials, road equipment, maintenance and operation, and external sources. Heavy metals, hydrocarbons, nutrients, particulates and de-icing salt are among the contaminants having received the greatest attention. Runoff, splash/spray and seepage through the road construction and the soil are major transport routes of pollutants from the road to the environment. During their downward transport through road materials and soils, contaminants in the aqueous phase interact with the solid phase. In saturated media, diffusion, advection and dispersion are the major processes of mass transport. In unsaturated soil, mass transport strongly depends on soil-moisture distribution inside the pores. Sorption/desorption, dissolution/precipitation and ion exchange reactions are the most significant chemical processes governing pollutant transport in soils. Redox conditions and acidity largely regulate heavy-metal mobility. Many heavy metals are more mobile under acidic conditions. Plants close to heavily trafficked roads accumulate traffic pollutants such as heavy metals. Heavy metals, organics, de-icing salt and other toxic substances disturb biological processes in plants, animals, micro-organisms and other biota and may contaminate water bodies and the groundwater. European legislation puts strong demands on the protection of water against pollution. Road operators are responsible for ensuring that the construction and use of roads is not detrimental to the quality of natural waters.

Keywords Contaminant · pollution · flux · soil process · pathway · chemistry · biota · biology · legislation

---

* *Co-ordinating Authors:*

✉ L. Folkeson
Statens väg- och transportforskningsinstitut/Swedish National Road and Transport Research Institute (VTI)
e-mail: lennart.folkeson@vti.se

✉ T. Bækken
Norsk Institutt for Vannforskning/Norwegian Institute for Water Research (NIVA)
e-mail: torleif.baekken@niva.no

A.R. Dawson (ed.), *Water in Road Structures,* DOI 10.1007/978-1-4020-8562-8_6,
© Springer Science+Business Media B.V. 2009

6.1 Context

Roads and road traffic influence the natural environment in a complex manner. At the same time as roads serve the transport of people and goods, roads take land and form barriers to the movement of people, animals and water in the landscape. A range of pollutants is emitted from roads and traffic and spread to the environment.

The pollutants are transferred away from the road mainly via road-surface runoff and aerial transport but also with percolation through the pavement. Runoff pollution is a much studied issue whereas much less is known about pollutants percolating through the pavement and embankment into the groundwater and surface waters.

The vast majority of the pollutants stay close to the road where they accumulate in vegetation, soil and also animals. To some extent, pollutants are transported further away mainly by aerial transport but also by water movement. In ecosystems receiving traffic pollutants, various ecosystem compartments and ecological processes will be affected.

Water is one of the most important transport media for the pollutants. Soil and water are the main targets of the pollutants (Fig. 6.1). Man, animals and plants are dependent on water of good quality, and legislation typically puts much emphasis on the protection of groundwater and surface water. Given the dense road network and rapidly increasing traffic, protecting the environment from road and traffic pollutants and securing a good water quality is an area of increasing concern to road planners and engineers.

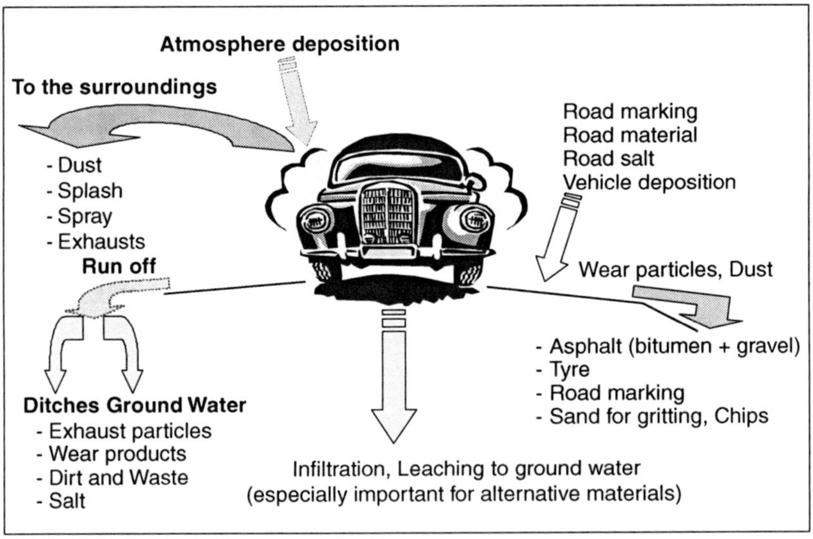

Fig. 6.1 Sources and routes of contaminants in the road environment. Reproduced by permission of the Swedish Road Administration

This chapter is devoted to sources, transport pathways and targets of road and traffic pollutants, as introduced in Chapter 1 (see Section 1.6 and Fig. 1.7). The domain dealt with is confined to the area vertically limited by the pavement surface and the groundwater table, and laterally by the outer drainage ditch at each side of the road (see Fig. 2.1). Pollutant sources such as traffic, cargoes, pavement and maintenance are briefly described. Knowledge of pathways and transport processes is important for the understanding of pollutant appearance in saturated or unsaturated porous media, and consequently for the understanding of effects on ecosystems and their compartments. Following a discussion of these issues, a concluding section briefly refers to EU legislation pertaining to the protection of waters as a natural resource, and the role of roads and traffic in that connection.

6.2 Sources

Pollution sources include five main groups: traffic and cargo, pavement and embankment materials, road equipment, maintenance and operation, and external sources. Road and traffic pollutants having received the greatest attention include heavy metals, hydrocarbons, nutrients (mainly nitrogen), particulates and de-icing salt (Table 6.1). Recently, precious metals worn from catalytic converters have also been given attention. In addition to these pollutants, a range of gaseous pollutants is emitted as a result of fuel combustion. These are to a large extent aerially transported away from the road area, and this issue is beyond the scope of this overview.

The amount of pollutants originating in road and traffic depends on several aspects related to road design, road materials, road maintenance and operation, types of fuel used and traffic characteristics such as volume of light and heavy vehicles, speed and driving behaviour (Pacyna & Nriagu, 1988; Legret & Pagotto, 1999; Sarkar, 2002; Warner et al., 2002; Bohemen & Janssen van de Laak, 2003).

To a great extent, heavy metals, polyaromatic hydrocarbons (PAH) and, to a varying extent, other pollutants (e.g. sodium and chloride from de-icing) emitted from road and traffic sources accumulate in the soil in the vicinity of the road (WHO, 1989; Münch, 1992; Zereini et al., 1997). This continuous accumulation poses a long-lasting stress to vegetation, animals, soil microflora and other compartments of the ecosystems close to roads but seldom gives rise to acute toxic effects. On the contrary, acute toxic effects may occur following the infrequent events of traffic accidents involving dangerous goods such as petrol and diesel as well as acids and other chemicals, sometimes in large quantities. It should be noted here that both concentrations and load are of importance – instantly high concentrations may cause acute damage or may be lethal whereas the long-term performance of the ecosystem (component) may be more influenced by the total load of pollutants over a period of time. Die-off of roadside trees or twigs due to the use of de-icing salt is an example of damage being caused either by instantly high concentrations or the load over time, or both (Bäckman & Folkeson, 1995).

Table 6.1 Sources of contaminants originating in different road and traffic sources

Source	Contaminant type	Common heavy metals	Platinum group elements	Sodium	Hydro-carbons (e.g. PAH, PCB)	Nutrients	Detergents	Organic matter	Particulate matter	Micro-organisms
Traffic and cargo	Car bodies	x			x		x		x	
	Tyres	x			x				x	
	Brake pads	x							x	
	Catalytic converters		x							
	Fuel, fuel additives	x			x		x	x	x	
	Lubricants	x			x					
	Cargo	x		x	x	x		x	x	x
	Spillage	x			x	x		x	x	
Pavement & embankment materials	Aggregate	x							x	
	Bitumen				x			x	x	
	Secondary (alternative) materials	x		x	x			x	x	
Road equipment	Crash barriers, signposts	x							x	
	Road markings							x	x	
Maintenance & operation	Winter maintenance			x			x	x	x	
	Summer maintenance			x		x	x	x	x	
	Painting	x							x	
	Vegetation control				x			x		
	Snow banks and heaps	x	x	x	x	x	x	x	x	x
External sources	Litter	x		x		x		x	x	x
	Excreta			x		x		x	x	x
	Long-range air pollution	x	x	x	x	x		x	x	x

"Cargo": spills and littering from cargoes as well as compounds released upon accidents involving dangerous goods. "Common heavy metals" here include iron, copper, zinc, cadmium, lead, chromium, nickel, cobalt and vanadium. "Platinum group elements" here include rhodium, palladium, iridium and platinum. Information from literature reviews including Sansalone & Buchberger (1997), James (1999), Leitão et al. (2000), Ek et al. (2004), Folkeson (2005).

6.2.1 Traffic and Cargo

Road traffic and cargo produce a range of compounds that pollute the environment. Corrosion of vehicle compartments is a source of heavy metals. Tyre wear gives rise to particles containing zinc, cadmium and iron (literature cited by Fergusson, 1990 [p. 420]; Landner & Lindeström, 1998; Sarkar, 2002). Brake pads and brake linings emit copper, zinc and lead (Weckwerth, 2001). Fuel, fuel additives and lubricants are sources of hydrocarbons. Lead (Pb) is no longer allowed in the EU states but in countries where leaded petrol is still used, e.g. many African countries, this metal is emitted in the exhausts. Wear of catalytic converters gives rise to emission of platinum, palladium and rhodium, though in minor amounts. Spills and littering from cargoes also release a wide range of contaminants. Car-polish and windscreen cleaning agents give rise to the spread of organic detergents. Snow banks along roads accumulate the pollutants over time and may become highly polluted. Through petrol and diesel spillages and other contamination, petrol-filling stations, often situated adjacent to roads, continuously contribute a range of contaminants, notably organic compounds from petrol products, to road runoff and the road environment.

6.2.2 Pavement and Embankment Materials

Pavement and embankment materials can be sources of contaminants that reach the environment either through leaching, runoff transport or aerial transport. The amount reaching the environment varies to a great extent with the type of material used in the various layers, the type, condition and wear resistance of the surface layer, the influence of water and traffic, and a range of other factors.

Pollutant leaching from modern types of bitumen used in asphalt pavements is usually low (Lindgren, 1998). As a substitute for or compliment to natural aggregates, various kinds of secondary materials may be used in road constructions. Some of the most commonly used secondary or manufactured materials include:

- crushed asphalt, concrete and brick (from old road pavements and demolished buildings);
- rock or soil associated with mining activities;
- by-products from metallurgical processes, such as slag;
- pulverised and bottom fuel ash – particularly "fly ash" from coal burning electricity generation; and
- other industrial by-products such as bottom ash from municipal solid waste incineration.

The re-use of materials can be considered advantageous from a natural resource-management point of view. The content of hazardous compounds must, however, be considered. A range of heavy metals and other pollutants such as oil and organic micro-contaminants (e.g. PAH, PCB) may be contained in such alternative materials. The concentrations and leaching ability vary greatly between materials and should be tested to ascertain feasibility for road-construction usage (Baldwin et al., 1997; Lindgren, 1998; Apul et al., 2003; Hill, 2004; Olsson, 2005; Dawson et al., 2006).

Pollutant leaching from road-construction materials containing potentially harmful chemicals has been subject to a Czech field study (Jandová, 2006). Water seeping down from the road surface through the pavement and embankment was collected 1.5 m beneath the road surface using the device described in Chapter 7 (Section 7.4.5 and Fig. 7.8), having passed through a pavement foundation formed of slag. The data of Legret et al. (2005), for water having passed through an asphalt containing recycled components, are given for comparison in Table 6.2. Significant PAH concentrations in the soil beneath the asphalt were observed also by Sadler et al. (1999) due to water entering the environment through leaching from asphalt surfaces. Results from leaching tests on standard hot-mix asphalt have been reported by Kriech (1990, 1991). Except for naphthalene, all PAH were below the detection limits. The same fact was observed for metals – only chromium was found in concentrations above the detection limit. Legret et al. (2005) analysed percolating water through two core samples containing 10% and 20% of reclaimed asphalt pavement. They also described leaching of selected heavy metals and PAH from reclaimed

Table 6.2 Leaching of pollutants from road construction materials containing recycled materials

Chemical characterization (μg/l except pH)	Jandová (2006) Slag under asphalt	Legret et al. (2005)	
		10% recycling	20% recycling
pH	6.99	6.9	6.9
Cu	9.8	20	21
Cr	14.9	5	8
Cd	<0.1	1.6	1.0
Ni	30.7	11	11
Zn	16.0	250	317
Pb	4.3	BDL	BDL
Anthracene	0.0001	BDL	BDL
Benzo(a)anthracene	0.0004	–	–
Benzo(a)pyrene	0.0012	BDL	BDL
Benzo(b)fluoranthene	0.0006	BDL	BDL
Benzo(ghi)perylene	0.0005	BDL	BDL
Benzo(k)fluoranthene	0.0009	BDL	BDL
Dibenzo(ah)anthracene	0.0001	BDL	BDL
Fluoranthene	0.0021	0.035	0.035
Indeno(123cd)pyrene	0.0007	BDL	BDL
Naphthalene	0.0006	–	–
Phenanthrene	0.0008	–	–
Pyrene	0.0019	–	–
Chrysene	0.0008	–	–
Acenaphthylene	0.0001	–	–
Acenaphthene	0.0002	–	–
Fluorene	0.0003	–	–

BDL = below detection limit; – = not analysed.

asphalt pavement in samples from an experimental site that were tested in both static batch tests and column leaching tests.

Where allowed, the use of studded tyres causes substantial pavement wear, typically in the range of 2–10 g/km/vehicle for modern pavements of high quality (Jacobson, 2005). The wear is higher from pavements of lower quality. Pavement wear results in high aerial concentrations of particles. Onto these particles, other pollutants such as heavy metals become adsorbed (Dahl et al., 2006; Lindbom et al., 2006). Aggregates of different mineralogical origin vary in their heavy-metal content. Granite/gneissic aggregates, e.g., have been shown to contain higher concentrations of heavy metals than does porphyry (Lindgren, 1996). This concentration difference in combination with lower resistance of granite/gneiss to studded-tyre wear results in higher release of Cu, Cr and Zn from this type of aggregate than from porphyry (Lindgren, 1996). The build-up of tyre-generated pavement-wear dust on the street surface and along streets during the winter often results in greatly elevated aerial particle concentrations during dry winter and spring days (Gustafsson, 2002). Dust generation from the unbound surface layers of gravel roads is a well-known problem (Oscarsson, 2007).

6.2.3 Road Equipment

Road equipment comprises crash barriers, road signs, sign-posts, lamp-posts, etc. Many of these structures are made of galvanized steel. Corrosion of these surfaces releases zinc to the environment (Barbosa & Hvitved-Jacobsen, 1999). Corrosion is promoted under moist conditions often prevailing as a result of splashing from the traffic during and after precipitation. Soiling and the use of de-icing salt further enhance the corrosion. Re-painting is usually preceded by the removal of old paint. The old paint may contain heavy metals. Regular washing of road equipment may contribute pollutants to the environment in cases where detergents are used (Folkeson, 2005).

6.2.4 Maintenance and Operation

Many measures taken within road maintenance and operation introduce pollutants into the highway environment. De-icing activities are among the most important of these. In countries with a cold climate, de-icing and snow clearing are important measures to reduce slipperiness and maintain the functionality of the road during periods with frost or snow. Ice and snow control is performed mechanically (ploughing) or with the use of chemicals. The most widely used chemical is sodium chloride (NaCl). As an anti-caking agent, a minute quantity of potassium-ferrocyanide is often added to the salt. At places, other chemicals than NaCl are used, e.g. urea on some bridges, or calcium chloride or calcium magnesium acetate (Ihs & Gustafson, 1996; Persson & Ihs, 1998). Winter operation of high-class roads, usually heavily trafficked highways, in cold regions is accompanied with the use of large quantities of salt. De-icing chemicals can thus be a considerable source of contamination of soil as well as groundwater and surface waters (Blomqvist, 1998; Johansson Thunqvist, 2003). Moreover, de-icing salt has been shown to mobilise heavy metals accumulated in roadside soils (Norrström & Jacks, 1998). Dust-binding chemicals used mainly on gravel roads include inorganic salts such as calcium chloride ($CaCl_2$) and magnesium chloride ($MgCl_2$) (Alzubaidi, 1999).

Roadside vegetation and its maintenance also influence the transport of pollutants having entered the road environment. Dense and tall vegetation close to the road will trap pollutants and diminish their spread away from the road (Folkeson, 2005). Upon decay of the plant litter, the pollutants trapped or taken up by the shoots will enter the soil and contribute to pollutant accumulation in the roadside ecosystem. If the mown vegetation is collected, pollutants in the cut material will be exported from the roadside ecosystem. Increasingly, around the world, especially in parts of Europe, the USA and Australasia, vegetated swales (Fig. 1.10) at the side of roads are being deliberately employed as a part of the environmental management of the highway runoff water. They aim to reduce the quantity and improve the quality of runoff that enters groundwater (see also Chapter 12, e.g. Fletcher et al., 2002).

In countries where chemical vegetation control is still not banned, herbicides are directly released into the roadside environment. Unless rapidly degraded into less harmful substances, these toxins may contribute to groundwater or surface-water contamination.

Ditch clearing involves the handling of soils that can be heavily polluted with organic pollutants and heavy metals. Displacement of the material to the outer slope will lead to the accumulation of pollutants in the road area and eventually to the leaching of, e.g., heavy metals to the groundwater or surface water bodies. Where rehabilitation of roads is planned, any spreading of pollutants having accumulated in the road body or the roadside should be avoided.

Road-runoff water carries large amounts of pollutants away from the road surface. The amounts so transported vary greatly depending on a range of factors, the most important being traffic volume and characteristics and amount of precipitation. Pollutant concentrations in runoff have been widely studied. Concentration ranges commonly reported are collected in Table 6.3.

Care must be taken both in road design and in road operation so as to avoid contamination of surface waters and the groundwater. The Water Framework Directive aims at securing good quality in all natural waters, not only where sensitive aquifers or drinking-water abstraction points could be at risk (see Section 6.5 below). Some national road authorities have handbooks for the treatment of highway runoff, e.g. Sweden (Vägdagvatten, 2004).

At some places, runoff water is diverted to retention ponds or other facilities for handling of pollutants (Hvitved-Jacobsen & Yousef, 1991). Facilities for protection of the environment from pollutants should be properly maintained so as to secure the continuous effectiveness of the facility. For instance, sediments in retention ponds accumulate large amounts of pollutants and must be treated or disposed in such a way that pollutants do not enter into the environment (Hvitved-Jacobsen & Yousef, 1991; Stead-Dexter & Ward, 2004).

Table 6.3 Illustrative values of highway run-off water quality obtained in various studies

Country, location, publication	AADT	pH		Conductivity (μS/cm)		Tot. susp'd solids (mg/l)		Pb (μg/l)		Zn (μg/l)		Cu (μg/l)		Cd (μg/l)		Cr (μg/l)	
		min	max	min	max	min	max	min	max	min	max	min	max	min	max	min	max
USA, Bellevue WA (Ebbert et al., 1983)[a]		3.4	7.9	12	1,480	1	2,740	4	1,800	–	–	–	–	–	–	–	–
USA, Ohio (Pitt, 1985)[b]		5.2	7.4	16	300	24	620	<100	820	30	370	–	–	–	–	–	–
Norway (Lygren et al., 1984)[b]	8,000	6.7	9.1	41	5,870	162	2,420	62	690	91	740	10	430	–	–	–	–
Germany (Stotz, 1987)[b]	41,000	–	–	–	–		137	–	202	–	360	–	97	–	–	–	–
	47,000	–	–	–	–		181	–	245	–	620	–	117	–	–	–	–
	40,600	–	–	–	–		252	–	163	–	320	–	58	–	–	–	–
UK (Revitt et al., 1987)[b]	37,600	–	–	–	–	2	192	–	181	–	–	–	63	–	–	–	–
UK (Hamilton et al., 1987)[b]	720	–	–	–	–	–	–	–	28.1	–	16.6	–	6.5	–	–	–	–
Germany (Dannecker et al., 1990)[b]	500	–	–	–	–	–	–	–	122	–	166	–	75.9	–	–	–	–
USA (Hvitved-Jacobsen & Yousef, 1991)[b]	–	5.9	7.8	45	175	–	–	30	379	13	173	10	101	–	–	–	–
UK (Hewitt & Rashed, 1992)[b]	150	–	–	–	–	–	–	1	151	0.7	65	0	14	–	–	–	–
France (Bardin et al., 1996)[b]	–	–	–	–	–	37	128	<5	90	177	681	9	49	–	–	–	–

Table 6.3 (continued)

Country, location, publication	AADT	pH		Conductivity (μS/cm)		Tot. susp'd solids (mg/l)		Pb (μg/l)		Zn (μg/l)		Cu (μg/l)		Cd (μg/l)		Cr (μg/l)	
		min	max	min	max	min	max	min	max	min	max	min	max	min	max	min	max
USA (Thomson et al., 1997)	–	–	–	–	–	–	116	–	–	–	169	–	–	–	–	–	–
USA, Texas (Barrett et al., 1998)	8,780	–	–	–	–	–	91	–	15	–	44	–	7	–	–	–	–
	47,200	–	–	–	–	–	19	–	3	–	24	–	12	–	–	–	–
	58,200	–	–	–	–	–	129	–	53	–	222	–	37	–	–	–	–
Portugal, Vila Real (Barbosa, 1999)	6,000	5.9	7.2	8.8	184	<8	147	<1	200	<50	1,460	<1	54	–	–	–	–
UK (Hares & Ward, 1999)	140,000	–	–	–	–	–	–	–	81	–	208	–	274	–	14.1	–	105
	120,000	–	–	–	–	–	–	–	70	–	188	–	248	–	11.9	–	86
UK (Moy et al., 2002)	71,900	–	–	–	–	–	88.6	–	–	–	8.6	–	–	–	–	–	–
	23,600	–	–	–	–	–	318	–	51.4	–	163	–	33.6	–	0.99	–	11.5
	36,100	–	–	–	–	–	101	–	50.4	–	66.8	–	23.3	–	0.56	–	9.08
	83,600	–	–	–	–	–	82.7	–	16.7	–	29.0	–	11.8	–	0.25	–	7.73
	65,000	–	–	–	–	–	45.8	–	15.4	–	55.7	–	17.6	–	0.43	–	4.82
	37,200	–	–	–	–	–	51.4	–	4.38	–	21.4	–	16.5	–	0.21	–	2.72
	All	–	–	–	–	15.2	1,350	0.00	178	0.00	536	0.00	90.0	0.00	5.40	0.00	49.0
USA (Kayhanian et al., 2003)	<30,000	–	7.0	–	–	–	168	–	1.2	–	35.3	–	6.5	–	–	–	1.7
	>30,000	–	7.4	–	–	–	145	–	6.1	–	79.1	–	14.7	–	0.3	–	2.6
	All	5.1	10.1	–	–	1	5,100	0.2	414	3	1,020	1	121	0.02	6.1	0.6	22
UK, Reading[c]	98,200	6.0	7.7	150	12,000	160	704	43	1,800	140	4,200	50	1,000	<1	13		<20
UK, Oxford[c]	77,700	6.5	6.7	72	2,000	70	134	<20	54	84	200	22	55		<1		<20
Netherlands, Nieuwegein#[c]	150,000	6.5	7.6	120	9,600	–	–	3	95	52	1,700	17	160	0	2	0	5
Netherlands, Spaarnwoude#[c]	90,000	5.7	7.8	90	3,500	–	–	0	88	28	290	13	61	0	3	0	20
Sweden, Svaneberg[c]	7,350	6.3	7.1	30	10,000	–	–	3	18	51	220	6	70	0	0	0	2
Sweden, Norsholm[c]	18,000	6.2	7.7	50	33,000	–	–	4	43	92	490	12	100	0	1	2	11

Table 6.3 (continued)

Country, location, publication	AADT	pH		Conductivity (μS/cm)		Tot. susp'd solids (mg/l)		Pb (μg/l)		Zn (μg/l)		Cu (μg/l)		Cd (μg/l)		Cr (μg/l)	
		min	max	min	max	min	max	min	max	min	max	min	max	min	max	min	max
Finland, Lohja[c]	13,700	6.8	7.6	59	5,100	<10	50	6	15	54	88	0	17	0.08	0.2	0	3
Finland, Utti[c]	8,000	6.9	7.1	57	2,400	<10	10	5	10	57	92	0	16	0.05	<0.3		<10
Denmark, Vejenbrod[c]	29,000	6.8	7.9	42	14,000	<10	40	8	46	47	330	3	95	<0.1	1	4	66
Denmark, Rud[c]	22,000	6.6	7.3	31	20,000	13	607	5	47	100	700	18	140	0.07	1	1	9
France, Erdre#[c]	24,000	6.7	7.8	41	5,300	6	507	5	41	130	460	<2	32	<0.10	2	<0.5	2
France, Houdan#[c]	21,000	7.0	7.9	91	1,300	0	114	10	76	<10	300	8	48	0.10	1	1	6
Portugal, Recta do Cabo[c]	21,800	7.5	8.3	120	1,400	18	1,560		<100	<100	170	2	130		<10		<100
Portugal, Vila Real[c]	8,500	6.6	7.5	<50	<110	<3	316		<100	1,100	2,000	1	<100		<10		<100

AADT = annual average daily traffic (vehicles/day); # porous asphalt.

[a] From Matos et al. (1999). [b] From Barbosa (1999). [c] From Folkeson (2000) and TRL (2002).

Likewise, water used for the washing of road tunnels (pavement, walls and roof) must be treated in a way that prevents the pollutants in the rinsing water from reaching the environment (Cordt et al., 1992; Barbosa et al., 2006).

6.2.5 Snow and Ice

In regions with a cold climate, snow and ice may cover the road surface for a period. Various machinery is used to clear roads of snow. On icy surfaces, sand or grit may be used to increase the friction. For de-icing purposes, road salt is used, mostly NaCl. The salt makes the road wet, thus keeping more of the pollutants on the road surface with potential to leak into cracks in the road surface and along the road shoulder.

If let lying for an extended period of time, snow deposited along roads often becomes heavily loaded with traffic pollutants via splash and spray. The deposition rates of pollutants to the snow banks along heavily trafficked roads may be high (Table 6.4). The resulting concentrations in the snow banks may also be high but depend on the amount of snowfall (Fig. 6.2). Many heavy metals increase their solubility in the presence of ions, e.g. resulting from de-icing with NaCl. Often occurring without a coinciding heavy rainfall which would have diluted the solution, the first flush following the snow melt has high concentrations of most water-soluble pollutants. This flush mobilises considerable amounts of pollutants, often over a short period of time.

Table 6.4 Rates of deposition on snow banks for a selection of traffic pollutants from streets in two cities of Norway (Bækken, 1994b; Bækken & Tjomsland, 2001)

AADT		2,000	6,000–7,000	15,000–38,000	88,000
Cd	mg/m^2/week	0.002–0.03	0.03–0.06	0.10	0.09
Cr	mg/m^2/week	0.15–2.2	0.52	3.06	1.10
Cu	mg/m^2/week	0.1–9	0.7–25	4.82–20	2.78
Fe	mg/m^2/week	11	188	672	485
Ni	mg/m^2/week	0.02–1.4	0.2–5.8	0.9–4.9	0.60
Pb	mg/m^2/week	0.2–1.1	1.3–1.4	4.9–8.1	5.40
Zn	mg/m^2/week	0.3–6	2.6–34	16–31	14
Sum PAH16	μg/m^2/week	165–242	293–1,940	1,390–1,680	5,520
Sum cPAH	μg/m^2/week	20–34	46–263	98–172	166
HCB	μg/m^2/week	0.007–2.6	0.9–7.8	3.00	1.00

AADT = annual average daily traffic. PAH16 = a selection of 16 internationally agreed standard polyaromatic hydrocarbons (PAH) congeners. cPAH = potentially carcinogenic PAH congeners. HCB = hexachlorobenzene.

6.2.6 External Sources

To some extent, contaminants occurring on the road surface or in the road area have other sources than the traffic or the road. Such sources may be either local or remote.

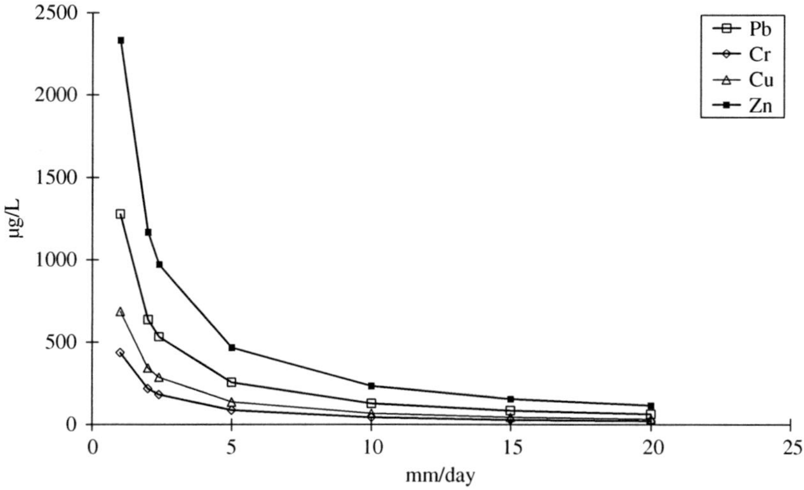

Fig. 6.2 Estimates of pollution concentrations in snow banks along a highly trafficked city road (AADT 40 000) as a function of the intensity of snowfall (expressed as mm of water) (Bækken, 1994b)

Local sources may include agricultural and industrial activities, dust and runoff water from buildings, e.g. copper-plated roofs, and heating by oil, coal and wood. Pollutants include particles, heavy metals, micro-organic pollutants, pesticides, organic carbon and compounds containing nutrients. At places, excreta from birds and other animals (mainly in built-up areas), as well as animal carcasses, may contribute nitrogen, phosphorus, organic compounds and micro-organisms (Murozumi et al., 1969; Elgmork et al., 1973; Wiman et al., 1990; Zereini et al., 2001).

Remote sources of long-range transported pollutants are mainly associated with industry, heating and traffic. These pollutants represent a wide variety of compounds including particles, heavy metals, nitrogen- and sulphur-containing compounds, micro-organic pollutants such as PAH and chloro-organic compounds (e.g. PCB, HCB). An important observation made by Landner & Reuther (2004), in a review study, is that long-range transported contaminants arriving in the road area will be of minor importance compared to the pollution originating from the road and traffic in the immediate vicinity.

6.3 Flow, Transport and Transformation Processes

Road and traffic pollutants are emitted in gaseous, solid or liquid form.

Materials used for the construction of pavements and embankments contain pollutants mainly in the solid state. However, pollutants initially in the solid state can be released into water. Two processes are at work:

- desorption – chemicals are detached from the solids to which they are loosely bound, and
- dissolution – chemicals are dissolved by adjacent water. Together these are known as leaching.

Pollutants will move from the solid phase to the dissolved phase until:

- The water cannot hold more ("solubility limit"); or
- There is no more solid phase to be desorbed or dissolved ("source limit"); or
- There is insufficient contact time for the processes of desorption or dissolution to complete ("availability limit").

Road construction materials may contain a variety of potentially harmful chemicals. The quantity of pollutants leached depends on factors including the surface area exposed to leaching, the material history and the pH, redox potential and other chemical and physical characteristics of the leachate.

A presentation of a conceptual model of water fluxes from the road construction is presented in Fig. 2.1.

6.3.1 Physical Processes

Many physical processes in the pavement, the embankment and the road environment influence the flow of water away from the road surface. Pollution transport is heavily influenced by the physical and chemical characteristics of the specific pollutants. It is also strongly influenced by the interaction between pollutants, and with materials making up the pavement and embankment. During their transport, pollutants interact with materials in the solid, liquid and gaseous form.

The physical processes of the movement of pollutants in and by fluids in roads and their environment can be described in terms of pollutant mass transport. Mass transport can take place in solution, in suspension or in the form of particulate matter. There are great differences between these processes, and they also vary between the unsaturated and the saturated part of the road construction. The majority of pollutant transport from the road surface is via surface runoff towards the soil, surface water and groundwater. Part of the precipitation falling on the pavement surface is infiltrated through the pavement or the soil adjacent to the pavement. In the beginning, the infiltrated water flows more or less vertically through the unsaturated zone. The nature of this flow mainly depends on the road geometry and the materials used in the road construction. Once having reached the groundwater, the infiltrated water will follow the direction of the groundwater flow that is usually more or less horizontal.

In porous pavements, in the embankment and in the soil adjacent to the road, the transport of pollutants is usually in water solution. Pavement cracks also allow transport of pollutants in the particulate form. Due to the clogging of pore spaces with particulates, however, this kind of transport is stopped some ten centimetres below the pavement surface (Brenčič, 2007 pers. comm.; cf. Khilar & Fogler, 1998).

6.3.1.1 Mass Transport in Saturated Media

Transport in saturated soil takes place in that part of soil where pores are completely saturated by water. In the road construction, this usually occurs in the subgrade but rarely in the sub-base. Three principal transport processes are defined:

- Diffusion – pollutants move from compartments with higher concentrations to compartments with lower concentrations, even if the fluid is not moving;
- Advection – pollutants are carried with the flow of the water;
- Dispersion – the pollutants are locally redistributed due to local variations in fluid flow in the pores of the soil or pavement material.

Diffusion

Diffusion will occur as long as a concentration gradient exists. The diffusing mass in the water is proportional to the concentration gradient, which can be expressed as Fick's first law. In one dimension it is defined as

$$F = -D_d \frac{dC}{dx} \tag{6.1}$$

where F = mass flux of solute (units of M/L^{2}T); D_d = diffusion coefficient (units of L^2/T); C = solute concentration (units of M/L^3) and dC/dx = concentration gradient (units of M/L^4). The negative sign indicates that movement is from areas of higher concentration to areas of lower concentration. In the case where concentrations change with time, Fick's second law applies. In one dimension it is defined as:

$$\frac{\partial C}{\partial t} = D_d \frac{\partial^2 C}{\partial x^2} \tag{6.2}$$

where $\partial C/\partial t$ denotes change of concentration with time.

Diffusion in pores cannot proceed as fast as it can in open water because the ions must follow longer pathways as they travel around grains of road material. To account for this, an effective diffusion coefficient D^* is introduced. It is defined as

$$D^* = \omega D_d \tag{6.3}$$

where ω is a dimensionless coefficient that is related to the tortuosity. Tortuosity is defined as the ratio between the linear distance between the starting and ending points of particle flow and the actual flow path of the flowing water particle through the pore space. The value of ω is always less than 1 and is usually defined by diffusion experiments.

Advection

In the road construction, a dissolved contaminant may be carried along with flowing water in pores. This process is called advective transport, or convection. The amount of solute that is being transported is a function of the solute concentration in the water and the flux of water infiltrating from the pavement surface. For one-dimensional flow normal to a unit area of the porous media, the quantity of flowing water is equal to the average linear velocity times the effective porosity and is defined as

$$v = \frac{K}{n_e} \frac{dh}{dl} \tag{6.4}$$

where v = average linear velocity (L/T); K = coefficient of permeability (i.e. hydraulic conductivity) (L/T); n_e = effective porosity (no units) and dh/dl = hydraulic gradient (no units).

Due to advection, the one-dimensional mass flux, F, is equal to the quantity of water flowing times the concentration of dissolved solids and is given as

$$F = v n_e C \tag{6.5}$$

One-dimensional advection in the x-direction is, then, defined as

$$\frac{\partial C}{\partial t} = -v_x \frac{\partial C}{\partial x} \tag{6.6}$$

where v_x is the velocity of flow in the x-direction. According to this one-dimensional advection equation, the mass transport in homogeneous porous media is represented with a sharp front.

Dispersion

Water in porous media is moving at rates that are both greater and less than the average linear velocity. In a sufficient volume where individual pores are averaged, three phenomena of mass transport in pores are present:

- As a fluid moves through the pores, it will move faster in the centre of pores than along the edges;
- In porous media, some of the particles in the fluid will travel along longer flow paths than other particles to travel the same linear distance;
- Some pores are larger than others, allowing faster movement.

Due to different velocities of water inside the pores, the invading pollutant dissolved in the water does not travel at the same velocity, and mixing will occur along the flow path. This mixing is called mechanical dispersion, and it results in a dilution of the solute at the advancing edge of flow. The mixing that occurs along the direction of the flow path is called longitudinal dispersion. An advancing solute front will also tend to spread in directions normal to the direction of flow because at the pore scale the flow paths can diverge. The result is transverse dispersion which is mixing in the direction normal to the flow path. In the road environment, the dispersal of a pollutant having penetrated into the sub-base will usually occur perpendicularly to the road course.

If we assume that mechanical dispersion can be described by Fick's laws for diffusion and the amount of mechanical dispersion is a function of the average linear velocity, a coefficient of mechanical dispersion can be introduced. This is equal to a property of the medium called dynamic dispersivity, being α times the average linear velocity, v_x.

In water flowing through porous media, the process of molecular diffusion cannot be separated from mechanical dispersion. The two are combined to define a parameter called the hydrodynamic dispersion coefficient, D:

$$D_l = \alpha_l v + D^* \tag{6.7}$$

$$D_t = \alpha_t v + D^* \tag{6.8}$$

where D_l = hydrodynamic dispersion coefficient parallel to the principal direction of flow (longitudinal) (with units of L^2/T) and D_t = hydrodynamic dispersion coefficient perpendicular to the principal direction of flow (transversal) (also units of L^2/T). α_l = longitudinal dynamic dispersivity and α_t = transversal dynamic dispersivity (both with units of L).

By the combination of the equations above and with proper initial and boundary conditions, the total mass transport of a non-reactive pollutant in two-dimensional saturated porous media can be described by an advection-dispersion equation defined as follows with v_x being the velocity of flow in the x-direction, as above:

$$D_l \frac{\partial^2 C}{\partial x^2} + D_t \frac{\partial^2 C}{\partial x^2} - v_x \frac{\partial C}{\partial x} = \frac{\partial C}{\partial t} \tag{6.9}$$

Often the dispersion and diffusion terms are combined with a "hydrodynamic dispersion coefficient", $D_h = D_l + D_t$, being used to combine the effects of diffusion and dispersion. Various analytical and numerical solutions of the equation are possible (see, e.g., Fetter, 1993) dependent on the boundary conditions, but will generally involve a distribution of contaminants, with distance from the source and with time, according to a probability function. In practice, the advection-dispersion equation is usually solved by numerical or analytical computer methods such as Hydrus or Stanmod.

6.3.1.2 Mass Transport in Unsaturated Soil

Mass transport in the unsaturated part of the road construction (the sub-base and upper part of the subgrade) strongly depends on the soil moisture distribution inside the pores. Where the mass transport is principally by advection then the water movement direction will control the contaminant flux direction. As the principal fluxes in the vadose zone are those due to evaporation and percolation, it follows that the direction of the mass transport will then be essentially vertical, upwards or towards the lower part of the subgrade.

Soil moisture travelling through the unsaturated part of the road construction moves at different velocities in different pores due to the fact that saturated pores through which the moisture moves have different-sized pore throats and different thickness of the water film on the mineral grains of the soil. The theory of mass transport inside of unsaturated soil is much more complicated than in saturated media. The processes are described in standard textbooks (e.g. Fetter, 1993; Hillel, 2004).

In general, the structure of the equations for mass transport in unsaturated soil is similar to the equations for saturated soil. They differ in that the diffusion and dispersion coefficients and flow velocities for unsaturated soil depend on the water content.

The rate of change of the total pollutant mass present inside the unsaturated part of the road construction must be equal to the difference between the pollutant flux going into the road pavement and that leaving it and going into the saturated subgrade. Due to the complex processes inside the unsaturated road layers, several sources and sinks of pollutants can exist. These processes can be associated with biological decay (for organic contaminants) as well as chemical transformations and precipitation.

If occurring in large quantities, organic compounds originating in petroleum products form a special case of great concern in connection with roads. Spillages of petroleum products from traffic accidents or from petrol-filling stations, often situated adjacent to roads, may result in large quantities of organic compounds entering the road surface of roadside soils. The different interaction of these organic fluids with the soil's chemistry will frequently increase the effective permeability and enhance the flux of the contaminants through the soil – behaviour referred to as incompatibility. Such situations are largely undesirable from an environmental and from other points of view.

Diffusion

The steady-state diffusion of solute in soil moisture is given by

$$F = -D^{*}(\theta)\frac{dC}{dx} \tag{6.10}$$

where F = mass flux of solute (units of M/L^{2}T); $D^*(\theta)$ = soil diffusion coefficient which is a function of the water content, the tortuosity of the soil, and other factors related to the water film on grains (units of L^2/T); C = the concentration of the contaminant (units of M/L^3), x = the distance in the direction of travel (units of L) and dC/dx = the concentration gradient in the soil moisture.

The second-order diffusion equation for transient diffusion of solutes in soil water is defined as

$$\frac{\partial C}{\partial t} = \frac{\partial}{\partial x}\left[D^*(\theta)\frac{\partial C}{\partial x}\right] \tag{6.11}$$

Advection

In road aggregates that are between saturated and residual saturation, some advection can occur with higher saturation allowing advective transport to increase. In terms of contaminant destination, the influence of advection is important. Diffusion will occur evenly in all directions in which the differences in contaminant concentrations exist, while advection will transport contaminants wherever the water is draining, either into a fin drain, or down the vadose zone towards the phreatic surface.

In an unsaturated soil, some of the void space is filled with gas. Due to evaporation, some contaminants will pass from the liquid phase into the gas phase both by volatilisation and by transport in water vapour. Contaminants within the gas will then be transported by diffusion and advection within the gas phase. Exchange processes transferring contaminants between the gas and liquid phases in the road construction are very complex. The transport in the gas phase inside a road construction can be substantial, especially when incidental spills appear on the road surface. However, the extent to which these processes occur inside the road construction is not well known and may be small for most metals. Certainly, as a soil or aggregate becomes less saturated the opportunity for advective transport reduces markedly, particularly because of the substantial reduction in permeability as described in Chapter 2, Section 2.8.

Dispersion

Rather as in Eq. 6.7, the soil-moisture dispersion coefficient, $D(\theta)$, is defined as the sum of the mechanical and diffusion mixing and is now expressed as:

$$D(\theta) = \xi\,|v| + D^*(\theta) \tag{6.12}$$

where ξ = an empirical dispersivity measurement (units L) that depends on the soil moisture and ν = the average linear soil moisture velocity. This definition of $D(\theta)$ may be contrasted with that for D_l given above for saturated conditions (see Eq. 6.7) which includes a soil tortuosity term, α, in place of the ξ term which is also controlled by the water content.

In a road aggregate where pores are only partially saturated, especially during dry periods, the capillary suction of the aggregate can increase very significantly, moving the aggregate towards its residual saturation condition and hindering advective contaminant transport. In these conditions, contaminant transport will, therefore, be very slow and primarily occur by diffusion.

6.3.1.3 Mass Transport in Surface Runoff

Where precipitation falls mainly as storm events, the majority of mass transport in surface runoff is connected with the start of the storm water runoff. This so-called first flush will mobilise pollutants having accumulated on the pavement surface since the previous storm event (Barbosa & Hvitved-Jacobsen, 1999). Concentrations and masses decrease with time, and the relationship between the mass and the contamination pulse depends on many factors (Sansalone & Cristina, 2004). The amount of pollutant in the storm runoff depends on several conditions before the rain. The consideration of the first-flush phenomenon, inclusive of contaminant fluxes, in stormwater treatment is of much concern among practitioners (Hager, 2001).

The transport of pollutants accumulated during dry weather can be described using the theory of sediment transport with water combined with semi-empirical equations. The wash-off rate of pollutants is directly proportional to the amount of material remaining on the surface. During a storm event, the mass of pollution present on the pavement is decreasing exponentially with time (Hall & Hamilton, 1991). The relation can be described as:

$$M(t) = M_0 e^{-JR\rho_w t} \tag{6.13}$$

where $M(t)$ = pollutant mass on the pavement surface (M/L^2) at time t (T); M_0 = pollutant mass on the pavement surface at the beginning of the storm hydrograph (M/L^2); J = rate coefficient (L^2/M); R = runoff (L/T) and ρ_w = density of water (M/L^3).

At the beginning of the storm runoff event, various particles from dry deposition are remobilised. As a consequence of the interaction between water and dry sediment during the storm, the concentrations in the diluted phase are also changing with time.

Part of the runoff water is mobilised by the traffic to form splash and aerosols which will be wind-transported away from the road. The vast majority of the pollutants so mobilised will be deposited close to the road but the smaller particles will be wind transported further away from the road, at least some hundred metres (Blomqvist & Johansson, 1999; Folkeson, 2005). To what extent pollutants are transported in the form of splash/spray or in the form of pavement-surface runoff is governed by factors such as traffic characteristics, weather conditions, topography and the type and condition of the pavement surface. For instance, aerial transport is more limited where porous asphalt is used as compared to conventional asphalt (Legret & Colandini, 1999; Legret et al., 1999; Pagotto et al., 2000).

6.3.1.4 Retardation and Enhancement

In most saturated soils, advection and diffusion/dispersion do not transport contaminants as fast as might be expected from a consideration only of these processes. Often, there is a movement of contaminant from the liquid phase to the solid phase due to various physio-chemical processes (see Section 6.3.2). Together, these processes act to *retard* the contaminant flux. Where the soil solids are, in effect, clean with respect to the contaminant prior to the contaminant's arrival, this retardation may be expressed using a very simple equation:

$$f_R = 1 + \rho_d k_d \tag{6.14}$$

where f_R is the retardation factor (no units), k_d (almost invariably expressed in units of l/kg = mL/g) is the partition factor which is discussed in the next paragraph and ρ_d is the dry density of the soil (for which units of Mg/m^3 will allow Eq. 6.14 to be used directly if k_d is expressed in units of l/kg).

The rate of contaminant flux is slowed by a factor of $1/R$ from that which would be expected assuming only advection, diffusion and dispersion have an effect. This approach allows the effects of physio-chemical processes to be simply modelled by adapting the advection-dispersion Eq. 6.9 as follows:

$$\frac{D_l}{f_R}\frac{\partial^2 C}{\partial x^2} + \frac{D_t}{f_R}\frac{\partial^2 C}{\partial x^2} - \frac{v_x}{f_R}\frac{\partial C}{\partial x} = \frac{\partial C}{\partial t} \tag{6.15}$$

The partition factor, k_d, is a very simple means of describing the concentration of a contaminant in the solid phase to the concentration of a contaminant in the fluid phase at equilibrium conditions. At low concentrations such as those normally experienced in the highway environment (except, perhaps, after certain spillages from vehicle accidents), a linear "isotherm" (relationship between the two concentrations) may not be too inaccurate and is a commonly used characterisation having the benefit of simplicity. Therefore, in such situations, a constant value of k_d is used.

Values of k_d are highly dependent on soil type, fluid and contaminant species. Values vary by several orders of magnitude for apparently small changes in some of these factors. Even with specific laboratory testing, the natural variability of ground conditions, mineral composition, particle size, etc. from place to place in a soil profile means that prediction of the retardation effect is very imprecise. Accordingly it is common to use published values and to compute the most and the least likely retardations that are credible. Values of k_d are available from many sources, notably from the US EPA (EPA, 1999).

It is possible for *enhancement*, the inverse of retardation, to occur, for example when a spillage changes the fluid chemistry causing leaching of contaminants previously bound into the soil or aggregate. When enhancement takes place, the contaminant flux is higher than would otherwise be anticipated. This can be modelled by a value of f_R of less than 1.0, although Eq. 6.15 will not be directly applicable as it assumes that the soil is initially clean as far as the contaminant of interest is concerned.

6.3.1.5 Movement of Non Aqueous Phase Fluids

Non-aqueous liquids, such as petroleum-based fluids, are not, in general, soluble in water so their movement must be considered separately. Although some of the liquid may be soluble or miscible in groundwater to such an extent that it is, thereby, subject to advection, diffusion and dispersion processes as described above, much may remain separate due to its different density and chemistry. These are termed non-aqueous phase liquids (NAPLs). Such fluids with densities less than that of water (light NAPLs) will float on top of groundwater in unconfined situations and their movement will, therefore, be controlled by the gradient of the top of the groundwater – which will act as the stimulus for movement – and the non-hydraulic permeability coefficient for that fluid and soil combination. Fluids with densities greater than that of water (dense NAPLs) will tend to flow vertically or sub-vertically through the groundwater until arrested by a soil stratum which is essentially impermeable to that fluid. Its movement will then be largely controlled by the gradient of the top of that stratum and the non-hydraulic permeability coefficient for that fluid and the soil in which it is contained.

Two of the more common sources of NAPLs in the road environment are spills from (e.g.) tankers and leaking storage tanks. It can be difficult to remove the NAPL from the ground by flow methods as the poor miscibility of the NAPL in water and the particular wettability characteristics between soil particles and the NAPL often means that small droplets are left behind in the soil pores from which the bulk of the NAPL has departed. These small droplets may present a continuing source of low-level contamination over long periods given their low miscibility with/solubility in the surrounding groundwater. A schematic of a light NAPL flow following a spill is illustrated in Fig. 6.3.

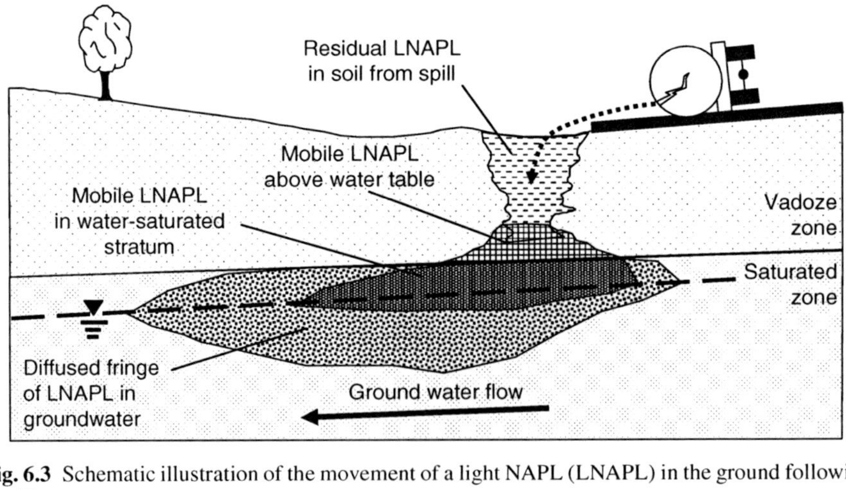

Fig. 6.3 Schematic illustration of the movement of a light NAPL (LNAPL) in the ground following a spill

6.3.2 Chemical Processes

The road construction is a multi-component system which is not isolated but open to physical, chemical and biological interaction with its surroundings. Reactions taking place in the road construction thus influence and are influenced by adjacent systems. For instance, the washing of the road surface by run-off brings organic and inorganic compounds (from sources mentioned in Section 6.2) to road shoulder materials and to neighbouring soils with which they may interact when water infiltrates. Seepage from the road surface into the road structure will also lead to chemical reactions with materials in the various road layers and the underlying soil.

Chemical reactions occurring in the road construction and adjacent soil systems commonly involve the solid and the liquid phases, but the gas phase can also play a role. The most significant chemical processes are sorption/desorption, dissolution/precipitation and ion exchange reactions.

Any transformation occurring during the chemical reaction induces a decrease in the total energy of the system. Under constant conditions, systems tend to evolve more or less quickly (depending on the chemical kinetics) towards a lower energy level. Chemical processes will occur as long as an equilibrium state is not reached or as long as the system is modified. Modifications can be induced by inputs and outputs of material or energy.

In natural waters, metals occur in various forms, so-called species. The speciation (the distribution of the various forms of a metal in a solution) depends on a wide range of factors. One of the most important factors is the presence of compounds capable of forming complexes. Other important factors are the acidity and the redox potential. The speciation greatly influences the solubility and mobility of heavy metals in soils.

In water, metals mainly occur either in ionic form or are associated with particulate matter. For practical reasons, filtering with a mesh size of 0.45 μm is often used to define a limit between ions and particulates.

Organic chemicals may also be present in natural waters, sometimes from natural sources, for example animal carcasses and excreta, decaying vegetation, etc., but often from the consequences of human actions – deliberate or accidental. The concentration of organic solids in the porous media greatly affects the partitioning of organic compounds between the aqueous and the solid phase as dissolved organic substances are, usually, preferentially sorbed to (or released from) organic solids.

The brief introduction given in this sub-section can only provide a few brief pointers to the complex description that would be required to fully explain the interaction between chemicals carried in groundwater and each other and their interaction with the solid structure through which they travel. It is sufficient, here, to make readers aware of the complexity of these and to be aware that both inorganic and organic chemicals, too, can undergo a wide range of reactions and transformations that can result in unexpected outcomes, both good and bad from an environmental point of view.

6.3.2.1 Adsorption/Desorption

In the road context, adsorption/desorption phenomena greatly influence the fate of pollutants entering the road construction, present therein or transported through road-construction layers and further down. Sorption phenomena are also of importance regarding pollutants possibly leached (dissolved) from some road materials (e.g. alternative materials) under the effect of infiltration, and adsorbed on a surface downstream. Sorption/desorption sequences (under the effect of surface characteristics and seepage pH, for example) can lead to a progressive downward transfer of substances.

Adsorption can be defined as the attachment or adhesion of a molecule or an ion in the gaseous or liquid phase to the surface of another substance (an adsorbent) in the solid phase or to the surface of a soil particle. Desorption describes the process by which molecules or ions move in the opposite direction. Adsorption/desorption is a universal surface phenomenon. It can occur at any surface, e.g. surfaces formed by any type of opening, capillary, crack, depression or other type of physical irregularity. The nature of the adsorbing surface plays an essential role in the process. The smaller the size of the soil particles, or the greater the porosity, the more efficiently the adsorption will occur because of the increase in surface area provided. Road pollutants are therefore leached much more quickly through a coarse-textured soil than through a clayey soil (Brenčič, 2006). This feature is of particular relevance when traffic accidents involve cargoes of harmful or toxic compounds.

The adsorption/desorption of substances between the liquid form and the surface of solid-state materials, such as soil particles, is one of the processes of greatest importance for the behaviour of inorganic and organic substances in the soil. The degree of adsorption increases with the concentration of the substance in the solution outside the adsorbent until a maximum is gradually approached. As the reaction kinetics depend on temperature (adsorption decreases with higher temperature because the molecules are more energetic and less easily held by their potential sorbent), the quantitative assessment of adsorption is done by means of so-called isotherms. Various models can be used to interpret isotherms, e.g. Langmuir, Freundlich or Brunauer-Emmet-Teller (BET) (Fig. 6.4) a variant of which is given in Eq. 6.16.

$$S = Q_T \beta C/(1 + \beta C) \tag{6.16}$$

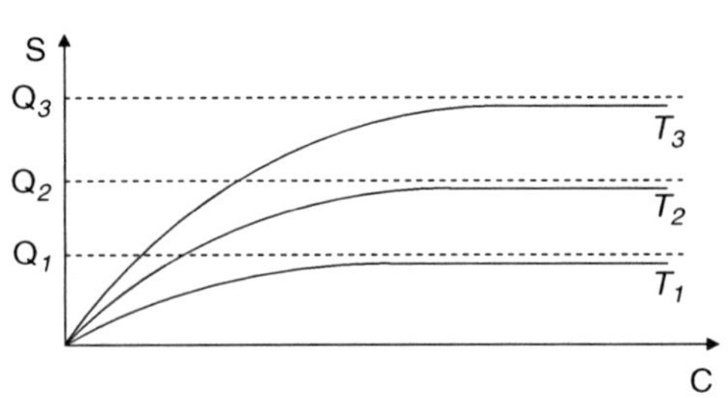

Fig. 6.4 Variation of the sorbed quantity (S) as a function of the concentration of sorbate (C) for different temperatures (T1>T2>T3) – Langmuir isotherm (adapted from Bontoux, 1993 and Selim & Sparks, 2001)

where S = mass of sorbate sorbed per mass of sorbent (typically in units of mg/kg); Q_T = maximum sorption capacity of the sorbent at temperature, $T(°)$, β = a variable that is only a function of the temperature, T, and C = aqueous concentration of sorbate (typically in units of mg/l).

Desorption can occur when a "new" ion (or other chemical) arrives at a sorption site and is sorbed, preferentially, over a previously sorbed ion of a different type. Less readily, sorbed species can be desorbed if the concentration of that species decreases in the groundwater around the sorbent.

Time is required for sorption/desorption reactions to become complete. Therefore the approach adopted both in analysis and in testing is to allow sufficient time for equilibrium to develop. Often this will take hours, perhaps days, to complete. Care is required when the input or output condition is changing due, for example, to flow bringing more contaminant. Then, true equilibrium may not be possible. The use of an isotherm approach necessitates the assumption of equilibrium conditions.

More important, though, is that the adsorption is also pH dependent; cations such as most metal ions are more strongly adsorbed at increasing pH. The degree of adsorption rises sharply in a short interval of increasing pH. This is due to the fact that the charge of the particle surfaces is greatly pH dependent. The pH of the soil thus largely regulates the mobility of heavy metals occurring in the soil. With the exception of some amphoteric compounds (e.g. some metal hydroxides) and some oxyanions (e.g. MoO_4^{2-}, AsO_4^{3-}), the general rule is that many heavy metals are more mobile at lower pH (Berggren Kleja et al., 2006).

6.3.2.2 Dissolution/Precipitation

Rainwater is able to dissolve gas present in the atmosphere (leading to acid rain, for example; see Section 6.2.6). Rainwater is also able to dissolve chemicals present at the road surface (e.g. metals, salts and some organics). Road materials are of course selected for not being soluble but trace elements present in natural and alternative materials can be released by dissolution when leached by seepage. The rise in a water table can also bring about dissolution. Dissolved elements can precipitate downstream where hydrous, pH and/or redox conditions differ from those upstream. Dissolution/precipitation sequences are also part of the circulation of chemicals. Dissolution of CO_2, whether from the air or biological activity, is of great importance to the pH of the soil solution, also in the road context.

Dissolution is the process by which a solution is formed when a soluble substance (a solute) is dissolved in a liquid (a solvent). A true solution is a uniform molecular or ionic mixture of one or more solutes in a solvent, as distinguished from a colloidal solution or dispersion in which the dispersed material is in the form of extremely small particles, 1 μm or less. The solute can be a solid or a gas.

As a polar molecule, water can dissolve ionic substances such as salts but also substances consisting of polar molecules with which the water molecule forms hydrogen bonds. The solubility (i.e. the maximal quantity of a chemical compound that can be dissolved per litre of solvent) is dependent on temperature, pH and activity coefficient.

Whether a substance is present in the dissolved or in the precipitated form is of crucial importance to its mobility and transport in the soil. This is especially true of heavy metals and other micro-pollutants (Ramade, 1998). Depending on variations in the chemical and physical properties of its environment, a given pollutant present in the soil can repeatedly change from being dissolved to being precipitated, and thus from being mobile to being less mobile. Some salts (ionic solids) are very soluble, for instance NaCl and $CaCl_2$ which are used for de-icing and dust-binding, respectively.

The degree of solution of any salt M_pX_q is governed by the dissolution equilibrium:

$$M_pX_{q(s)} <-> pM^{q+} + qX^{p-} \tag{6.17}$$

where the solubility product $K_s = (M^{q+})^p \cdot (X^{p-})^q$. Thus, the greater the solubility product, the greater the solubility of the salt. This equilibrium can be coupled with other equilibria, e.g. acid-base, redox or complexation equilibria. The solubility of a salt is, e.g., dependent also on the pH and the redox status of the soil.

Carbonates are important to the mobility of heavy metals. Carbonates are dissolved upon the interaction with water and with the carbon dioxide present in air, water and soil. Calcium carbonate ($CaCO_3$) is a major constituent of calcareous rock. Where enough free carbonate ions ($CO_3{}^{2-}$) are present, they will react with heavy metal ions to form immobile precipitates, e.g. lead carbonate. The mobility of many heavy metals is low in calcareous soils. On the contrary, heavy metals are often more mobile in acidic soils where carbonates are largely absent (Selim & Sparks, 2001).

Hydroxides of Fe and Mn also play a major role in natural waters and soils. The solubility of hydroxides depends on the acidity (pH) of the water or the soil solution. The solubility of hydroxides decreases when pH increases, passes through a minimum and then increases at higher pH.

Organic molecules that contain polar groups or create hydrogen bonds are to a great extent soluble in soil solution and water. This is the case for organic molecules with groups such as hydroxyl, amine, carboxylic acid, carbonyl, ester or ether.

In the road situation, adsorption and desorption will happen routinely whereas precipitation will depend on the ion concentration. At low concentrations, many metals are under-saturated with respect to their associated mineral phases so that their mobility/retardation is governed by adsorption/desorption. At higher concentrations, both adsorption and precipitation may be occurring to take ions out of solution, but it is the dissolution/precipitation processes that will determine the aqueous concentration of the metal.

6.3.2.3 Exchange Reactions

Exchange reactions take place between two reactants, usually meaning that both are in the liquid phase (although some surface complexation reactions may involve an exchange reaction, too). They include electron exchanges (reactions between oxidizers and reducers), proton exchanges (reactions between acids and bases) and "particle"[1] exchanges (formation of complexes from ions or molecules) (Stumm & Morgan 1996).

On its way from the road surface downwards, the infiltrating seepage (carrying chemicals accumulated during rainfall and runoff) will encounter and interact with varying redox-potential and acidity conditions in the various layers of the road construction and soil layers beneath. The resulting more or less steady conditions will govern the equilibria of chemical reactions. More or less oxidizing or reducing road/soil materials will, through dissolution, create more or less oxidizing or reducing conditions. This will influence the toxicity of some chemicals (chromium for example). In like manner, road/soil materials will influence the acidity/alkalinity of the medium and its buffering capacity. Under special conditions, e.g. where very alkaline man-made road materials are present, percolating water can reach very high pH levels followed by more neutral conditions in subsequent layers. In this way, the buffer capacity of the road/soil materials can mitigate the influence of an acid or base spillage, should it occur.

Reactions Between Oxidizers and Reducers (Electron Exchange)

Many chemical reactions imply the transfer of electrons from one chemical species to another. These reactions are called redox reactions and they are usually rather slow. In soil and water, redox reactions involve hydrogen ions and are thus greatly pH dependent. The most important redox reactions involve oxygen, carbon, nitrogen, sulphur, manganese and iron. In polluted soils, arsenic and mercury can also participate.

The redox potential, and changes thereof, play a crucial role in the behaviour of metals in soils. For instance, iron oxides are formed at high redox potentials. Iron oxides and hydroxides are capable of adsorbing heavy metals onto their surfaces, which will greatly reduce the mobility of the heavy metals. When the redox potential is lowered, the iron oxides dissolve and the adsorbed heavy metals are released and will be available for leaching further down the soil profile. Many redox reactions in nature are speeded up by certain bacteria, however. The bacteria utilise the energy released from redox reactions (Berggren Kleja et al., 2006).

Reactions Between Acids and Bases (Proton Exchange)

The amount of protons (H^+) in solution greatly influences most chemical reactions. Proton transfer reactions are usually very fast. According to the Brønsted definition, protons are provided by an acid and captured by a base. To each acid Ac there is a corresponding base Ba:

$$Ac_1 \leftrightarrow Ba_1 + H^+ \tag{6.18}$$

---

[1] The use of the word 'particle' here is to differentiate the type of chemical component being exchanged from protons and electrons which are much smaller. It is not used to indicate a solid and visible particle (e.g. of sand) as elsewhere in the book.

The acid and the corresponding base constitute an acid/base couple (Ac_1/Ba_1).

In most natural waters, the pH lies within the range from 5 to 8. All the substances dissolved into water (gases, mineral and organic compounds) contribute to the acid-base equilibrium of water. All components of the carbonate system make a major contribution to the acid neutralizing capacity (called alkalinity) of the water and to its base neutralizing capacity (acidity). The buffering capacity of water (the ability of the water to maintain its pH despite any addition of H^+ or OH^-) is also largely determined by the carbonate system. However, dissolved silicates, ammonia, organic bases, sulphides and phosphates also contribute to the alkalinity. In like manner, non-carbonic acids, polyvalent metal ions and organic acids contribute to the acidity.

Rainwater often contains strong acids originating in atmospheric pollutants (dissolution of gases leading to HCl, HNO_3, H_2SO_4). Acid rain may increase the heavy-metal solubility in soils. The pH effect of strong acids on soil and water will depend on the buffering capacity of the soil or water, however. Oxidation reactions lead to a decrease in pH whereas reduction tends to increase the pH.

Formation of Aqueous Complexes ("Particle"[2] Exchange)

Complexes are chemical compounds consisting of a central atom (metal) and ligands (consisting of a group, molecule or ion) tied to the central atom with at least one co-ordination bond. A chelate is a special form of complex where the ligand is attached to the central atom by at least two bonds. The most common ligand in water solutions is the water molecule itself but anions such as hydroxide, carbonate, hydrocarbonate, sulphate and organic acids also form ligands. The formation of a complex from a metal and a ligand is a balanced reaction characterized by a constant (K_c) that is often pH dependent. Some complexations can be considered as "surface complexation" reactions (e.g. of a metal with an iron oxide) as opposed to "aqueous complexation" reactions.

Organic and inorganic complexes are present in all natural waters. Organic acids such as humic acids (originating in humus formation upon decay of plant litter) make up one of the most important types of ligands in natural waters. Humic acids and other types of humic substances greatly affect the solubility and thus the availability of heavy metals to biota. In soil water, humic substances occur in dissolved form and in more or less insoluble aggregates. Compared to heavy metals occurring as insoluble aggregates, heavy metals occurring in the dissolved form are much more mobile and available and therefore more toxic to biota (Berggren Kleja et al., 2006).

Among the inorganic complexes, hydroxides of Fe and Mn are common in natural soils. From a pollution point of view, it is of great importance whether the hydroxides are present in dissolved or precipitated form because hydroxides regulate the mobility of heavy metals. The stability of the hydroxide complexes is greatly

[2] See previous footnote

governed by the pH. Depending on the soil type, but also on the degree and characteristics of the pollution load, roadside soils vary greatly in pH. In many cases, pH is higher close to the road than further away (James, 1999).

Components present in the road/soil environment and likely to form complexes with heavy metals include hydroxides, carbonates, hydrocarbonates, sulphates and organic acids. They originate from deposition, road materials and infiltrating water.

6.3.3 Biological Processes

Usually, roadside soils are or become covered by vegetation. Especially where the plant cover is large or the vegetation dense, the vegetation as a physical body influences the air-borne transport of pollutants from the road and traffic to the surrounding environment. Usually, however, the vegetation is kept low by mowing and bush cutting. To some extent, pollutants deposited on leafy surfaces enter the interior of the plant.

Under good growing conditions, plants will produce a more or less dense root system. Their root mass will greatly influence the movement not only of water but also of pollutants in the soil. Root uptake can withdraw large quantities of water from percolation. Root uptake also forms an important pathway of pollutants into the plant. The tendency to be taken up by roots differs greatly between contaminants and also between plant species. Once absorbed, the pollutants become trapped within the plant and they are therefore removed from the soil system until either the plant is consumed or decomposed.

The vegetation is also a producer of organic matter. Upon death, the plant with its shoot and root parts will form plant litter which will eventually be decomposed to form soil-organic matter. Soil-organic matter is an important factor in a range of biological, chemical and physical processes in the soil.

The ability of plants to take up pollutants, especially through their roots, is actively or passively utilised for run-off treatment. This form of bio-remediation can be an efficient means of treating pollutants accumulating in the road environment. Ditches are often vegetated, and plants such as tall-growing grass or sedge species often absorb and retain heavy metals and other pollutants to a considerable degree. In the case of organic pollutants, the plants help degrade at least some of the compounds. In the case of heavy metals, the pollutants will stay in the roadside environment unless cut vegetation (or the ditch mass) is removed and transported elsewhere. Heavy metals in themselves are not degradable. The use of plants for either absorption or biodegradation (organism-mediated breakdown of substances) of contaminants in soil is known as phytoremediation.

Roadsides are inhabited or otherwise utilised by a variety of animals. Through grazing, animals will ingest pollutants present in or on the biomass. Likewise, animals of prey will ingest any pollutants present in their prey. In the case of mobile animals, this will form a pathway of pollutants to the environment away from the roadside.

Roadside soils also accommodate a range of animals exploring the soil resources. Burrowing organisms such as earthworms and arthropods ingest large quantities of soil. Soil ingestion and excretion is an important means of contaminant transport within the soil. This process may also mobilise contaminants that had previously been bound to soil particles by sorption processes. Tunnelling will also create channels for water flow, which increases soil permeability to water. This will result in any future intrusion of contaminated water passing through the soil more rapidly, which reduces the ability of the soil to adsorb the contaminants.

Every soil is also inhabited by micro-organisms. Micro-organisms are highly involved in the turnover of organic matter in the soil. In natural soils, a wide range of complicated microbial processes involving enzymes participate in the prolonged process in which organic substances from plant, animal and microbial matter are decomposed into simple compounds. Some of these constitute nutrients necessary for biomass build-up with the help of photosynthesis. Organic exudates produced by micro-organisms also greatly influence soil structure.

Bacteria, algae and fungi are highly involved also in the transformation of soil pollutants. Many organic pollutants are gradually degraded to less harmful compounds by the action of micro-organisms. Also heavy-metal pollutants are influenced by micro-organisms. Chelating agents exuded by micro-organisms greatly influence the chemical form and mobility of heavy metals in the soil. Unlike organics, heavy metals, which in themselves are elements, are not decomposed, even if they are transformed into chemical compounds which may render them either less or more available to plant and animal life. The availability and toxicity of heavy metals to plants, animals and micro-organisms is greatly influenced by the heavy-metal speciation. Often, the free hydrated form is the most prevailing form, and also the most available and toxic to biota.

In the vicinity of roads, road pollutants accumulate in soil, water and other ecosystem compartments. There is a wealth of literature documenting various types of detrimental effects of road and traffic pollutants on plants, animals and micro-organisms (see, e.g., Scanlon, 1991). Even if micro-organisms are especially sensitive to toxic substances, plants and animals are also sensitive. The sensitivity differs greatly between various plant, animal and microbial groups, and between toxic substances.

Of the substances occurring in elevated concentrations in road environments, heavy metals, PAH and de-icing salt are the most relevant and most studied. Generally, biological processes involving enzymes are known to be especially prone to disturbance from heavy-metal pollutants. Micro-organism-mediated processes such as organic-matter breakdown, humification, and nitrogen and phosphorus mineralization are largely susceptible to disturbance from elevated heavy-metal concentrations. Such effects have been reported from the vicinity of roads (Tyler, 1974). Reduced photosynthesis rate, growth and reproductive ability are among the most commonly reported effects of heavy-metal exposure to plants and animals (Bazzaz et al., 1974; Rolfe & Bazzaz, 1975; Sprague, 1987; Holdway, 1988; Weis & Weis, 1991; Sarkar, 2002).

Contaminants, especially those with high mobility, often reach surface waters and the groundwater. De-icing-salt contamination of groundwater and surface water bodies is often a problem in countries using de-icing salt (Johansson Thunqvist, 2003).

6.4 Pathways and Targets

Once having entered the road area, pollutants may start their transport to other ecosystem compartments. Any ecosystem compartment that may be affected by a pollutant can be considered a target. The pollutants will not be permanently trapped at these destinations but may stay there for a prolonged period of time.

Pollutants in the solid and in the liquid form are transported to the environment in various ways (see Fig. 6.5):

- infiltration into the road structure and further transport to the groundwater;
- pavement runoff;
- splashing to the road shoulders and ditches;
- spray.

The relationship between different sources of pollution and different targets will be a function of

- the "strength" of the source (i.e. the rate of the emission);
- the pollutant pathway from the source to the target;
- the physical and chemical processes affecting the pollutant during the transport; and
- the "vulnerability" of the target.

Water-borne transport of pollutants occurs on the road surface and on top of the adjacent soil but also in the interior of the road structure. Even if there is good knowledge of parts of these processes, there is insufficient knowledge to provide a quantitative appraisal of these pollutant fluxes on top of, inside and around the road structure towards the different targets. Therefore only a qualitative description of the possibly impacted targets can be provided.

The wearing course of a road is not an impervious layer. Under the influence of rainfall infiltration, pollutants previously settled on the surface course can infiltrate into the road structure. Pollutants included in the matrix of road materials can eventually be made soluble. Then, the first target (A1[3]) is the soil underlying the road structure (the vadose zone). As the "road leachate" can go on percolating towards the saturated zone of the subsoil, the second possible target (B) is groundwater. The distribution of pollutants between targets A1 and B will vary depending on the pollutant, the nature of the underlying soil, the prevailing physical and chemical conditions in the soil, the thickness of the vadose zone and the dynamics of the aquifer.

Another part of the rainfall will be transported on the surface of the road. Runoff water can infiltrate into the road shoulder that is usually made of permeable material. In cold climates the winter precipitation as snow melts in one or more short periods of time during winter and spring. The polluted melt water may infiltrate into the road shoulder/ditches, or run away on the top of a still frozen soil. Even when

[3] Alphanumeric codes are as used in Fig. 6.5

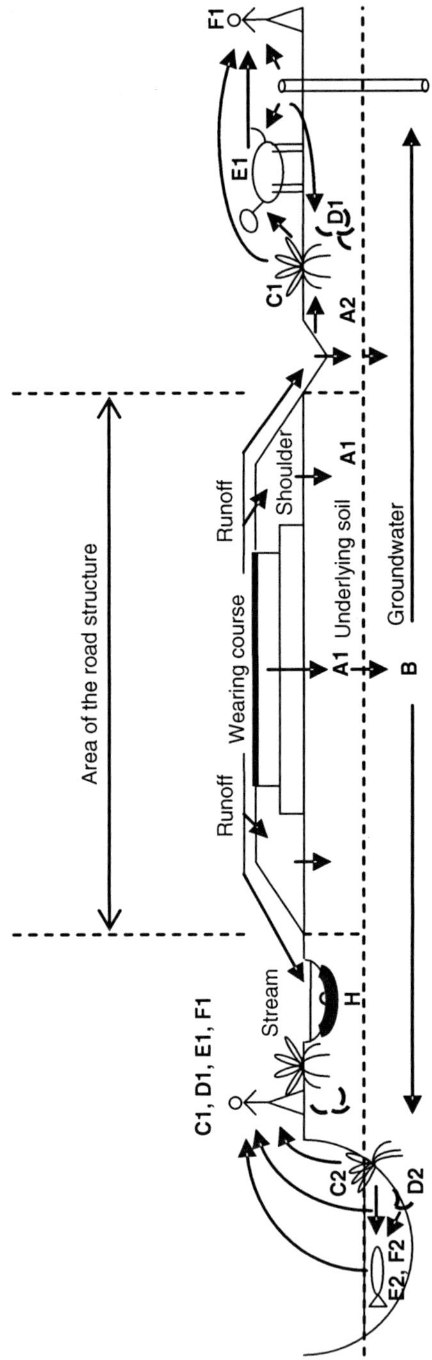

Fig. 6.5 Possible contaminant targets (letters – refer to text) and their relations (*arrows*)

the runoff pathway is somewhat different from those described above, the targets remain the same. Where roads are equipped with an impervious collection system, the runoff water can be transported to a permeable ditch or to more sophisticated water treatment facilities such as infiltration basins and settling basins (retention basins) – see Chapter 13, Section 13.4.8. In the first case, the soil adjacent to the road structure (A2) is a target. In the second case (infiltration down to the saturated zone), the groundwater (B) will also be a target.

From roadside soil, pollutants can become available for plants (C1) or soil-inhabiting animals (D1). These plants and animals can act as sources of contamination of herbivorous (E1) and carnivorous (F1) organisms. Some hazardous substances can accumulate in the organisms and further be biomagnified in the food chain.

As groundwater (B) can be used for drinking water supply or for irrigation, the plant and animal targets can also be impacted through this target.

Runoff water, sometimes collected in treatment facilities, is eventually discharged into natural surface waters. The water in streams, lakes and ponds can thus be a target of pollution (G). In cases of heavy use of road de-icing salt, lake targets may become permanently stratified due to high-density salt water concentrating in the deep water layer. The result is stagnant hypolimnion water (the lower part of the lake volume) with oxygen depletion and biologically dead areas.

As runoff-water pollutants are often adsorbed to particles, the bottom sediments of lakes and slow flowing streams become significant targets (H). Lake sediments may become almost permanent traps for the pollutants. Eventually the water and sediments become sources of contamination of aquatic organisms being plants (C2), decomposers (D2), herbivores (E2) or carnivores (F2) (Bækken, 1994a; Bækken & Færøvig, 2004). As humans are users of water resources, and often the top predator in the food chains, they are the ultimate target (F1).

Similar targets (C2, D2, E2 and F2) can be reached in cases where it is the impacted groundwater (target B) that feeds a surface water body. In streams and lakes, herbivorous (target E1) or carnivorous (target F1) terrestrial consumers can be impacted through the consumption of targets C2, E2 and F2. And finally, similarly to groundwater, surface water bodies can be used for drinking water supply and for irrigation and can therefore impact targets C1, D1, E1 and F1.

6.5 European Legislation

Across Europe, the legislation on the influence of road and road traffic on water and water bodies and associated ecosystems is wide and complex. European legislation in general prohibits water pollution and limits influences on the water biotopes. These general rules are transferred into national legislation in very different ways. Realization of these rules depends on the country's prevailing natural conditions (e.g. climatic regime), uses of water and technical regulations concerning road planning, design, construction and maintenance. In general, water pollution from roads is regulated in two main groups of legislation: environmental law and construction law. This section refers to environmental law.

Water is one of the most comprehensively regulated areas of the EU environmental legislation with directives regulating quality and standards for, e.g., dangerous substances in water, fishing water, drinking water and groundwater. The Water Framework Directive (WFD) of 2000 (EU, 2000) is the most important directive under the group of environmental law that regulates water pollution. The Groundwater Directive (EU, 2006), on the protection of groundwater against pollution and deterioration, is also a feature of the WFD.

The purpose of the WFD is to establish a framework for the protection of inland surface waters, transitional waters, coastal waters as well as groundwater. It aims at enhanced protection and improvement of the aquatic environment, and ensures the progressive reduction of pollution of water, based on a long-term protection and prevention of further pollution. Common environmental quality standards and emission limit values for certain groups or families of pollutants should be laid down as minimum requirements in Community legislation.

At latest 15 years after the date of entry into force of the WFD, i.e. 2015, Member States shall have protected all their water bodies with the aim of having a good water status. Good water quality is such that the concentrations of pollutants do not exceed the quality standards applicable under other relevant Community legislations.

Furthermore, the WFD presents an indicative list of what in general is considered the main groups of pollutants in water. Some of these are toxic while others are nutrient salts or substances causing oxygen depletion. In particular a number of priority substances have been listed and given special attention (the List of Priority Substances in the field of water policy). The List contains 33 substances [http://europa.eu.int/comm/environment/water/water-framework/priority_substances.htm]. Some of these are typical traffic and road pollutants.

The substances on the list are already controlled, to varying degrees, by EU and national legislation. Further controls, independent of the WFD, are expected for a number of substances as a result of European and other international regulations. The European Parliament and the European Council will adopt specific measures against pollution of water by individual pollutants or groups of pollutants presenting a significant risk to or via the aquatic environment, including such risks to waters used for the abstraction of drinking water. For those pollutants, measures will be aimed at the progressive reduction and, for priority hazardous substances, at the cessation or phasing out of discharges, emissions and losses.

6.6 Concluding Remarks

Road-related pollution sources include traffic and cargo, pavement and embankment materials, road equipment, maintenance and operation, and external sources. Road and traffic pollutants having received the greatest attention include heavy metals (e.g. from vehicle corrosion, cargo spills and road equipment), hydrocarbons (from fuels, lubricants and bitumen), nutrients (generated from motor exhausts), particulates (from pavement and exhausts) and de-icing salt. Runoff, splash/spray and seepage through the road construction and the soil are major transport routes of pollutants from the road to the environment.

Pollutant transport through road materials and soils in the road environment is governed by the same physical processes as those occurring in soils elsewhere. During their downward transport, contaminants in the aqueous phase interact with the solid phase. For mass transport in saturated media, diffusion, advection and dispersion are the major processes. Mass transport in unsaturated soil strongly depends on soil-moisture distribution inside the pores. After prolonged dry periods, the first flush of runoff often contains large quantities of pollutants accumulated on the road surface. Long-lying snow close to roads accumulates traffic pollutants.

In road soils, like elsewhere, the most significant chemical processes governing the transport of substances including pollutants are sorption/desorption, dissolution/precipitation and exchange reactions. Sorption of substances in the liquid form on soil particles greatly influences pollutant solubility and transport in soils. Redox conditions and acidity (pH) largely regulate the solubility and thus the mobility of heavy metals. Many heavy metals are more mobile under acidic conditions.

Roadside vegetation influences the transport of traffic contaminants through air, water and soil. Plants close to heavily trafficked roads accumulate traffic pollutants such as heavy metals. Heavy metals, organics, de-icing salt and other toxic substances disturb biological processes in plants, animals, micro-organisms and other biota and may contaminate water bodies and the groundwater.

European legislation puts increasingly strong demands on the protection of water against pollution. Road-keepers are responsible for ensuring that the construction and use of roads is not detrimental to the quality of natural waters.

Strategies for the protection of the environment from road and traffic pollutants should primarily be directed towards limiting the generation of pollutants. As a complement to source-based measures, mitigation measures aim at reducing the dispersal of pollutants to the roadside environment and detrimental effects on soil, water and biota. Principles of road and traffic pollution prevention and mitigation include both technical and biological methods some of which are briefly outlined in Chapter 12.

Including consideration of measures for environmental protection at an early planning stage is much more cost efficient than retrofitting measures and installations afterwards. To judge the need for prevention and mitigation measures, chemical and biological characterization of soil and water is often required. Principles for the sampling and analysis are briefly described in Chapter 7.

The issue of contaminants in the environment is a very large subject and it is not possible within a few chapters to fully address the issues, even limiting the coverage to highway-related topics. Readers who want to explore further will find no shortage of reading material and can readily study the underlying science in much more detail than has been possible in this chapter (e.g. Fetter, 1993; Rand & Petrocelli, 1995; Charbeneau, 1999).

References

Alzubaidi, H., 1999, "*Drift och underhåll av grusvägar, Litteraturstudie*" (In Swedish with a summary in English; *Rehabilitation and maintenance of gravel roads, State-of-the-art*), VTI meddelande **852**, Linköping.

Apul, D.S., Gardner, K.H., Eighmy, T.T., Brannaka, L. & Benoit, J., 2003, "A review of water movement in the highway environment: Implications for the recycled materials use", In: Eighmy, T.T. (ed.), *Beneficial Use of Recycled Materials in Transportation Applications*, Air and Waste Management Association Press, pp. 195–204.

Bäckman, L. & Folkeson, L., 1995, "*The influence of de-icing salt on vegetation, groundwater and soil along Highways E20 and 48 in Skaraborg County during 1994*", VTI meddelande **775A**, Linköping.

Bækken,T., 1994a, "Effects of highway pollution on a small Norwegian lake", *Science of the Total Environment*, **146/147**, pp. 131–139.

Bækken, T., 1994b, "*Trafikkforurenset snø i Oslo*", (in Norwegian; *Traffic polluted snow in Oslo*) – NIVA Rapport LNR 3131-1994, Oslo.

Bækken, T. & Færøvig, P.J., 2004, "*Effekter av vegforurensninger på vannkvalitet og biologi i Padderudvann*" (In Norwegian; *Effects of highway pollutants on water quality and biology in Lake Padderudvann*), Vegdirektoratet, Teknologiavdelingen, Publikasjon **106**, Oslo.

Bækken, T. & Tjomsland, T., 2001, "*Trafikkforurenset snø i Drammen sentrum. Konsekvenser av snødumping for vann- og sedimentkvalitet i Drammenselva*" (In Norwegian; *Traffic-polluted snow in Drammen city centre. Effects of snow disposal on water and sediment quality in River Drammenselva*), NIVA Rapport LNR 4460-2001, Oslo.

Baldwin, G., Addis, R., Clark, J. & Rosevear, A., 1997, "*Industrial by-products in road construction water quality effects*", CIRIA Report **167**, London, CIRIA.

Barbosa, A.E., 1999, "*Highway runoff pollution and design of infiltration ponds for pollutant retention in semi-arid climates*", Ph.D. Dissertation Series, The Environmental Engineering Laboratory, Aalborg University, Denmark.

Barbosa, A.E. & Hvitved-Jacobsen, T., 1999, "Highway runoff and potential for removal of heavy metals in an infiltration pond in Portugal", *Science of the Total Environment*, **235**, pp. 151–159.

Barbosa, A.E., Saraiva, J. & Leitão, T., 2006, "*Evaluation of the runoff water quality from a tunnel washing*", 8th Highway & Urban Environment Symposium, Nicosia, Cyprus, 12–14 June, 2006, 10pp.

Barrett, M.E., Irish, L.B., Malina, J.F. & Charbeneau, R.J., 1998, "Characterization of highway runoff in Austin, Texas, Area", *Journal of Environmental Engineering*, **124**, pp. 131–137.

Bazzaz, F.A., Carlson, R.W. & Rolfe, G.L., 1974, "The effect of heavy metals on plants: Part I. Inhibition of gas exchange in sunflower by Pb, Cd, Ni and Tl", *Environmental Pollution*, **7**, pp. 241–246.

Berggren Kleja, D., Elert, M., Gustafsson, J.P., Jarvis, N. & Norrström, A.-C., 2006, "*Metallers mobilitet i mark*" (In Swedish; *Metal mobility in soil.*), Naturvårdsverket Rapport **5536**, Stockholm.

Blomqvist, G., 1998, "*Impact of de-icing salt on roadside vegetation—A literature review*", VTI rapport **427A**, Linköping.

Blomqvist, G. & Johansson, E-.L., 1999, "Airborne spreading and deposition of de-icing salt—a case study", *Science of the Total Environment*, **235**, pp. 161–168.

Bohemen, H.D. van & Janssen van de Laak, W.H., 2003, "The influence of road infrastructure and traffic on soil, water, and air quality", *Environmental Management*, **31**, pp. 50–68.

Bontoux, J., 1993, "*Introduction à l'étude des eaux douces, eaux naturelles, eaux usées, eaux de boisson*", (In French; *Introduction to the study of fresh water, natural water, waste water, drinking water.*), Cébédoc, Liège, Belgium.

Brenčič, M., 2006, "Groundwater and highways interaction: past and present experiences of highway construction in Slovenia", *Environmental Geology (Berl.)*, **49**, pp. 804–813.

Brenčič, Mihael, 2007, personal communication.

Charbeneau, R.J., 1999, "*Groundwater Hydraulics and Pollutant Transport*", New Jersey, Prentice Hall, 593pp.

Cordt, G., Pacher, F. & Wach, R., 1992, "*Schlußbericht zum Forschungsvorhaben über Straßentunnelabwässer und Gewässerschutz*", Bundesministerium für wirtschaftliche Angelegenheiten, Straßenforschung, Heft **408**.

Dahl, A., Gharibi, A., Swietlicki, E., Gudmundsson, A., Bohgard, M., Ljungman, A., Blomqvist, G. & Gustafsson, M., 2006, "Traffic-generated emissions of ultrafine particles from pavement–tire interface", *Atmospheric Environment*, **40**, pp. 1314–1323.

Dawson, A.R, Hill, A. & Boothroyd, P., 2006, "The hydraulic and geo-environmental behavior of the sub-pavement environment", Geotech. Special Publ'n., No. 154, Am. Soc. Civil Eng'rs., ed. Huang, B., Meier, R., Prozzi, J. & Tutumluer, E., pp 287–295.

Ek, K.H., Morrison, G.M. & Rauch, S., 2004, "Environmental routes for platinum group elements to biological materials—a review", *Science of the Total Environment*, **334/335**, pp. 21–38.

Elgmork, K., Hagen, A. & Langeland, A., 1973, "Polluted snow in southern Norway during the winters 1968–1971", *Environmental Pollution*, **4**, pp. 41–52.

EPA, 1999, "*Understanding Variation in Partition Coefficient, kd, Values*", EPA Report 402-R-99-004A, US Environmental Protection Agency.

EU, 2000, "*The Water Framework Directive*", 2000/60/EC.

EU, 2006, "*The Groundwater Directive*", 2006/118/EC.

Fergusson, J.E., 1990, "*The Heavy Elements. Chemistry, Environmental Impact and Health Effects*", Pergamon Press, Oxford, 614pp.

Fetter, C.W., 1993, "*Contaminant Hydrogeology*", Macmillan Publishing Company, 458pp.

Fletcher, TD., Peljo, L., Fielding, J., Wong, T.H.F. & Weber, T., 2002, "*The Performance of Vegetated Swales for Urban Stormwater Pollution Control*", Proc. 9th Int. Conf. Urban Drainage, Portland, OR, paper **51**, on CD.

Folkeson, L. (ed.), 2000, "*Pollution from roads and vehicles and dispersal to the local environment: Monitoring of 14 case studies*", POLMIT (Pollution of Groundwater and Soil by Road and Traffic Sources: Dispersal Mechanisms, Pathways and Mitigation Measures), RO-97-SC.1027, EC Transport RTD Programme of the 4th Framework Programme, Draft report, Deliverable D3.

Folkeson. L., 2005, "*Spridning och effekter av tungmetaller från vägar och vägtrafik. Litteraturöversikt*" (In Swedish; *Dispersal and effects of heavy metals from roads and road traffic. Literature survey*), VTI Rapport **512**, Linköping.

Gustafsson, M., 2002, "*Väg-och gaturengöring som åtgärd mot höga partikelhalter orsakade av vägdamm*" (In Swedish with a summary in English; *Road- and street sweeping as a measure against high particle concentrations caused by road dust*), VTI meddelande **938**, Linköping.

Hager, M.C., 2001, "Evaluating first flush", *Stormwater*, http://www.forester.net/sw_0109_evaluating.html (visited 20 Dec. 2007).

Hall, M.J. & Hamilton, R.S, 1991, "Highway runoff transport modelling", In: Hamilton, R.S. & Harrison, R.M. (eds.), *Highway Pollution*, pp.131–164, Elsevier.

Hares, R.J. & Ward, N.I., 1999, "Comparison of the heavy metal content of motorway stormwater following discharge into wet biofiltration and dry detention ponds along the London Orbital (M25) motorway", *Science of the Total Environment*, **235**, pp. 169–178.

Hill, A.R., 2004, "*Leaching of Alternative Pavement Materials*", Ph.D. Thesis, Univ. Nottingham.

Hillel, D., 2004, "*Introduction to environmental soil physics*", Elsevier, 494pp.

Holdway, D.A., 1988, "The toxicity of chromium to fish". In: Nriagu, J.O. & Nieboer, E. (eds.), *Chromium in the natural and human environments*, pp. 369–97, Advances in Environmental Science and Technology, **20**, John Wiley & Sons, New York.

Hvitved-Jacobsen, T. & Yousef, Y.A., 1991, "Highway runoff quality, environmental impacts and control", In: Hamilton, R.S. & Harrison, R.M. (eds.), *Highway Pollution*, Studies in Environmental Science, **44**, pp.165–208, Elsevier.

Ihs, A. & Gustafson, K., 1996, "*Calcium magnesium acetate (CMA)—an alternative deicing agent, A review of the literature*",VTI meddelande **789A**, Linköping.

Jacobson, T., 2005, "*Undersökning av dubbslitaget vintern 2004/2005 och validering av VTI:s slitagemodell*" (In Swedish; *Study of the pavement wear by studded tires during the winter 2004/2005 and validation of the VTI pavement-wear model*), VTI Notat **36-2005**, Linköping.

James, C., 1999, "*The impact of pollutants from roads and vehicles on the local natural environment*", POLMIT Project Report, PR/SE/603/99, RO-97-SC.1027, Transport Research Laboratory, 131pp.

Jandová, V., 2006, "*Kontaminace vod polutanty obsaženými v konstrukčních vrstvách vozovek*" *(Průběžná zpráva o realizaci projektu 1P050C003 za rok 2005)* (In Czech; *Water contamination with pollutants contained in the constructional layers of the roads. Report of the project No. 1P050C003 for year 2005)*, CDV, Brno, 20pp.

Johansson Thunqvist, E.-L., 2003, "*Estimating chloride concentration in surface water and groundwater due to deicing salt application*, Ph.D. Thesis, Department of Land and Water Resources Engineering, Royal Institute of Technology, TRITA-LWR PhD **1006**, Stockholm.

Kayhanian, M., Singh, A., Suverkropp, C. & Borroum, S., 2003, "Impact of annual average daily traffic on highway runoff pollutant concentrations", *Journal of Environmental Engineering*, **129**, pp. 976–990.

Khilar, K.C, & Fogler, H.S., 1998, "*Migration of Fines in Porous Media*", Kluwer Academic Publishers. 171pp.

Kriech, A.J., 1990, "*Evaluation of hot mix asphalt for leachability*". Asphalt Institute. Heritage research group, Indianapolis, Indiana, USA, http://www.asphaltinstitute.org/ai_pages/Technical_Focus_Areas/Tech_Issues_Environmental.asp.

Kriech, A.J., 1991, "*Leachability of asphalt and concrete pavement*". Asphalt Institute. Heritage research group, Indianapolis, Indiana, USA. http://www.asphaltinstitute.org/ai_pages/Technical_Focus_Areas/Tech_Issues_Environmental.asp.

Landner, L, & Lindeström, L., 1998, "*Zinc in society and in the environment. An account of the facts on fluxes, amounts and effects of zinc in Sweden*", Swedish Environmental Research Group (MFG). Kil, 160pp.

Landner, L. & Reuther, R., 2004, "*Metals in Society and in the Environment. A Critical Review of Current Knowledge on Fluxes, Speciation, Bioavailability and Risk of Adverse Effects of Copper, Chromium, Nickel and Zinc*", Kluwer Academic Publishers.

Legret, M. & Colandini, V., 1999, "Effects of a porous pavement with reservoir structure on runoff water: Water quality and fate of heavy metals", *Water Science and Technology*, **39**, pp. 111–117.

Legret, M., Nicollet, M., Miloda, P., Colandini, V. & Raimbault, G., 1999, "Simulation of heavy metal pollution from stormwater infiltration through a porous pavement with reservoir structure", *Water Science and Technology*, **39**, pp. 119–125.

Legret, M. & Pagotto, C., 1999, "Evaluation of pollutant loadings in the runoff waters from a major rural highway", Science of the Total Environment, **235**, pp. 143–50.

Legret, M., Odie, L., Demare, D, & Jullien, A., 2005, "Leaching of heavy metals and polycyclic aromatic hydrocarbons from reclaimed asphalt pavement", *Water Research*, **39**, pp. 3675–3685.

Leitão, T.E., Lehmann, N., Smets, S., Lobo Ferreira, J.P.C. & Holm, P., 2000, "*Pollution from Roads and Vehicles and Dispersal to the Local Environment: Mass Flux and Mass Balance Calculations; Assessment of Pollution of Groundwater and Soils by Road and Traffic Sources*", WP2/4, LNEC and VKI, 47pp.

Lindbom, J., Gustafsson, M., Blomqvist, G., Dahl, A., Gudmundsson, A., Swietlicki, E. & Ljungman, A.G., 2006, "Exposure to wear particles generated from studded tires and pavement induces inflammatory cytokine release from human macrophages", *Chemical Research in Toxicology*, **19**, pp. 521–530.

Lindgren, Å., 1996, "Asphalt wear and pollution transport", *Science of the Total Environment*, **189/190**, pp. 281–286.

Lindgren, Å., 1998, "*Road construction materials as a source of pollutants*", Doctoral thesis, Department of Environmental Engineering, Division of Traffic Engineering, Luleå University of Technology, Luleå.

Matos, J., Portela, M. & Mourato, S., 1999, *"Reabilitação da Bacia de Decantação de Fátima"* (In Portuguese; *Rehabilitation of Fátima Basin*), Parecer Técnico, IST- BRISA, Lisbon, 58pp.

Moy, F., Crabtree R.W. & Simms, T., 2002, *"Long Term Monitoring of Pollution from Highway Runoff: Final Repor"*, Report by WRc plc for Highways Agency and Environment Agency.

Münch, D., 1992, "Soil contamination beneath asphalt roads by polynuclear aromatic hydrocarbons, zinc, lead and cadmium", Science of the Total Environment, **126**, pp. 49–60.

Murozumi, M., Chow, T.J. & Patterson, C., 1969, "Chemical concentrations of pollutant lead aerosols, terrestrial dusts and sea salts in Greenland and Antarctic snow strata", *Geochimica et Cosmochimica Acta*, **33**, pp. 1247–1294.

Norrström, A.-C. & Jacks, G., 1998, "Concentration and fractionation of heavy metals in roadside soils receiving deicing salts", *Science of the Total Environment* **218**, pp. 161–174.

Olsson, S., 2005, *"Environmental assessment of municipal solid waste incinerator bottom ash in road constructions"*, Licentiate Thesis in Land and Water Resource Sciences, Royal Institute of Technology, Stockholm, TRITA-LWR LIC **2030**, Stockholm.

Oscarsson, K., 2007, *"Dust suppressants for Nordic gravel roads"*, Licentiate Thesis in Highway Engineering, Department of Civil and Architectural Engineering, Royal Institute of Technology, TRITA-VT FR **07:02**, Stockholm.

Pacyna, J. M. & Nriagu, J. O., 1988, "Atmospheric emissions of chromium from natural and anthropogenic sources", in: Nriagu, J. O. & Nieboer, E. (eds.), *Chromium in the natural and human environments*, Advances in Environmental Science and Technology, **20**, John Wiley & Sons. New York, pp. 105–23.

Pagotto, C., Legret, M. & Le Cloirec, P., 2000, "Comparison of the hydraulic behaviour and the quality of highway runoff water according to the type of pavement", *Water Research*, **34**, pp. 4446–4454.

Persson, K. & Ihs. A., 1998, *"Kalciumklorid i vinterväghållningen. Litteraturstudie"* (In Swedish, with a summary in English: *"Calcium chloride in winter maintenance, A literature review"*), VTI meddelande **829**, Linköping.

Ramade, F., 1998, *"Dictionnaire encyclopédique des sciences de l'eau – Biogéochimie et écologie des eaux continentals et littorals"*, (In French; *"Encyclopedia of water sciences—Biogeochemistry and ecology of continental and litoral waters"*), Edisciences International, Paris.

Rand, G.M. & Petrocelli, S.R., 1995, *"Fundamentals of Aquatic Toxicology: Effects, Environmental Fate and Risk Assessment"*, 2nd edition, Taylor & Francis, 1148pp.

Rolfe, G.L. & Bazzaz, F.A., 1975, "Effects of lead contamination on transpiration and photosynthesis of loblolly pine and autumn olive", *Forest Science*, **21**, pp. 33–35.

Sadler, R., Delamont, C., White, P. & Connell, D., 1999, "Contaminants in soils as a result of leaching from asphalt", *Toxicological and Environmental Chemistry*, **68**, pp. 71–81.

Sansalone, J.J. & Buchberger, S.G., 1997, "Partitioning and first flush of metals in urban roadway storm water", *Journal of Environmental Engineering*, **123**, pp. 134–143.

Sansalone, J.J. & Cristina, C.M., 2004, "First flush concepts for suspended and dissolved solids in small impervious watersheds", *Journal of Environmental Engineering, American Society of Civil Engineers*, **130**, pp. 1301–1314.

Sarkar, B. (ed.), 2002, *"Heavy Metals in the Environment"*, Marcel Dekker, New York, 725pp.

Scanlon, P.F., 1991, "Effects of highway pollutants upon terrestrial ecosystems", In: Hamilton, R.S. & Harrison, R.M. (eds.), *Highway Pollution*, Studies in Environmental Science **44**, pp. 281–338, Elsevier.

Selim, H.M. & Sparks, D.L., 2001, *"Heavy Metals Release in Soils"*, Lewis Publishers, Boca Raton, USA.

Sprague, J.B., 1987, "Effecs of cadmium on freshwater fish", In: Nriagu, J.O. & Sprague, J.B. (eds.), *Cadmium in the aquatic environment*, Advances in Environmental Science and Technology, **19**, pp. 139–169, John Wiley & Sons, New York.

Stead-Dexter, K. & Ward, N.I., 2004, "Mobility of heavy metals within freshwater sediments affected by motorway stormwater", *Science of the Total Environment*, **334/335**, pp. 271–277.

Stumm, W. & Morgan J.J., 1996, "*Aquatic chemistry, chemical equilibria and rates in natural waters*", 3rd edition, John Wiley & Sons, 1022pp.

Thomson, N.R., McBean, E.A., Snodgrass, W. & Mostrenko, I., 1997, "Sample size needs for characterizing pollutant concentrations in highway runoff", *Journal of Environmental Engineering*, **123**, pp. 1061–1065.

TRL., 2002, "*Pollution from Roads and Vehicles and Dispersal to the Local Environment: Final Report and Handbook*", Deliverable D5, POLMIT, Pollution of Groundwater and Soil by Road and Traffic Sources: dispersal mechanisms, pathways and mitigation measures, RO-97-SC.1027, Transport Research Laboratory, (electronic publication).

Tyler, G., 1974, "Heavy metal pollution and soil enzymatic activity", *Plant and Soil*, **41**, pp. 303–311.

Vägverket, 2004, "*Vägdagvatten – Råd och rekommendationer för val av miljöåtgärder*", (In Swedish; "*Road runoff. Advice and recommendations for choosing environmental measures*"), Vägverket publikation 2004:**195**, Borlänge.

Warner, L.R., Sokhi, R.S., Luhana, L., Boulter, P.G. & McCrae, I., 2002, "Non-exhaust particle emissions from road transport", In: International Conference "*Transport and Air Pollution*", 19–21 June 2002, Graz, pp. 265–272.

WHO, 1989, "*Lead – Environmental Aspects*", Environmental Health Criteria, **85**, World Health Organisation, Genève, 106pp.

Weckwerth, G., 2001, "Verification of traffic emitted aerosol components in the ambient air of Cologne (Germany)", Atmospheric Environment **35**, pp. 5525–5536.

Weis, P. & Weis, J. S., 1991, "The developmental toxicity of metals and metalloids in fish", in: Newman, M. C. & McIntosh, A. W. (eds.), *Metal Ecotoxicology. Concepts & Applications*, pp. 145–169. Lewis Publishers. Chelsea, Michigan.

Zereini, F., Alt, F., Rankenburg, K., Beyer, J.-M. & Artelt, S., 1997, "Verteilung von Platingruppenelementen (PGE) in den Umweltkompartimenten Boden, Schlamm, Straßenstaub, Straßenkehrgut und Wasser", *Umweltwissenschaften und Schadstoff-Forschung*, **9**, pp. 193–200.

Chapter 7
Contaminant Sampling and Analysis

Teresa Leitão*, Andrew Dawson*, Torleif Bækken, Mihael Brenčič, Lennart Folkeson, Denis François, Petra Kuřímská, Roman Ličbinský and Martin Vojtěšek

Abstract This chapter presents a general overview of procedures and methods for sampling and analysis of contaminants in water and soil in the road environment. The chapter concerns the water and seepage in road structures under the influence of traffic loading, and in the adjacent ground extending to the water table where contaminant seepage is of concern. The text gives an introduction to this subject and guides the reader to relevant literature with detailed information about practices of sampling and analysis. The chapter in divided into five main sections: principles of data collection and storage, sampling design, water and soil sampling procedures, and in-situ and laboratory measurements and analyses.

Keywords Road contaminants · data collection and storage · sampling design · in-situ and laboratory analysis · water · soil

7.1 Introduction

The purpose of contaminant sampling and analysis is mainly to characterize a specific road in terms of its runoff characteristics and the water percolating vertically through the road structure, as well as the existing state of quality of the adjacent water and soil. Sampling and analysis can also be performed to identify a specific pollution episode. The environmental compartments, usually considered as being potentially affected, comprise surface waters, groundwater, soil and soil water. Together they can give a global picture of contaminant dispersion and pathways after entering the soil (see Chapter 6).

---

* *Co-ordinating Authors:*
✉ T. Leitão
Laboratório Nacional de Engenharia Civil / National Laboratory for Civil Engineering (LNEC), Portugal
e-mail: tleitao@lnec.pt

✉ A.R. Dawson
University of Nottingham, Nottingham, UK
e-mail: andrew.dawson@nottingham.ac.uk

A.R. Dawson (ed.), *Water in Road Structures*, DOI 10.1007/978-1-4020-8562-8_7,
© Springer Science+Business Media B.V. 2009

7.2 Principles of Data Collection and Storage

7.2.1 Data Collection

It is a general rule that data must always be collected with a specific purpose in mind. This rule is as much true concerning contaminant levels in soils and groundwaters as it is in any other field. Therefore, before any sampling regime is contemplated and before any specimens are analysed, it is essential to decide the purpose for which samples and analyses might be required. Then a systematic sampling programme and network needs to be designed and planned with the purpose and constraints clearly in view. It is usually a non-routine task to identify the questions that need answering by the sampling and analysis programme. Perhaps these will be suggested because of standards set by an environmental regulator. Perhaps they will arise because of a client's desire to establish or maintain a reputation for environmentally responsible behaviour. Perhaps they will be set as a consequence of the desire to establish a benchmark for future work. Probably there will be some combination of these prompts together with others. It is the monitoring agency's job to ensure that the data collected is that which is really needed and not that habitually collected. There is no point in collecting data that is never used, nor will it usually be possible to go back and collect missing data.

If a road is to be constructed of an unusual material, then a programme of laboratory assessments of the leaching potential of the material might be established long before construction is planned. The purpose here would probably be to understand the possible yields of contaminants in time, and in concentration, under conditions that are expected to pertain in-situ once construction commences and, also, after construction is completed. Initially a wide range of elements might be studied to determine what species are of concern within the runoff contaminants (see Chapter 6). Later studies might take place on a reduced number, only addressing those elements previously identified as potentially giving concern. Probably some kind of computer modelling will be intended to extend the in-isolation laboratory results to in-situ conditions. In that case, it is necessary to obtain data in a form suitable for use in the selected modelling program. Once construction commences, data collection in-situ will most probably be required. At that time the aim would be to confirm that the interpretation made from the laboratory assessments and numerical modelling are, indeed, valid.

In examples where a potentially contaminating material is to be used, for which construction experience and contaminant behaviour is already understood, laboratory studies might be restricted to confirming that the material is, indeed, similar to previously used examples. In-situ investigations might then be limited to confirming similar behaviour to that previously experienced. In practice, no two installations are identical, so some sample collection and analyses are likely to be required to explore the specific application.

For many studies of contaminants in the road environment, a "base-line" study is required. This investigation has the aim of characterizing the in-situ ground and water quality condition *prior* to a planned action. Thus, the naturally occurring conditions of a "greenfield" site would be recorded before a new road was constructed in the vicinity, or a contaminated site would be characterised prior to activity designed to improve water quality. In both cases the aims are, first, to be able to determine the effects of the activity on the surrounding ground and water quality and, secondly, to assess whether some intervention is, therefore, needed.

7.2.2 Data Storage and Retrieval

An appropriate database / record keeping system must be provided (or constituted):

- To hold the data.
- To have data extracted/interpreted in a manner that has meaning. There is no point in collecting data that cannot be successfully accessed.
- To allow it to be interrogated in a way that permits the likely users (owners, regulators, researchers, etc.) to apply the method of interpretation that meets their needs. As most data storage is archived electronically, consideration should also be given to providing secure network access.
- To be easily maintained and amended.
- To have some semi-automatic processing capability that will alert the database owner to take some investigative action if the data contained seems to indicate a problem. There are more than a few examples of a record system holding the data that would have indicated a potential problem long before it became an issue .... if only someone had looked at the data!
- The database / system must be properly documented and backed-up. Accepted archival systems are required for both electronic and paper records.
- The database should hold all the data that passes certain pre-defined quality levels. The quality levels should not be set too high otherwise too many useful data points will be excluded. The disadvantage is that some invalid or unreliable readings will be stored. Therefore, sufficient data points should be stored in the database so that later data analyses can differentiate genuinely high or low values from those readings that are unreliably high or low.

An ongoing budget should be secured to enable monitoring and database maintenance to continue over the full time-scale required by the probable contaminant transport behaviour. If it cannot be ensured, a sustainable "fall-back" programme should be incorporated into the plan.

Modern software systems are readily, and economically, available to provide secure, accessible data storage and retrieval capability. Data entry to these systems is also considerably more user-friendly than in the past. The standardized data format of the Association of Geotechnical and Geoenvironmental Specialists (AGS, 2004) is one such system that has the great advantage of being non-proprietary. Thus, data stored in this manner is readily interchangeable between different users. Field data can be collected by "palmtop" ("PDA") computer and combined with applications developed using open source software (e.g. Walthall & Waterman, 2006; Chandler et al., 2006).

7.3 Sampling Design

In establishing a monitoring programme and data collection schedule, the following points will need addressing. This is not a comprehensive list for every eventuality, but most monitoring programmes will need to consider this list as a minimum:

- The equipment, skills, storage and transportation facilities (and, if required, power to the site):
  - Sampling, transport and storage protocols should be obtained or prepared to promote good practice and consequently to yield reliability of readings obtained from tests.
  - A responsible person should be identified for sample collection and instrument readings and made available over the life of the programme.
  - Health and safety plans are required for personnel engaged in the sampling.
- Timing of sample collection:
  - Sometimes a frequency that reduces over time will be satisfactory if adequate behaviour is demonstrated by early analysis, especially where some new material or construction is being tried or where a singular event has occurred (such as a cargo spillage). Otherwise regular sampling will usually be preferred. There are some cases where only a single sample or one "before" and one "after" sample are necessary at any particular point.
  - Sampling needs to take account of the weather conditions in which the media will be sampled / measured (rain, wind, heat, etc.). Sometimes collection should be at a fixed time of day, a certain temperature or in a certain season when, otherwise, there wouldn't be (or would be doubts about) comparability between specimens.
  - Sometimes samples are to be collected from a discrete, sometimes from a continuous, source. In some low flow situations, sampling may exhaust the source until slow seepage provides the next specimen. This may have an influence on sampling frequency and/or volume.
- Location of sampling:
  - There is usually a conflict between the desired number of points to be sampled and the budget available. As far as possible, the locations should aim to ensure spatial reliability and representativeness. In particular the design should give confidence that all "hot-spots" will be located and that invalid and unreliable readings can be easily distinguished from genuine extreme values.
  - Vertical and horizontal positions in the ground should be chosen depending on the expected source and route of contaminant flow and the receiving media. For example, where traffic factors are thought to be influential, closeness to the wheel path may be important; where runoff is important, measurements in the verge may be important; where a developing pollutant front is to be detected (or refuted) then positions laterally and beneath the source point may be required. There are similar issues for water bodies – at what depth(s)

and water velocity(ies) should specimens be collected? In the context of road construction, such a question will be irrelevant if water is collected from seepages.
  - Consideration should be given to the accessibility of sampling locations once the road is open to traffic. Access through the carriageway will often require lane closures that are expensive or difficult to arrange. The access point may also be difficult to keep sealed under traffic loading thereby compromising the quality of specimens collected.

- Protection:
  - Sample collection points and instruments should be designed to be replaceable if they become damaged or aged. Appropriate protection of collection points and instruments against vandalism, traffic over-run, grass-cutting and other maintenance/rehabilitation will often be needed.

- Materials to be sampled:
  - The sampling programme needs to assess all relevant material, i.e., groundwater, surface water, soils and aggregates as well as these media in their reference condition(s).
  - Adjacent surface water bodies may need assessment if an affect by polluted water from the road is suspected.
  - Soil or construction material may need monitoring as well as groundwater. If the subgrade or a construction material will (or may) transport or sorb contaminants, it may need to be sampled from time-to-time to check for any alteration (e.g. permeability value or sorbed contaminant level). Similarly, if construction material, in-place, is thought to be a source of ongoing contamination, the plan should consider sampling it and testing its leachability as use continues.
  - While water samples are only tested for their chemical properties, soil specimens may need testing for total solids make-up, organic content, mineralogy, particle size distribution and specific surface area. The last two are important in understanding sorption behaviour. Water can also be extracted from soil samples, e.g. by using a centrifuge.

- Selecting parameters for analysis:
  - It is expensive to analyse for all species all of the time. Therefore, it may be necessary to identify key analytes that may act as indicators of change in the seepage/transport process, and to concentrate monitoring on these.
  - Often, a regular frequency of assessment can be maintained at modest cost if full chemical analysis is sometimes replaced by surrogates, e.g. pH, Eh, electrical conductivity, etc. by electrical means.
  - Where "trigger" values have been set for some intervention, the concentrations of the "trigger" species will require specific, ongoing, analysis.
  - It can be a false economy only to analyse collected specimens for the analytes of known concern. A record of the contamination of other species may yield

important information on an underlying chemical process (e.g. ion exchange) or may give rise to unexpected values that will lead to the identification of some unexpected problem or benefit.
- The water or soil should be sampled in amounts large enough to permit all the desired testing to be performed on it.

Some duplicated sampling will be needed, usually from the same location. The aims are:

- to ensure evidence of a reading's representativeness;
- to adequately define statistical scatter in readings;
- to monitor genuine fluctuations in source concentration (e.g. as a consequence of flow levels, season, traffic, etc.);
- to ensure that both mean and "worst-case" values are available; and
- to allow repeated analyses in case of dispute.

To ensure best practice in this area most environmental regulatory authorities issue guidance on sampling (see reference list at end of Section 7.4.1).

7.4 Water and Soil Sampling Procedures

7.4.1 Introduction

The procedure for sampling is primarily influenced by the source of the water (e.g. in a borehole, in a pipe) and by the equipment available with which to sample it. The equipment, itself, is largely controlled by the sampling location. Once collected by the sampling device, water samples must be quickly processed before changes in make-up occur due to various physical and biological processes. The following sections discuss the collection and immediate treatment of surface and sub-surface waters and of soil. They provide a general overview of the techniques and procedures for application in pavements and the ground around highways. There is insufficient space to cover all aspects of sampling techniques and practice, so interested readers are directed to other texts where these aspects are fully described. In particular, reference may be made to the following sources :

- for runoff: FHWA (1987, 1996), Hamilton et al. (1991), Hvitved-Jacobsen & Yousef (1991), and Wanielsita & Yousef (1993);
- for surface water: Ruttner (1952), Krajca (1989), Environment Agency (1998);
- for groundwater, soil water and soil: Barcelona et al. (1985), Canter et al. (1987), Nielsen (1991), Clark (1993) and Boulding (1995).

7.4.2 Sampling of Runoff

Sampling can aim at documenting contaminant concentrations and fluxes during and after storm events (or other rain or snow melt), at mirroring the load and flux of contaminants over an extended period of time, or at characterizing an accidental discharge.

Road runoff presents specific characteristics that change significantly from place to place, depending on site characteristics and many other factors (Barbosa & Hvitved-Jacobsen, 2001). Therefore, the monitoring programme should also include the characterization of the most important factors such as: traffic characteristics (volume, speed, type of vehicles, fuel types, etc.), geographic location, climate, topography, drainage area and road design, pavement characteristics, right-of-way characteristics and adjacent land use. Although the quality of water is being influenced by several external characteristics, it is the duration and intensity of the pavement washing that determines the degree of dilution and transport of pollutants. The site selected for runoff sampling should be a drainage area (with significant and representative size) where the runoff from the paved and unpaved areas can be isolated from other sources within the selected highway system.

Runoff sampling may be used to characterize the influence of the mean daily traffic of a road section. For that purpose samples must be collected in a discrete way, in a surface drainage pipe (e.g. Fig. 7.1), during rain or snow thawing events, and be associated with a specific road drainage area. The runoff flow measurement devices are usually installed transversally in the outflow pipe from the system draining the area of pavement that is of interest and must include a calibrated component that links the observed water level to the corresponding runoff volume by knowing the shape and dimension of the pipe. Runoff sampling depends on the occurrence of precipitation, and therefore, sampling is facilitated by the use of sampling equipment that is activated by a sensing device which automatically starts the sampling whenever the flow rate exceeds a previously defined value. Such sampling equipment also has the benefit of allowing the sampling of the first flush of a storm event.

This type of monitoring gives discrete information about the changes in runoff pollution during a certain period and allows the determination of the Event Mean Concentration (EMC) and the Site Mean Concentration (SMC). An EMC is

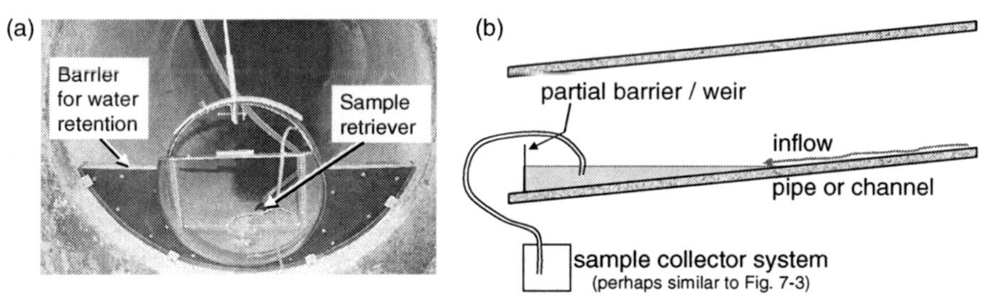

Fig. 7.1 Device for runoff sampling in a (**a**) road ditch section or drainage pipe and (**b**) principle of implementation

calculated for an individual storm event as the total mass load of a pollutant parameter (CV) divided by the total runoff water volume (V) discharged during the storm (Hvitved-Jacobsen & Vollertsen, 2003):

$$\text{EMC} = \Sigma CV/\Sigma V \tag{7.1}$$

A SMC is a characteristic runoff annual pollution loads for a specific site, typically defined by the arithmetic mean value of the EMC's measured at one site (Hvitved-Jacobsen & Vollertsen, 2003).

To evaluate the EMC it is advisable to sample the whole event in a way that is time constant and volume proportional. In this method, discrete samples are collected at equal time increments and composed proportional to the varying flow rate during the sampling period (FHWA, 1987, 1988). In order to mirror seasonal variations, a monitoring programme should, ideally, include several discrete storm events for a one-year period. Authors differ in their opinions about the minimum number of storm events needed. Some authors consider 10, others 6 storm event episodes, each one characterized by a minimum of 5 samples (Barbosa, 1999, Burton & Pitt, 2002). Probably, the choice is influenced by storm severity and other climatological factors that may vary widely between countries and regions.

To allow the characterization of road runoff throughout a specific period, runoff monitoring devices should be:

- automated;
- include a rainfall device;
- include a runoff flow measurement device; and
- include sampling and recording equipment.

All these devices should operate in phase with each other in order to allow the determination of lag time from rainfall to runoff and the determination of pollutant load in each sample.

Figure 7.2 illustrates an alternative approach, common for measuring lower flow volumes than observed in runoff collection pipes – e.g. rainfall and seepage water. Water is fed into a bucket (Fig. 7.2a) that tips alternately to left and right when it fills. The number of tips may be counted electronically. Figure 7.2b shows a typical installation, in this case with the possibility of some of the water being sampled for later analysis. By integrating the output of the tipping buckets with the automatic water sampling equipment (as shown in Fig. 7.3) it is possible to sample water on a volumetric basis rather than a time basis.

Where discrete sampling is not feasible, runoff sampling can be accumulative. For that it is possible to arrange special sampling devices, in the form of a gutter installed along the sealed pavement's edge to capture runoff water and then to conduct it to a type of tank, such as a rainwater settling tank or a retention tank installed close to or away from the road. Afterwards the tank can be sampled by abstracting a proportion into a vessel installed in a chamber dug into the roadside soil. These devices can give a broad idea about the road pollution but are not appropriated for load calculations if their draining section is not easily delineable. These sampling

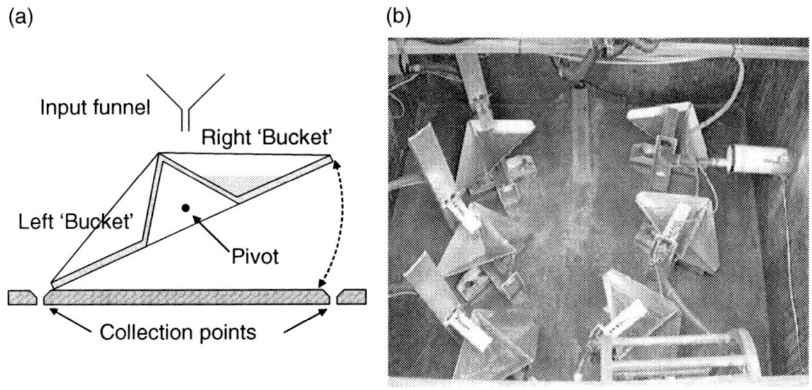

Fig. 7.2 Tipping bucket system to measure flow: (**a**) principle of operation; (**b**) with adjacent, periodic, sampling system (from Hendry et al., 2004). Photo. credit: Lincoln University, New Zealand

methods are less informative since they give information about the road pollution integrated over a period of time. Also, other sources than the road and its traffic might contribute to the contaminants collected in the tanks. Furthermore, change in water condition can take place between the time that water arrives at the tank and the time at which a specimen is collected. Sampling from tanks, therefore, only gives a qualitative overview of road pollution. In the case of an overall sample taken in

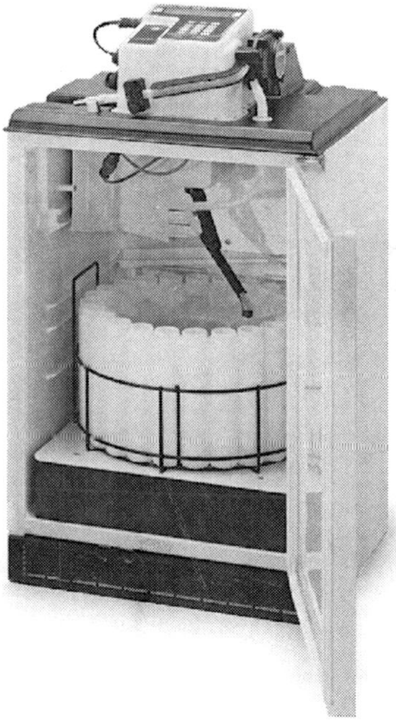

Fig. 7.3 Automatic water sampler with refrigerated cabinet (Courtesy of Hach. ©Hach Company, 2007)

a settling tank or retention tank, water collection must be made at several locations and depths of the tank so as to provide a composite sample representative of the water in the entire tank.

7.4.3 Sampling from Surface Water Bodies

Water samples should be collected from surface water bodies (lakes, rivers, streams, ponds, etc.), taking into account the velocity field in flowing water and any possible stratification in standing water. Once the location and frequency of water sampling are selected there are a set of procedures for sampling water in the natural environment (from surface water bodies) that need to be considered. Figures 7.4 and 7.5 give two examples of bottles for surface water sampling.

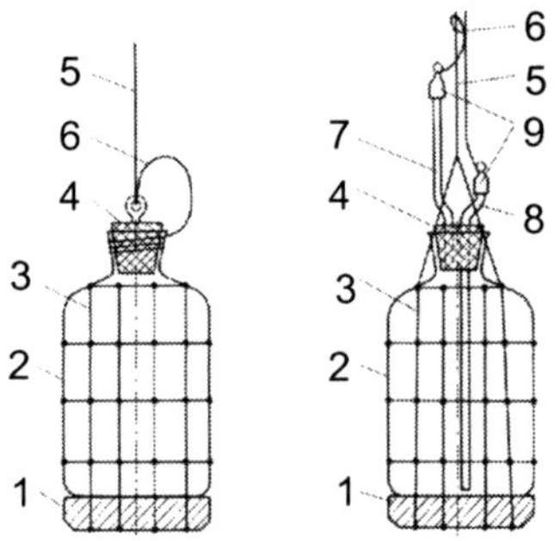

Fig. 7.4 Simple bottle type samplers. Left: Meyer's submersible bottle; Right: Submersible bottle based on Dussart's principle

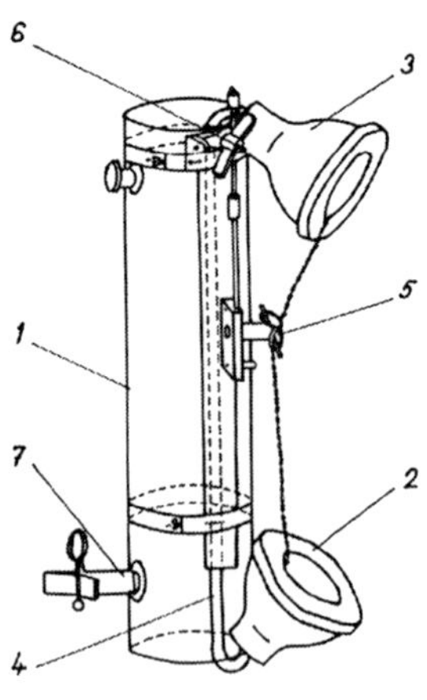

Fig. 7.5 Van Dorn's water bottle (or flushed sampler), vertical sampling model

7.4.4 Sampling of Groundwater

Groundwater sampling can be performed in existing facilities (wells, piezometers, springs, etc.) or in new ones that need to be built. In this latter case, it is more frequent to install piezometers of a small diameter, just enough to be compatible with the monitoring equipment to be subsequently used. Most groundwater sampling installations must be located downgradient to the road discharges in terms of the local groundwater flow if they are to look for road-induced contamination. A few upstream locations may also be chosen to allow a reference water condition to be established. Their depth should be at least 2–3 m below the minimum annual groundwater level in order to avoid having a dry piezometer. The necessary number of piezometers depends on the dynamics of groundwater (i.e. varying condition with time and place) (Leitão, 2003). Higher permeability and hydraulic gradient implies monitoring in more locations and more frequently.

To obtain a sample that is representative of the water in the well or piezometer in question, there should be a purging operation until electrical conductivity, pH and temperature of the outflow water have stabilized (Aller et al., 1989). In low permeability soils the purging operation should try to minimize the water displaced during purging, otherwise recharge of the sampling location will take too long to allow realistic sampling. The water can then be sampled directly or using a device similar to Van Dorn's water bottle, but with a size compatible with the well diameter (Fig. 7.6).

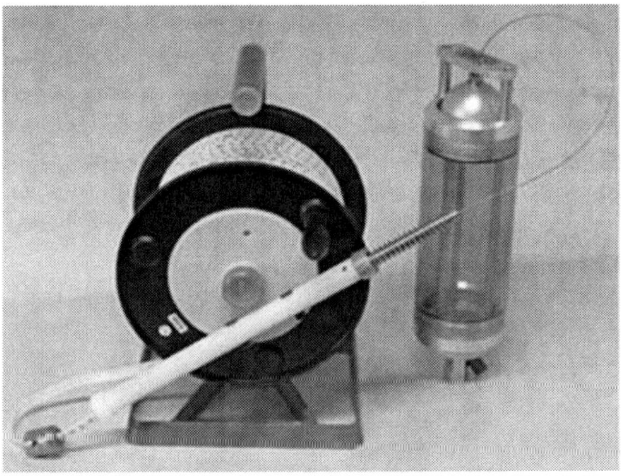

Fig. 7.6 Example of a groundwater sampler: a bailer filled by lowering beneath water surface so that tube fills from the top. Reproduced by permission of Hydrokit

7.4.5 Sampling of Soil and Soil Water

Having entered the soil environment near roads, contaminants will either be retained in the soil or transported through the soil. Depending on soil characteristics and other environmental conditions, different contaminants are transported with the soil water through the soil at varying rate. Mobile compounds (such as chloride) move rapidly whereas many heavy metals and organic contaminants move much slower. Often, contaminant concentrations are much higher in the upper soil layers than further down the soil profile.

Sampling of soil water gives a picture of the rate of transportation of contaminants down a soil profile whereas sampling of soil gives a picture of the contaminant quantities having accumulated in the various soil layers over a long period of time.

Seepage and soil water (pore-retained water) can be sampled, although with less ease than groundwater below the water table. There, suction lysimeters (also called tension lysimeters) may be used. In principle, the soil water is sucked out of the soil through a lysimeter body that acts as a membrane or filter (Fig. 7.7). These devices include a high air-entry porous tip inside which a partial vacuum may be applied via a flexible pipe connected to an external vacuum pump. By this means water is pulled into the tip and, after collection, is sampled by gravity when possible or by gas displacement. For heavy-metal sampling, lysimeters should be made of Teflon, glass, PET or other material unable to sorb the metals. The flux of contaminants down the soil profile is often of interest. Since tension lysimeters give information on concentrations only, water volumes have to be measured separately or modelled so as to make calculations of pollutant fluxes possible.

Besides the more classical methods of soil water sampling, alternative road surface infiltration samplers have been designed in which water seeps from the road surface down and goes through separate layers of pavement and embankment towards a circular "funnel" (Sytchev, 1988). The device is installed during pavement construction with the layers of pavement being placed over the top of the sampling inlet. Two layers of siliceous sand of different grain sizes are situated there on fine-mesh screens to prevent entrance of solids to a sampling bottle where the seeping water is collected. The water amount in the sampling bottle is detected by measuring resistance (conduction sensor). There are two small metal pieces in the bottle; resistance between them is different when there is air or water. When a sufficient

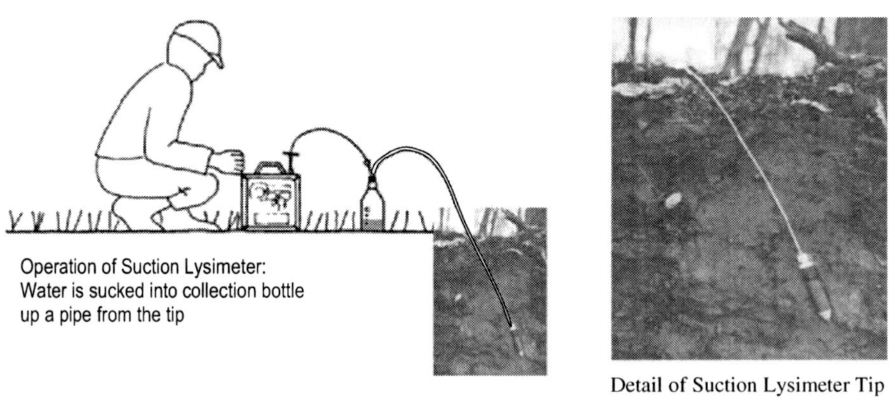

Fig. 7.7 Example of a lysimeter. Reproduced by permission of Prenart Equipment ApS

amount of water has been collected in the sampling bottle, a gas (commonly N_2) is injected into the bottle so as to close the valve under the funnel. The water specimen is then forced back to the surface and runs into the sampling bottle (Fig. 7.8a,b).

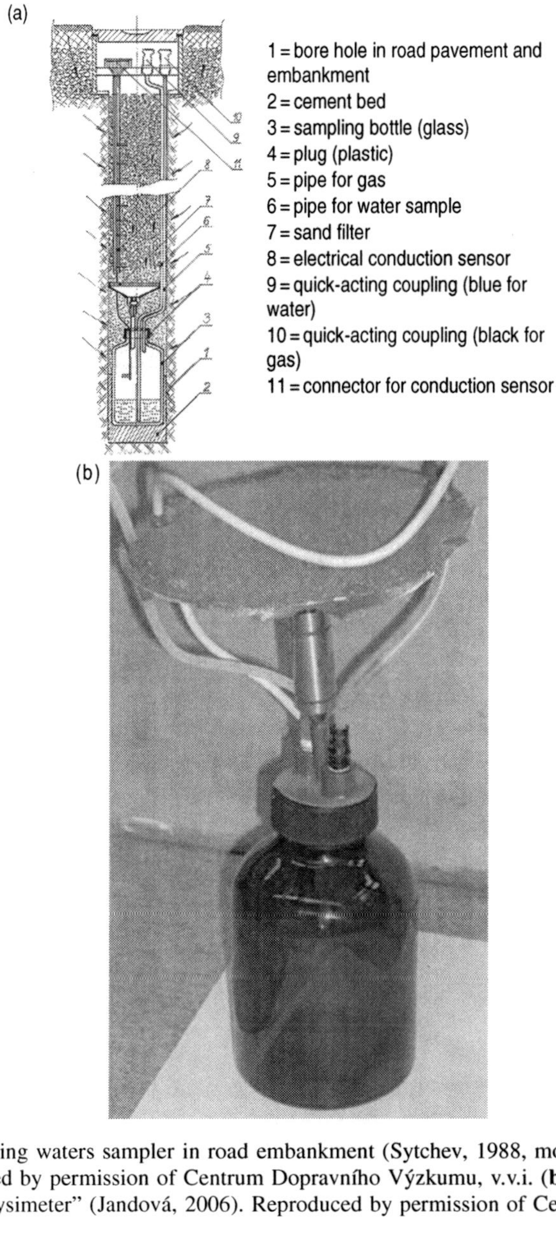

Fig. 7.8 (**a**) Seeping waters sampler in road embankment (Sytchev, 1988, modified in Jandová, 2006). Reproduced by permission of Centrum Dopravního Výzkumu, v.v.i. (**b**) Sampling bottle, valve and "area lysimeter" (Jandová, 2006). Reproduced by permission of Centrum Dopravního Výzkumu, v.v.i.

Soil sampling can be performed in any season, except in periods of frost. Sampling from consecutive soil depths gives information on the displacement of accumulated pollutants down the soil profile. Natural upper soil profiles are usually richer in organic matter content favouring the retention of several pollutants, namely heavy metals and organic pollutants. This pattern should be analysed in order to observe differences in contaminant content and behaviour across a soil profile. Usually, soil samples are taken out using a steel cylinder of a given volume so as to allow volume-related physical and chemical analyses. Just as for water, soil samples should be transported without delay and kept cool.

To collect samples beneath pavements, a core hole will usually be required in the pavement surface. Drilling conventionally uses a water-cooled core cutter, but these should not be used when abstracting samples for chemical assessments as the water will be likely to change the chemical conditions in the underlying ground by introducing contaminants and/or diluting what was already there.

7.4.6 Water and Soil Storage

As physico-chemical and biological reactions occur in the soil and the water, sampling periods of short duration are recommended. Some chemical variables should be measured in-situ in a sub-sample (temperature, pH, redox potential, and electrical conductivity) whereas the main water sample is preserved to prevent reduction or loss of target analytes, and transported to the laboratory without delay and kept cool until further treatment. Preservation stabilizes analyte concentrations for a limited period of time. Some samples have a very short holding time (from few hours to some days).

Each analytical method available will have its own requirements for specimen preparation. The most appropriate sampling method specifications for each parameter can be found in many textbooks (e.g. SMEWW, 1998). They should consider, for each chemical parameter, the bottle type (glass/plastic, dark), preservative (acidification, cold), typical sample volume, the need of filtration, and maximum storage time.

7.5 In-situ Measurements

7.5.1 Introduction

When collecting a sample of water, certain principal variables that are prone to more or less rapid change upon sample storage must be measured in-situ. These variables characterize the status of the water at the time of sampling. These variables commonly include electrical conductivity, pH, temperature, redox potential, and sometimes also total hardness, turbidity, salinity and dissolved oxygen. Among the in-situ variables, electrical conductivity determination gives the most important information about water quality since it gives an indication of the salts dissolved in water.

Ion selective electrodes provide, in principle, a method for users to determine the concentrations of many ions. However, the instruments need careful (and often repeated) calibration to reference concentrations and washing in a buffer solution between successive readings. This makes their routine use on-site somewhat problematic and prevents their sensible use as remote instrumentation. For this reason further details of these instruments are given in Section 7.6.2.

7.5.2 pH

To measure pH in-situ, so-called pH testers (for a rough estimate of the pH value) and pocket (portable) pH meters are used (Fig. 7.9). Periodic calibration of the instrument is required.

The determination of pH is very fast and reliable when a combined glass electrode is used. It enables an automatic measurement over long time intervals with the accuracy of $\pm$ 0.01 pH units. The glass electrode can be used even in strongly acidic and alkaline solutions, and also in the presence of oxidizing or reducing substances. It must be constantly immersed in water. With time, all glass electrodes deteriorate due to alkali leaching from the surface layers.

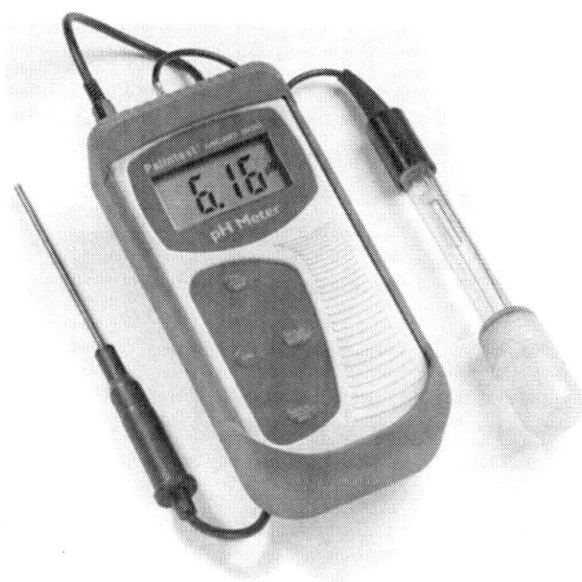

Fig. 7.9 Pocket pH meter. Reproduced by permission of Palintest

7.5.3 Redox Potential (in-situ)

The redox potential, i.e. a measure (in volts) of the affinity of a substance for electrons compared with hydrogen, may also be determined in the field using electrical, hand-held equipment, this time employing an inert oxidation-reduction electrode.

7.5.4 Electrical Conductivity

Electrical conductivity is typically measured in-situ, being an important, yet simple, indicator of pollution since the ability of water to conduct electricity increases as the proportion of dissolved ions increases. It can be measured directly through the insertion of probes and the resistance (or conductivity recorded) or indirectly through air-coupled "aerials". However, because there are so many factors that affect electrical conductivity (e.g. presence of metals, saturation), it is normally best to use conductivity techniques as means of locating areas of anomalous response. These can then be investigated by alternative techniques to discover whether pollution is the cause and, if so, its degree and type.

7.6 Laboratory Measurements

Contaminants may be held both in pore water and on/in the solids fraction of soil samples. Often it is desirable to know how much contaminant could be released from the sample. Simple separation of the pore water (e.g. by a centrifuge method) will not enable us to know how much contaminant might be released from the solids by desorption and leaching. To find this information, extraction tests of some kind need to be performed in which the contaminant is encouraged to move from the solids into the liquid phase by the arrangement of the tests. This is the subject of the first part of this section. Once the liquid phase has been extracted, chemical tests can be performed on the contaminated water – this is described in the second part of this section.

7.6.1 Extraction Methods

7.6.1.1 Introduction

Soils are complex matrices made up of numerous constituents with different and variable physical and chemical properties. Such constituents present variable capacities of interactions with pollutants, which drives the partitioning between the liquid and the solid phases. Pollutants of soils can thus be dissolved in the solution, can be adsorbed or make complexes with organic or inorganic constituents, or can be partially or totally transformed (bio-geo-chemical dynamic) (ADEME, 1999). All these forms are in relation and can, according to the type of matrix and the properties of the pollutant or external factors, induce an increase or a decrease in the mobility and (bio) availability of pollutants. These different forms can be extracted selectively from the matrix thanks to laboratory extraction methods using appropriate chemical reactants (Tessier et al., 1979; ADEME, 1999).

The environmental performance of a material is rather based on release than on total content of potentially dangerous constituents (van der Sloot & Dijkstra, 2004). Selective extraction procedures allow the assessment of the geo-chemical distribution of pollutants in the solid matrix (Colandini, 1997), and therefore the choice of extraction methods depends on the purpose of the investigation. This section essentially deals with inorganic pollutants.

Organic pollutants are often insoluble in water, though may be miscible by surfactants (e.g. detergents) or may exist in water as emulsions. Alternatively, the water and the organic chemical may be self-segregating leading to layered "oil" and other fluids, their relative positions dependent on relative density. Interaction of the organic fluid and solid is complex, depending largely on surface chemistry effects which will not be explained here (see e.g., Yong et al. (1992) for further information). Sorption of organic fluid is largely limited to organic solids.

7.6.1.2 Extraction Through Leaching and Percolation Methods

Leaching can be defined as the process by which soluble constituents are dissolved and filtered through the soil by a percolating fluid, while percolation can be described as the movement of water downward and radially through subsurface soil layers, continuing downward to groundwater (US EPA, 1997 in ADEME, 1999). This led Tas & Van Leeuwen (1995) to define leachate as water or wastewater that has percolated through a column of soil or solid waste in the environment (in ADEME, 1999). The laboratory terminology describes the leaching test as a technique of leaching of solid products by an appropriate solvent in order to extract its soluble fraction (ADEME, 1999). Leaching tests are a kind of extraction technique.

Extraction may be achieved in a number of ways that may be usefully classified as follows:

- Static tests in which the solid specimen is placed in a container with a fixed volume of fluid (leachant) for a certain period of time during which a static equilibrium is reached between the solid and the solution. Such tests are carried out in few hours. Among them one can distinguish:
    - Single batch tests in which the leaching solution is unique. Depending on the test method this may, or may not, involved agitation of some form to quickly reach steady-state conditions. Agitated tests focus on measuring the chemical properties of a material-leachant system rather than the physical, rate-limiting mechanisms. In non-agitated extraction tests the material and leachant are mingled but not agitated: these tests measure the physical, rate-limiting mechanisms.
    - Serial batch tests in which a series of single leaching tests is carried out on the same solid specimen. Such an approach, by means of a succession of steady states, is intended to exhaust the total amount of removable pollutant or, at least, to monitor change in leaching with volume of water passing. Leaching tests are generally carried out in few hours. They are simpler to apply but are less realistic than the percolation tests described next.

- Dynamic tests in which, in a column, a continuous supply of fresh leachant is passed through the specimen and withdrawn after contact with the solid fraction. Contrary to static tests, they allow assessment of the release as a function of time. After a while a dynamic equilibrium can be reached generating a continuous release. Such tests can last up to several dozens of days. Among them one can distinguish:
    - Up-flow percolation tests in which the column is fed from the bottom. This method implies saturated conditions and avoids preferential flows in the column. It may induce pressure migration. The flow through the column is easier to control than in the down-flow percolation test and this means that it is more often used despite being less representative of usual flow conditions in soils.
    - Down-flow percolation tests in which the leachant flows under gravity through a partially saturated column. These tests are especially useful in studying biochemical activity in the vadose zone (Fig. 7.10).

As many soils are rather impermeable, the test is often accelerated by applying large pressure differentials across the specimen of soil, but this reduces the contact time of the water with the solids, so careful interpretation of results is then necessary to ensure that the laboratory result can be applied to the in-situ conditions with meaning.

In each of these tests, the fluid can be water (often distilled and deionised) or it may seek to be representative of in-situ or "worst-case" groundwater (e.g. a weak acid). For static leaching tests, when the water content of the material is too high (sediments, sludge), interstitial water can be recovered by means of a centrifuge.

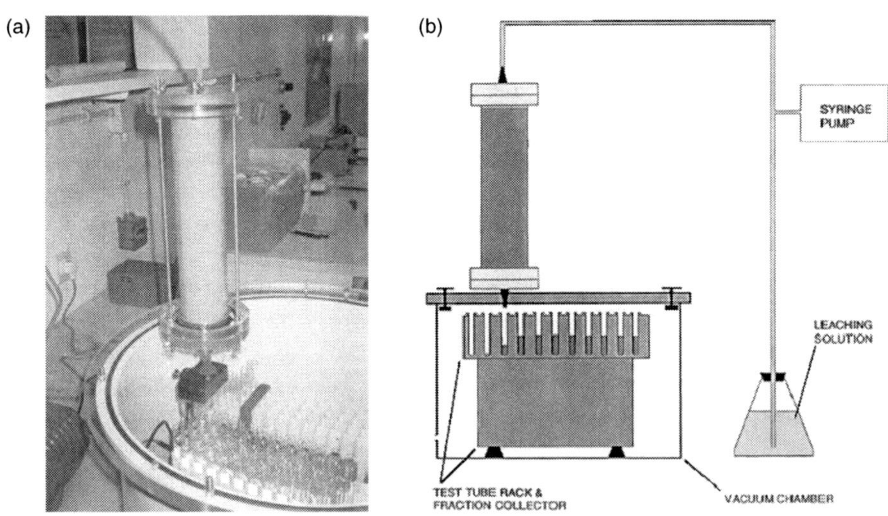

Fig. 7.10 Down-flow percolation test device; (**a**) photograph, (**b**) line diagram. Diagram reproduced by permission of Soil Measurement Systems.com

The centrifuged pellet is then used to carry out leaching. The centrifuge supernatant may also be subjected to testing.

Chemical mechanisms controlling the release of pollutants are dissolution, sorption and diffusion. Diffusion will be controlled by concentration differences and by the total available contaminant content (which can be far lower than the total content). The pH of the material and its environment (in the laboratory: the solvent) are most important as dissolution of most minerals and sorption processes are pH dependent. The oxidation/reduction state of the material and its environment influences the chemical form of the contaminant and its solubility. Complexed forms are generally more soluble than non-complexed ones. The presence of solid and dissolved organic matter or humic substances can enhance the leaching. High ion strength of the solution in the material or in its environment generally increases the leaching of contaminants. Temperature increase leads to higher solubility. Lastly, time of contact is an important factor for the release amount (van der Sloot & Dijkstra, 2004).

The form of the material (granular, monolithic or cemented) is an important physical factor influencing the transport of a contaminant from the material to the liquid phase. Indeed, the release behaviour of granular materials is percolation (advection) dominated, while for monolithic materials it is diffusion dominated. For granular materials, the particle size determines the distance between the centre of the particle and the surface area of exchange and also, for a given amount of material, the total exchange surface. The latter factor is also important for monolithic materials, considering the shape of the monolith. For granular (in column) and monolithic materials, the porosity and the permeability are important factors on release (van der Sloot & Dijkstra, 2004).

Several parameters can be controlled in leaching test protocols in order to highlight different leaching behaviours:

- the relative amount of solvent in contact with the material (expressed in litres/kilogram of dry material, or sometime in litres/sq. metre for monoliths) or the flow through columns;
- the nature of the solvent (generally de-ionised water);
- the time of contact;
- the pH of the solvent (natural or controlled in order to maintain specific values);
- the granular or the monolithic form of the material;
- the crushing of the material to a certain particle size;
- the porosity and the permeability of compacted granular materials implemented into columns; and
- temperature (which generally is ambient temperature or controlled at 20°C).

Table 7.1 presents some examples of typical leach test methods that can be found in the literature. Also listed, for completeness, are speciation tests that aim to separate out different leaching species.

7.6.1.3 Sequential Extraction Methods

Selective extraction can be considered as an "operational speciation" as it corresponds to the quantification of elements bound to specific phases of the soil, rather

Table 7.1 Examples of leaching (and speciation) tests from around the world (adapted from van der Sloot et al. (1997) and from Hill (2004))

LEACHING TESTS FOR GRANULAR MATERIALS

Single Batch Leaching Tests (equilibrium based)

pH Domain 4 - 5	**pH 5 - 6**	**Material Dictated**	**Complexation**	**Low L/S**
TCLP	Swiss TVA	DIN 38414 S4	MBLP (Synth)	MBLP
EPtox		NF X-31-210	(California WET test)	EN 12457–3 (at L/S =2 & 10)
Availability test (NEN 7341)		Ö-norm S2072		Wisconsin SLT
California WET		EN 12457		
Ontario LEP		Canada EE MCC-3C		
Quebec QRsQ		ASTM D 3987		
Soil HAc		Soil – $NaNO_3$		
		Soil – $CaCl_2$		

Multiple Batch and Percolation Tests (mostly based on local equilibrium)

Serial Batch (low L/S)	**Serial Batch L/S>10**	**Percolation or Flow Through Tests**
UHHamburg	NF-X 31-210	NEN 7343 (NVN 25008) column up
WRU	WRU	ASTM column upward
EN 12457–1 (at L/S = 2)	ASTM D4793-88	Column German (pH static)
	NEN 7349 (NVN 2508)	
	MEP method 1320	
	Sweden ENA	
	MWEP	
	EN 12457-2 & -4 (at L/S = 10)	

Static Methods	**Speciation Methods**
Pacific Northwest Lab. MCC-1	Sequential chemical extraction
Pacific Northwest Lab. MCC-2	pH static test procedures
Compacted granular tank leaching test (Rutgers/ECN)	

LEACHING TESTS FOR MONOLITHIC MATERIALS

ANSI/ANS 16.1
Tank leaching test NEN 7345 (a static test, non-agitated)
Spray test (impregnated wood)
Swedish MULP
EN 1744-3 (static test, agitated)

The codes in this table refer to various standards or standards originating organisations. Readers who are uncertain of their meaning are referred to the original sources. L/S = Liquid:solid ratio.

than to an exhaustive analysis of the chemical species in the material. (Tessier et al., 1979; Quevauviller et al., 1993). Selective extraction procedures of pollutants from soils can be simple or can be organised according to a sequential or parallel extraction pattern.

Simple extraction procedures are not much used to determine the operational speciation of metals in materials but are used in soil sciences in order to quantify their potential availability for plants.

Sequential and parallel extractions follow the same principle: to submit the material to a series of reactants in order to identify associations between the different components of the material and the pollutant. Such procedures are more informative than simple extractions as they allow study of the geo-chemical partitioning of pollutants.

Parallel extractions involve different test portions of the same sample subjected to different reactants, while sequential extractions aim to submit the same sample to a well-ordered series of reactants with increasing aggressiveness. Different parallel extraction procedures have been proposed by Serne (1975), Förstner & Patchineelam (1976) and Cazenave (1994) (as cited in Lara-Cazenave, 1994) and also different sequential extraction procedures e.g. Gupta & Chen (1972), Engler et al. (1974), Tessier et al. (1979), Salomons & Förtsner (1980), Meguellati (1982), Welté et al. (1983) and Morrison & Revitt (1987) (as cited in Flores-Rodrigues, 1992).

Associations that are usually studied in most selective extraction protocols are (Colandini, 1997):

- the exchangeable fraction: pollutants are removed from clayey minerals and amorphous materials by simple ion exchange (neutral salts such as $MgCl_2$, $BaCl_2$ or $CH_3CO_2NH_4$ are used);
- the fraction associated to carbonates: metals (co-)precipitated with natural carbonates are easily dissolved by a pH decrease (a weak acid as CH_3COOH is enough to dissolve calcite and dolomite);
- the fraction associated to metal oxides: metals associated to oxides of Fe, Al and Mn are extracted by means of a reducing agent (as hydroxylamine hydrochloride – $NH_2OH.HCl$);
- the organic fraction: under oxidizing conditions (H_2O_2 is used under acidic conditions) organic compounds are mineralized and metals are released; and
- the residual fraction: includes elements that are naturally present into the matrix of minerals.

In principle the different mineral phases can be quite precisely isolated thanks to the use of a series of extractions. However, the chemical attack on a phase does not always lead to a complete dissolution of pollutants contained in that phase, and for sequential extractions this can result in the dissolution of metals contained in other phases of the sample. Moreover, pollutants released by mineralization of a phase can be reincorporated by remaining phases. Thus, during sequential extraction measuring errors can accrue through the different steps. Despite these drawbacks, there remains a key benefit: that this method requires less material than parallel extraction.

As a conclusion of a European research programme (Ure et al., 1993) a 4-step harmonized sequential extraction method of heavy metals from soils and sediments was proposed by the Bureau Communautaire de Référence:

- extract metals of the exchangeable and acid extractable fraction (using CH_3COOH 0.11 M);
- extract metals of the reducible fraction (using NH2OH, HCl 0.1 M; pH 2);

- extract metals of the oxidizable fraction (using H_2O_2 8.8 M; CH_3COONH_4 1 M; pH 2); and
- extract the residual fraction (using HF + HCl 15.5 M).

7.6.2 Chemical Analysis

7.6.2.1 Introduction

Chemical analysis allows determination of the chemical composition of collected samples and, therefore, to identify specific compounds in the chosen environment. Each chemical compound has one or more analytical methods, from the many different methods available, that are more suitable for obtaining an accurate determination of concentration. This section of this chapter presents a brief summary of the analytical methods most used at present for chemical composition identification. It includes coverage of toxicity tests that properly supplement chemical analyses when used to assess the possible impact on living organisms.

7.6.2.2 Selective Ion Measurement

Ion selective electrodes (ISE) are membrane electrodes that respond selectively to specified ions in the presence of other ions. ISE include probes that measure specific ions and gasses in solution. ISE are most commonly used to determine cations and anions. An ISE (with its internal reference electrode, Fig. 7.11) is immersed in an aqueous solution containing the ions to be measured, together with a separate, external reference electrode.

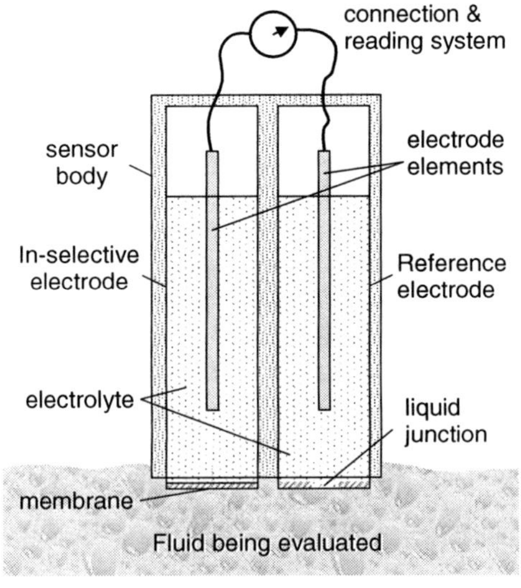

Fig. 7.11 Ion selective electrode – main constituent parts

Table 7.2 Examples of ion-selective electrodes and measurement ranges

Ion	Concentration range ($mol.l^{-1}$)
Ag^{+}/S^{2-}	$10^{-7} \leftrightarrow 1$
Ca^{2+}	$5 \times 10^{-7} \leftrightarrow 1$
Cd^{2+}	$10^{-7} \leftrightarrow 1$
Cl^{-}	$5 \times 10^{-5} \leftrightarrow 1$
CN^{-}	$10^{-6} \leftrightarrow 10^{-2}$
Cu^{2+}	$10^{-8} \leftrightarrow 1$
F^{-}	$10^{-6} \leftrightarrow 1$
H^{+}	$0 < pH < 14$
K^{+}	$10^{-6} \leftrightarrow 1$
NH_3	$10^{-6} \leftrightarrow 1$
NO_3^{-}	$6 \times 10^{-6} \leftrightarrow 1$
Pb^{2+}	$10^{-7} \leftrightarrow 1$

The most commonly used ISE is the pH probe (see Section 7.5.2). Other commonly used ISEs measure electrical conductivity, metals (see Table 7.2) and gases in solution such as ammonia, carbon dioxide, nitrogen oxide and oxygen.

The principle of the measurement is ion exchange between the ion which is dissolved in the solution being monitored and the ions behind the membrane Fig. 7.11. The electro-chemical membrane permits the desired ions to cross it, resulting in a charge on the fluid inside the membrane. At the same time the same amount of charge is passed from the reference electrode to the sample solution, thus maintaining electrical equilibrium. ISEs are normally available as pen-sized probes that can be lowered into the fluid to be assessed – see Fig. 7.12. An excellent guide to ISEs

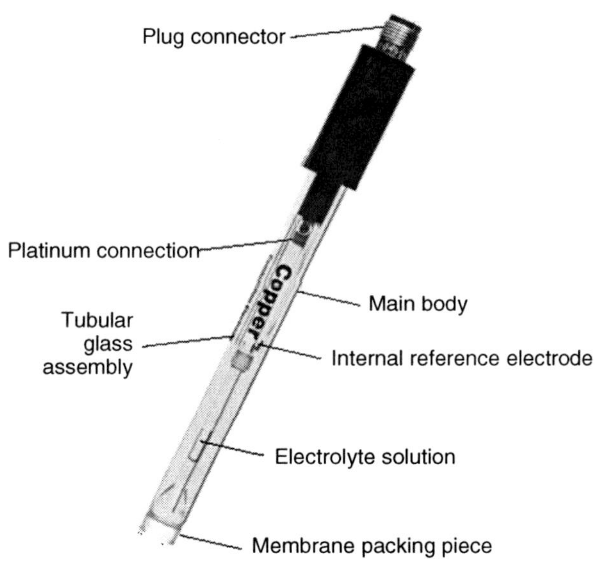

Fig. 7.12 Ion selective electrode (Medek, 1993, Zýka, 1991). In this version, adapted from, and reproduced by permission of van London – pHoenix company

and their use is available on-line (Rundle, 2000). Ions commonly analysed using ISEs are listed in Table 7.2.

Chemical analyses are able to give more precise figures but ISEs can be useful to give an approximate value and also indicate a need for more advanced analyses.

7.6.2.3 Quantitative Analysis

Historically, chemical analysis of water was achieved by titration methods. It is, practically, impossible to use these on water containing a pollutant at a low concentration due to the need to collect a very large volume of water that can be concentrated to permit a weighable amount of chemical to be obtained at the end of the procedure. Also, such procedures are very time-consuming and operator sensitive. Therefore, modern analysis is based on electrical, atomic and spectrographic techniques.

The most common analytical techniques suitable for determination of the presence of selected pollutants are synthesised in Table 7.3.

Table 7.3 Analytical techniques suitable for determination of the presence of selected pollutants

Technique	Detection limit level	Basic cations*	Common heavy metals‡	Platinum group elements†	Organic compounds (e.g. PAH and their derivates, HCB, PCB)
Gas Chromatography (GC)	Depends on sample preparation and detector used				X
Liquid Chromatography (LC)	Depends on sample preparation and detector used				X
Atomic Absorption Spectrometry (AAS)	$mg.l^{-1}$ or $\mu g.l^{-1}$ (ppm or ppb)	X	X		
Inductively Coupled Plasma (ICP)	$\mu g.l^{-1}$ or $ng.l^{-1}$ (ppb or ppt)		X	X	
Molecule Absorption spectrometry in the UV – VIS environment (UV – VIS)	$mg.l^{-1}$ or $\mu g.l^{-1}$ (ppm or ppb)	X	X		
Ion Exchange Chromatography (IEC)	Depends on sample preparation and detector used	X	X		

* = include sodium, potassium, calcium, barium and magnesium.
‡ = include iron, copper, zinc, cadmium, lead, chromium, nickel, cobalt and vanadium.
† = here include rhodium, palladium, iridium and platinum. PAH = polyaromatic hydrocarbons, HCB = hexachlorobenzene, PCB = polychlorinated biphenyls. ppm = parts per million (10^{-6}). ppb = parts per billion (10^{-9}). ppt = parts per trillion (10^{-12}).

7.6.2.4 Eco-toxicity Tests

Despite most regulatory constraints being based on physico-chemical analysis, the hazard toward the natural environment represented by a contaminated solution or matrix cannot simply be assessed on the basis of the single analytical approach. The latter supposes that the contaminants can all be identified and are not too numerous (which is not always the case), but moreover, the chemical concentration does not provide any information about phenomena of synergy or antagonism between pollutants, and does not provide information about the toxicity towards living organisms (criterion H14 of the European Directive 91/689). Biological methods can do so (ADEME, 1999). The purpose of these methods is to assess the eco-toxicological danger of solutions and matrix. They are carried out in vitro on biological species chosen for their sensitivity to pollution (Ramade, 2000).

Eco-toxicity tests can also be carried out from solids thanks to extraction techniques (see Section 7.6.1 on leaching and percolation, above).

Biological test analyses range between classical tests on organisms measuring survival to tests on cells and enzymatic activity.

Water from different parts of the road pavements and embankments and their surrounding environment may be analysed for toxicity to plants, animals, fish and humans. The methods used for collection of the water for this purpose will be as for collection for analyses for chemical compounds. It is, however, especially important that the water quality does not change during the toxicity test. Therefore, it must be kept cool and in dark, and quickly transported to the laboratory (see Section 7.4.6).

The classical tests for deciding toxicity, biological degradation and bioaccumulation are tests according to international standards (OECD Guidelines, ISO). The tests use living micro and macro-organisms (plants, animals) or cell cultures to characterise the toxicity of tested single chemical compounds or mixtures of compounds. In vitro methods use cells or enzymes and proteins for the testing of single compounds or complex mixtures.

When assessing the environmental effects, the test solution is often subjected to several test organisms such as algae, crustaceans and fish to search for differences in the sensitivity of organisms at different trophic levels of the ecosystem.

Table 7.4 Some examples on standard toxicity tests

Category	Description/species	Test code
Toxicity		
Algae	Growth inhibition, *Selenastrum capricornutum*	OECD210, ISO8692
Crustaceans	Immobilisatione, *Daphnia magna*	OECD202, ISO6341
Fish	Death, *Salmo trutta* 96 h	OECD203
Degradation		
Micro-organisms	Easy degradation	OECD301, A,D,F
Bioaccumulation		
Fish	Bioaccumulation, fish	OECD305

Codes refer to OECD (Organisation for Economic Co-operation & Development) and ISO (International Standards Organisation) test procedures

Eco-toxicity tests (Table 7.4) may be classified as acute tests or chronic tests. Acute tests are tests with effects showing within a short time. A classical acute test is the measurement of the survival of organisms. The results are recorded as the concentration at which half the number of test organisms survive/die during the test period (LC50, Lethal Concentration). If the test period is 96 h the concentration referred to will be 96 h LC50. The chronic tests are conducted during a longer period at lower test compound concentrations. The end point is not death, but some secondary sub-lethal effect.

7.7 Concluding Remarks

This chapter presents a general overview of water and soil sampling and analysis in the road environment. The main principles of data collection and storage, and methodologies for sampling design are presented. Furthermore, water and soil sampling procedures as well as in-situ and laboratory measurements and analyses methods are described, with an elucidation about their usefulness, potentialities and fields of application.

It is intended that the information presented in this chapter, as well as the bibliographic material that is referenced at its end, can provide a sufficient and valuable base from which the reader can consider the best choices for contaminant sampling and analysis methodologies, accordingly to the purpose of his/her investigation, and considering the abilities of available methods and tools as presented above.

References

ADEME, 1999, "*Ecotoxicité des Sols et des Déchets: Extraction des Polluants, Données et Références*", (Ecotoxicity of soils & wastes: extraction of pollutants: definitions and references), Agence de l'Environnement et de la Maîtrise de l'Energie, Book 3198, 138pp.

AGS, 2004, "*Electronic Transfer of Geotechnical and Geoenvironmental Data (Edition 3.1)*", Association of Geotechnical and Geoenvironmental Specialists, www.ags.org.uk.

Aller, L., Truman, W.B., Hackett, G., Petty, R., Lehr, J.H., Sedoris, H., Nielsen, D.M. & Denne, J.E., 1989, "*Handbook of Suggested Practices for the Design and Installation of Ground Water Monitoring Wells*", US Environmental Protection Agency,. EPA 600/4-89/034, Dublin, Ohio, 398pp.

Barbosa, A.E., 1999, "*Highway Runoff Pollution and Design of Infiltration Ponds for Pollutant Retention in Semi-Arid Climates*", The Environmental Engineering Laboratory PhD. Dissertation Series, Aalborg University, Denmark.

Barbosa, A.E. & Hvitved-Jacobsen, T., 2001, "Infiltration Pond Design for Highway Runoff Treatment in Semiarid Climates", *J. Environ. Eng., ASCE*, **127**,(11), pp. 1014–1022.

Barcelona, J.M., Gibb, J.P., Helfrich, J.A., & Garske, E.E., 1985, "*Practical Guide for Ground-Water Sampling*", Champaign, Illinois State Water Survey, 37pp.

Boulding, J.R., 1995, "*Practical Handbook of Soil, Vadose Zone, and Ground-Water Contamination. Assessment, Prevention, and Remediation*", Florida, Lewis Publishers, Inc., 1995, 948pp.

Burton, G.A. Jr., & Pitt, R., 2002, "*Stormwater Effects Handbook: A Tool Box for Watershed Managers, Scientists, and Engineers*", CRC Press, Boca Raton, 911pp.

Canter, L.W., Knox, R.C. & Fairchild, D.M., 1987, "*Ground Water Quality Protection*", Michigan, Lewis Publishers, Inc., 1987, 562pp.

Chandler, R.J., Beaumont, A., Evans, D., Toll, D. & Quinn P., 2006, "Combining the Power of AGS and XML – A Data Format for the Future", *Proceeding of GeoCongress* "Geotechnical Engineering in the Information Technology Age", ed. DeGroot, D.J., DeJong, J.T., Frost, D. & Baise, L.G., Atlanta, American Soc. Civil Eng., 6pp.

Clark, L., 1993, "Monitoring and Protecting Ground Water Quality", *Pollution Prevention*, **3**(4) pp. 51–56.

Colandini V., 1997, "*Effets des Structures Réservoirs à Revêtement Poreux sur les Eaux Pluviales: Qualité des Eaux et Devenir des Métaux Lourds*", (in French: "Effects of water storage structures beneath permeable surfaces on rainwater: quality of water and destiny of heavy metals"), PhD thesis, Université de Pau et des pays de l'Adour, 174pp. & annexes.

Environment Agency, 1998, "*Quality Management System for Environmental Sampling*", National Sampling Procedures Manual, Vol. **25**, Environment Agency. Bristol, UK.

FHWA, 1987, "*Methodology for Analysis of Pollutant Loadings from Highway Stormwater Runoff*", Federal Highway Administration, Report N°. FHWA/RD-87/086, U.S. Department of Transportation, Washington and Woodward-Clyde Consultants.

FHWA, 1988, "*Evaluation of Pollutant Impacts from Highway Stormwater Runoff. Design Procedure*", Federal Highway Administration, Report N°. FHWA/RD-88/006, U.S. Department of Transportation, Washington and Woodward-Clyde Consultants.

FHWA, 1996, "*Evaluation and Management of Highway Runoff Water Quality*", Federal Highway Administration, Report N°. FHWA-PD-96-032, U.S. Department of Transportation, Washington, 457pp.

Flores-Rodriguez, J., 1992, "*Les Métaux Toxiques dans les Eaux Pluviales en Milieu Urbain : Caractéristiques Physico-Chimiques*", (in French: "Toxic metals in rainwater in the urban environment : physio-chemical characteristics"), PhD thesis, Université de Paris XII – Val de Marne, 216pp. & annexes.

Hamilton, R.S., Harrison, R.M., McCrae, I.S. & Ward, N.I., 1991, "Highway Pollutant Monitoring", in: Hamilton, R.S. & Harrison, R.M. (eds:) *Highway Pollution, Studies in Environmental Science*, Vol. **44**, Elsevier Science Publishing, New York, pp. 48–99.

Hendry, T., Cameron, K.C., Di, H.J., Moir, J., Smith, N. & Beale, N., 2004, "Automatic Measurement of Drainage Flux and Nitrate Leaching Losses in a Heavy Clay Soil", 'SuperSoil 2004', *3rd Australian & New Zealand Soils Conf.*, 5–9 Dec., Univ. Sydney, Australia, 5pp, available at http://www.regional.org.au/au/asssi/supersoil2004/pdf/1733_hendryt.pdf.

Hill, A.R., 2004, "*Leaching of Alternative Pavement Materials*", PhD thesis, School of Civil Engineering, University of Nottingham.

Hvitved-Jacobsen T. & Vollertsen, J., 2003, "Pollution from Road Runoff. Concepts and Engineering", pp. 1–22, in "*Curso Sobre Características de Águas de Escorrência de Estradas em Portugal e Minimização dos seus Impactes*", 13–15 October, Laboratório Nacional de Engenharia Civil, 176pp.

Hvitved-Jacobsen T. & Yousef Y.A., 1991, "Highway Runoff Quality, Environmental Impacts and Control", in Hamilton RS & Harrison RM, (eds.), *Highway Pollution, Studies in Environmental Science*, Vol. **44**, Elsevier, Amsterdam pp 165–208.

Jandová, V., 2006, "*Kontaminace vod polutanty obsaženými v konstrukčních vrstvách vozovek*", Prubežná zpráva o realizaci projektu 1P050C003 za rok 2005. (In Czech. "*Water Contamination with Pollutants Contained in the Constructional Layers of the Roads*", Report of the project No. 1P050C003 for year 2005), CDV, Brno, 20pp.

Krajca, J., 1989, "*Water Sampling*", Ellis Horwood Limited, Chichester, 212 pp.

Lara-Cazenave M.B., 1994, "*Spéciation des Métaux Lourds (cadmium, cuivre, plomb et zinc) dans les Eaux de Ruissellement*", (in French. "*Speciation of Heavy Metals (cadmium, copper, lead & zinc) in Storm Water*"), PhD thesis, Université de Pau et des pays de l'Adour, 149pp.

Leitão, T.E., 2003, "Monitorização de Águas Subterrâneas e de Solos" (Groundwater and soil monitoring in road environment), in proceedings of the course "*Características das Águas de*

Escorrência de Estradas e Minimização dos seus Impactes", Barbosa, A.E., Leitão, T.L. & Hvitved-Jacobsen, T., & Bank, F. (eds.), pp. 149-158.

Medek, J., 1993, "*Experimental Methods*", PC-DIR Brno, Brno, 180pp. In Czech.

Nielsen, D.M., 1991, "*Practical Handbook of Ground-Water Monitoring*", Chelsea, Michigan, Lewis Publishers Inc., 1991, 660pp.

Quevauviller P., Ure A., Muntau H. & Griepink B., 1993, "Improvement of Analytical Measurements within the BCR Programme: Single and Sequential Procedures Applied to Soil and Sediment Analysis", *Int. J. Environ. Anal. Chem.*, **51**, N°. 14, pp. 129–134.

Ramade F., 2000, "*Dictionnaire Encyclopédique des Pollutions*", (in French. Encyclopaedic Dictionary of Pollutions), Ediscience international, Paris, 690pp.

Rundle, C., 2000, "*Beginners' Guide to ISE Measurement*", available at www.nico2000.net/Book/Guide1.html.

Ruttner, F., 1952, "*Fundamentals of Limnology*", University of Toronto Press, Toronto.

Smeww, 1998, "*Standard Methods for the Examination of Water and Wastewater*", 20th ed., American Public Health Association/American Water Works Association/Water Environment Federation, Washington DC, USA. http://www.ec.gc.ca/nopp/docs/rpt/2MM5/en/C6.cfm#table3.

Sytchev, K. I., 1988, "Water Management and the Geoenvironment", Vol. **1**, In *Geology and the Environment*, an International Manual in three volumes, Kozlovsky E. A. (ed.), UNESCO Paris/UNEP Nairobi, 1988.

Tas J.W. & Van Leeuwen C.J., 1995, "Glossary", in: van Leeuwen C..J & Hermens J.L.M. (eds.), *Risk Assessment of Chemicals: An Introduction*, Dordrect, Netherlands, Kluwer Academic Publishers, pp. 339–361.

Tessier A., Campbell P.G.C. & Bisson M, 1979, "Sequential Extraction Procedure for the Speciation of Particulate Trace Metals", *Anal. Chem.*, **51**(7), pp. 844–851.

Ure A., Quevauviller P., Muntau H., Griepink B., 1993, "*CEC/BCR Information Report EUR report 14763 EN*", European Commission, Brussels, 85pp.

US EPA, 1997, "*Terms of Environment: Glossary, Abbreviations, and Acronyms*", Revised Edition, Dec., available at http://www.epa.gov/trs/.

van der Sloot H.A. & Dijkstra J.J., 2004, "*Development of Horizontally Standardized Leaching Tests for Construction Materials: a Material Based or Release based Approach? Identical leaching mechanisms for different materials*", Energy Research Center of the Netherlands, Report No-. ECN-C-04-060, 44pp. & annexes.

van der Sloot, H.A., Heasman, L. & Quevauviller, Ph. (eds.), 1997, "Harmonization of Leaching/Extraction Tests", *Studies in Environmental Science* **70**, Elsevier Science B.V., Amsterdam, 281pp.

Walthall, S. & Waterman M.K., 2006, "The Collection and Management of Geotechnical & Geoenvironmental Data", *Proceedings GeoCongress* 'Geotechnical Engineering in the Information Technology Age', DeGroot, D.J., DeJong, J.T., Frost, D. & Baise, L.G. (eds.) Atlanta, American Soc. Civil Eng., 5pp.

Wanielsita, P.W. & Yousef, Y. A., 1993, "*Stormwater Management*", John Wiley & Sons, Inc., USA, Canada, 579pp.

Zýka, J, 1991, "*Instrumentation in Analytical Chemistry*", vol. **1**, Ellis Horwood, Chichester, 368pp.

Yong, R.N., Mohamed, A.M.O. & Warkentin, B.P, 1992, "*Principles of Contaminant Transport in Soils*", Elsevier, Holland, 327pp.

Chapter 8
Water Influence on Bearing Capacity and Pavement Performance: Field Observations

Robert Charlier*, **Pierre Hornych, Mate Sršen, Åke Hermansson, Gunnar Bjarnason, Sigurður Erlingsson and Primož Pavšič**

Abstract This chapter presents a mechanical behaviour study, i.e. the bearing capacity as a function of the moisture degree. The field point of view is expressed and the chapter summarises a number of observations on road behaviour, in relation to variations of moisture. First, the road structure is recalled with respect to the mechanical analysis point of view. Then some observations on field under temperate climate, humid, are given. In a second step, the specific case of frost and thawing are discussed.

Keywords Bearing capacity · moisture level · field measurement · stiffness · rutting · thawing

8.1 Introduction

This chapter introduces the study of mechanical behaviour, i.e. it seeks to describe bearing capacity as a function of the moisture degree. This description is based on a summary of road behaviour with respect to variations of water content as observed in-situ.

Initially, the road structure is presented from the point of view of a mechanical analysis. The mechanical and hydraulic specific behaviour of each subgrade layer is discussed. Next, the chapter briefly analyses water penetration and the water effect on the road structure layers before illustrating the water-induced mechanical effects. The change of water content over time is shown for specific locations that are subjected to a humid, temperate, climate. Then a consideration of the development of ruts, i.e. the strains and deformation, with time is given. A discussion of the stiffness of a road structure is also given as it relates to the moisture level. A clear decrease of the stiffness and an increase of the strain accumulation are observed when the moisture increases.

* *Co-ordinating Author:*
✉ R. Charlier
University of Liège, Belgium
e-mail: robert.charlier@ulg.ac.be

A.R. Dawson (ed.), *Water in Road Structures,* DOI 10.1007/978-1-4020-8562-8_8,
© Springer Science+Business Media B.V. 2009

In the second part of the chapter, specific cases of frost and thawing action are discussed. In particular, the stiffness decrease during and after thawing is described. Field measurements of temperature, of layer moisture and of deflectometer stiffness are presented.

8.2 Pavement Behaviour in Relation with Moisture: Water Influence on Bearing Capacity

8.2.1 Different Types of Road Structures Versus Sensitivity to Water

Road pavements are multilayer structures (see Fig. 1.5) generally comprising a surface course and one or more asphaltic or granular base layers, resting on a pavement foundation. Chapter 1 introduced the major pavement layers – the foundation, the sub-base, the pavement base and a surfacing (see Section 1.4.2).

Water permeability should normally increase from the top of the pavement (the asphalt or concrete layers) downward until about 0.7 m depth (see Chapter 5). Otherwise water would accumulate onto the low permeability layer and keep the upper layer wet; freezing of the accumulated water might then unbind the upper layer. This would decrease it's bearing capacity and service life. An exception to these rules are the porous asphalt surfaced pavements described in Chapter 5, Section 5.7, which are designed to carry water within their thickness.

Pavement structures can be divided in four main groups:

- **Thin bituminous pavements**, which consist of a relatively thin bituminous surface course, resting on one or more layers of unbound granular materials. These are typically used for carrying low traffic levels.
- **Thick bituminous pavements** consisting of a bituminous surfacing, over one or two bituminous layers/asphaltic concrete (AC) (base) then an aggregate (sub-base). Their application is, typically, for high traffic levels. They may be considered as flexible, but they are much stiffer then the preceding pavements.
- **Semi-rigid pavements** comprising a bituminous surfacing over one or two layers of materials treated with hydraulic binders (e.g. concrete). This type of pavement is also appropriate for high traffic levels.
- **Portland cement concrete (PCC) pavements** which consist of a Portland cement concrete slab (15–40 cm thick), possibly covered with a thin bituminous surfacing, resting on a sub-base (bound or unbound), or directly on the foundation. The concrete slab can be continuous with longitudinal reinforcement, or discontinuous. Once again, this type of pavement is also appropriate for high traffic levels.

These various pavement types present different types of mechanical behaviour, and different deterioration mechanisms. However, for all structures, water plays a major role in pavement deterioration.

In **thin bituminous pavements**, high stresses are transmitted to the unbound granular layers and to the subgrade; and lead to permanent deformations of these layers. Because unbound layers and subgrades are sensitive to water content, the performance of these pavements is strongly dependent on variations of moisture conditions. This can lead, in particular to "edge failures": water infiltrates from the pavement shoulders, under the edge of the pavement, leading to subsidence at road edges. As these pavements are very sensitive to moisture, impermeability of the surface course and good drainage are very important for their performance.

In **thick bituminous pavements**, the much lower flexibility of their bituminous base layers means that the stresses transmitted to the soil are much lower, and the risk of permanent deformations in the soil, as well as the sensitivity to the water content of the soil, are lower. The main mechanism of deterioration of these pavements is cracking due to the combined effects of traffic-induced strains and thermally-induced movements, causing high tensile stresses in the bound layers. Once the pavement is cracked, water infiltration accelerates the degradations, leading to weakening of the subgrade, attrition at the lips of the cracks, and material chipping away to form potholes (see Chapter 5). Without maintenance, deterioration can lead rapidly to total ruin.

In **pavements with layers treated with hydraulic binders**, the main deterioration process is generally due to reflective cracking. Thermal contraction and shrinkage in the cement treated materials create transversal cracks. These cracks generally rise up through the surfacing and appear on the pavement surface at fairly even spacing (5–15 m). These cracks tend to deteriorate and split under traffic loads. Then again, water infiltration is a major problem, leading to a deterioration of the bonding between bound layers, a decrease of the bearing capacity of the subgrade in the cracked area, thus decreasing load transfer and favouring attrition of the crack lips. On these pavements, protection against water infiltration, by using relatively thick surface courses and by sealing the shrinkage cracks, is essential;

In concrete pavements, due to the high modulus of elasticity of concrete, only very low stresses are transmitted to the foundation. Thermal cracking in concrete structures is generally controlled by transverse joints, or by the longitudinal reinforcement, producing only very fine micro-cracks. Two main types of damage are observed:

- Cracking created by excessive tensile stresses at the top or base of the slabs due to the combination of traffic loads and deformations of the slabs due to thermal gradients; and
- Reduction in bearing capacity around joints and cracks, leading to pumping phenomena. This reduction is essentially due to the presence of water at the interface between the slab and the sub-base. Under loading by vehicles, water at the interface is locally highly pressurised, high pressure gradients appear, inducing high water flow velocities which can erode the sub-base material (near a pavement edge crack or joint), reduce the bearing capacity of the support and reduce load transfer between the slabs. This is generally observed as edge or corner cracking.

For all pavements, **freezing and thawing** phenomena are also a major source of deterioration. In frost sensitive, fine grained soils, freezing leads to a concentration of water near the frozen zone (due to the so-called cryo-suction process – see

Section 4.6). This leads to heaving of the pavement, and then loss of bearing capacity during the thaw period. It should also be noticed that in cold climates, where winter tyres with studs are used, the wear of the surfacing of the pavement and the consecutive re-paving is often faster than fatigue or deformation damage.

8.2.2 Pavement Design and Climatic Effects

Most actual pavement design methods are based on the same principles. Linear elastic calculations are used to determine the stresses and strains in the pavement layers, under a reference traffic load, and then the calculated stresses and strains are compared with maximum allowable values, depending on the nature and characteristics of the pavement materials. Most usual design criteria are the following:

- For **bituminous layers**: a fatigue criterion, often limiting the tensile strain at the bottom of the bituminous layer to prevent upward cracking in thin bituminous layers. For thicker asphalt layers, where the cracking may be from the top-downwards due to aging, traffic and thermal effects, fatigue relationships will need correlating to actual cracking performance if this calculation route is selected;
- For **cement-treated layers**: a fatigue criterion, limiting the maximum tensile stress at the bottom of the treated layer; and
- For **subgrades**; a criterion limiting the vertical elastic strain at the top of the layer, to avoid risks of rutting.

Thus, pavement design is based on elastic calculations and on the application of a limited number of design criteria (mostly fatigue criteria). Calculations are generally performed with constant material properties (corresponding to the initial characteristics of the materials after construction) and a design traffic, defined by an equivalent number, NE, of standard axle loads (ESALs). For thicker asphalt pavements and pavements containing cement-treated layers, unless there has been a specific calibration between the causes of pavement deterioration, such as ageing and deterioration of materials and climatic effects (temperature and moisture variations, frost), then these factors will need assessment before the safety of the road pavement and embankment can be secured. The effect, either way, is likely to increases the road cost by over-design of the structure.

Design against frost is generally based on the evaluation of the frost sensitivity of the subgrade (by frost heave tests, swelling tests, or on the basis of empirical classifications). When the subgrade is sensitive to frost, a thermal propagation model is often used to determine the thickness of protective material needed to reduce penetration of frost into the subgrade. However, swelling of frost-sensitive soils during freezing, or loss of bearing capacity during the thaw period are generally not taken into account, except in some countries (some examples are given in the Annex).

8.2.3 Influence of Water Infiltration on Pavement Deterioration and Mechanical Degradation

Changes in water content, especially excess moisture, in pavement layers combined with traffic loads and freezing and thawing can significantly reduce pavement service life. Failures associated with moisture are detected on roads all over the Europe. There is some evidence to suggest that water has less impact on thick and well-construced pavements than it does on thinner ones (Hall & Crovetti, 2007). It appears that in thicker pavements the effect of water may be more indirect than in thinner ones, reducing material stiffness leading to later distress.

To minimize the negative effects of moisture on pavement performance, first we have to identify the sources of infiltration of water. There are many different possible sources of water infiltration in pavement systems. Mainly, the presence of water in pavement is due to infiltration of rainwater through the pavement surfaces through joints, cracks and other defects, especially in older, somewhat deteriorated, pavements and shoulders. An important source is also migration of liquid water upwards to the freezing front. Water may also seep upward from a high groundwater table due to capillary suction or vapour movements, or it may flow laterally from the pavement edges and side ditches.

The significance of the routes of infiltration depends on the materials, climate, and topography.

Many pavement failures are the direct result of water entering the pavement courses and/or the subgrade. Water entry in the compacted unsaturated material will increase water pressure or decrease suction, and in turn, reduce the effective stress (see Chapter 9). Hence, the strength and the elastic and plastic stiffnesses of the pavement material and the subgrade will be reduced. The rate of traffic-induced deterioration of the road will increase during this time. The loss of strength and stiffness can lead, in the extreme, to rutting and other forms of surface deformation or, more commonly, to pavement edge failures. The worst situation occurs with poorly-compacted granular material (as a result of shear strength reduction) with frost susceptible soils or with cohesive swelling clays (as a result of damaging volume changes).

Water seeping through the pavement can also transport soil particles and cause erosion and pumping (i.e. transport) of fines as well as leaching of many materials (see Chapter 6). Moisture entry can also affect the performance of the surface course by causing stripping of bitumen from aggregate, layer separation between bound courses, and pothole formation (see Section 5.5). A probable mechanism of pothole formation is illustrated in Fig. 8.1. A somewhat similar failure mechanism has also been observed in slabbed concrete pavements (Roy & Johnson, 1979):

- water entered through the joints between the slabs;
- the water softened the supporting layers allowing the slab to deflect under traffic;
- the increased dynamic movement of the slab when trafficked caused a "pumping" action by which water was rapidly displaced from the pores in the supporting material;

(i) Stage 1 Water is forced into a cracked pavement

(ii) Stage 2 Water travels along the interface between the asphalt (AC) and the compacted aggregate base

(iii) Stage 3 Asphalt lifts to allow pressure to dissipate, but also allows more water to be pumped into the crack

(iv) Stage 4 Asphalt is dislodged due to traffic action and water pressure

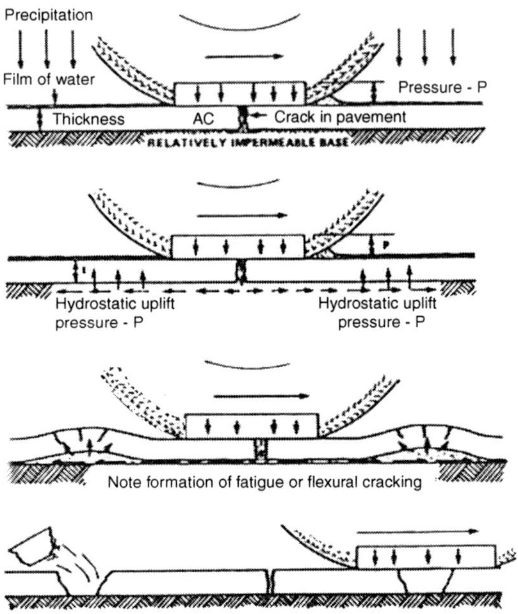

Fig. 8.1 Layer separation and pothole formation (from Gerke, R.J. (1979), cited by Lay (1986)). Reproduced by permission of M. G. Lay

- the fast-moving water eroded the finer materials;
- at their edges, the concrete slabs progressively lost support from the underlying layers as material was washed away;
- the deflections of the slab became greater and the erosion more rapid eventually leading to cracking of the unsupported concrete at edges and corners of slabs; and
- dirty water was seen to squirt from the joints in the pavement when the slabs were trafficked.

Pavement damage, associated with water, can be divided into moisture-caused and moisture-accelerated distresses. Moisture-caused distresses are those that are primarily induced by moisture, while moisture-accelerated are those that are initiated by different factors, but the rate of deterioration is accelerated by presence of water. Most commonly observed damages due to the moisture are:

- surface defects;
- surface deformations; and
- cracking.

Generally, water threatens the stability of soil and it is a particular problem for pavement structures since they are built through areas of changeable moisture quantities.

It has been generally established that the increased subgrade water content during spring results in increased deformability, i.e. a decreased bearing capacity of the pavement. These changes in bearing capacity are, in particular, obvious for silty and clayey materials.

Water contents contained in materials under flexible pavements are influenced by the amount and intensity of rainfall. Periods of long rainfall of low intensity can be more severe than concentrated periods of high intensity, since the amount of moisture absorbed by the soil is greatest under the former conditions. Also, the combined effects of rainfall and freezing temperatures determine, in part, the extent of pavement damage.

Water can penetrate into the pavement structure in several different ways as follows (Fig. 8.2):

- Seepage from the elevated surrounding soil, which depends on the hydraulic gradient and soil permeability coefficient.
- Rise and fall of the phreatic surface, which depends on the climatic circumstances and soil composition (e.g. following heavy rains there is an increase in the sub-surface water level in permeable strata, while water remaining on the surface of impermeable ground will drain away or evaporates and there is no risk of a rapid rise in sub-surface water level).
- The penetration of water through damaged pavement surfaces causes high local concentration of water in penetration areas. Under heavy traffic load, such a condition can result in significant damage if the subgrade made of changeable material is in contact with water. Even worse pavement damage can result from freezing of the structure and subgrade when soaked with water.
- The penetration of water through shoulders (if the shoulders are permeable or their surface is deformed in such a way as to allow the retention of water), which depends on the material permeability, compactness of the surface, inclination and the drainage from pavement surface. The effect is similar to the effect of water penetration through a damaged pavement surface.

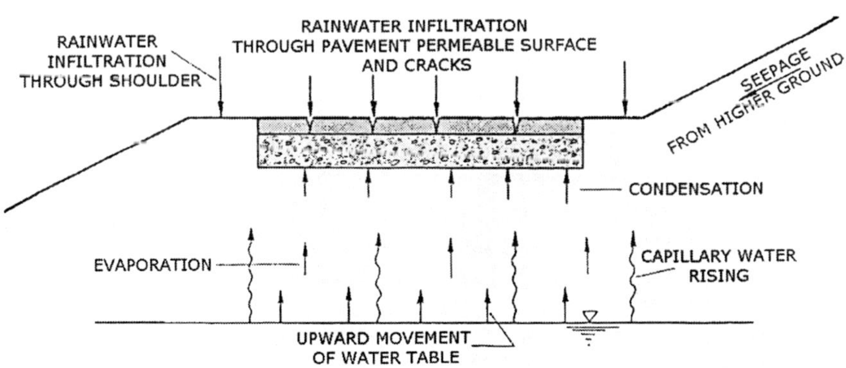

Fig. 8.2 Possibilities of water movements into the pavement zone (from Moris & Gray, (1976) cited by Lay (1986)). Reproduced by permission of M. G. Lay

- Capillary rise from the foundation soil. The water rises from the foundation soil through the fine-grained soil up to the pavement structure.
- Evaporation of water from the foundation soil and its condensation under the pavement structure if the pavement structure is colder than the soil.

Moisture behaviour in a pavement can be considered as occurring in three phases:

- an entry phase – which occurs quite rapidly;
- a redistribution phase when water moves within the material in response to suction and gravity; and
- an evaporative phase when water, as vapour, leaves a material or moves to other layers. Water vapour movements can occur under temperature gradients with the water vapour travelling from a warm to cool area when it then condenses.

Water content under road pavements will vary seasonally, annually and over longer periods. Seasonal variations in water content are commonly located in the upper 1–2 m, and in the metre or so of pavement width at the edge of the pavement surface (the outer wheel path is clearly the critical zone). Significant moisture changes will only occur immediately after rainfall if very permeable layers exist.

The behaviour of the pavement related to moisture should be considered in reference to the climate, type of soil, groundwater depth and moisture concentration in the soil.

8.2.4 Water Content Variations in Pavements

In the last ten years, significant progress has been made in the measurement of in-situ water contents in pavements, using in particular TDR probes (see Chapter 3, Section 3.2.2). These measurements have shown that, often, significant amounts of water infiltrate in to pavements through the pavement surface and from the shoulders.

Low traffic pavements are particularly exposed to water infiltration. Examples of moisture measurements on a typical flexible pavement (6 cm thick bituminous surfacing and granular base) are shown in Figs. 8.3 and 8.4. Figure 8.3 shows that the daily variations of water content in the granular base and in the clayey subgrade (near the pavement edge) are important and strongly related with the rainfall. Figure 8.4 shows average water contents measured in the granular base, at different locations, near the centreline of the pavement and near the edge. The critical zone is clearly the pavement edge where the water content is about 2 percentage points higher than near the centreline. In this pavement, subjected to a mild oceanic climate, seasonal variations of water content are low, but they can be more important with more continental climates.

Thick bituminous or cement-treated pavements are less permeable, and water infiltrates mainly when cracking develops, thus accelerating the deterioration. In such pavements, protection against water infiltration, by proper maintenance (crack sealing, renewal of the surface course) is one of the main concerns.

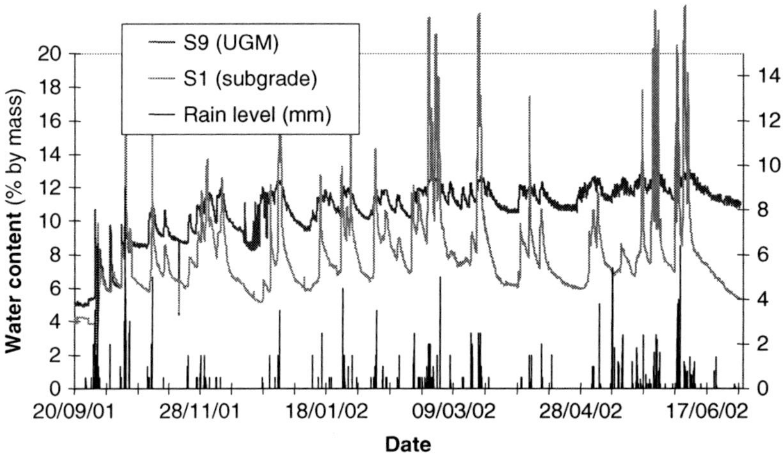

Fig. 8.3 Water content variations in the granular base and subgrade of a low traffic pavement (near the pavement edge)

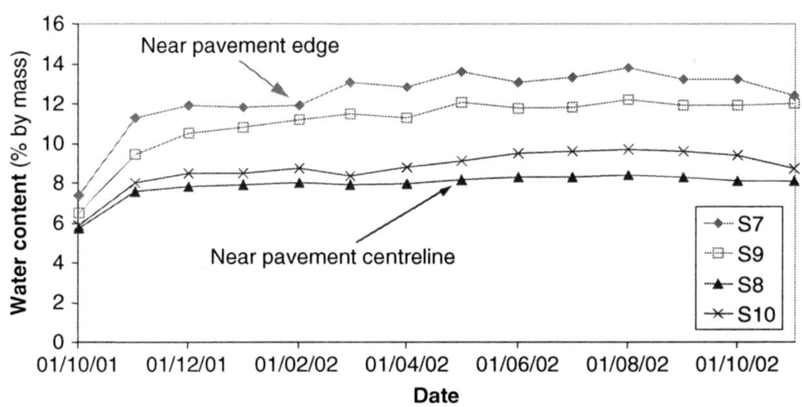

Fig. 8.4 Monthly average water contents in the granular base, at the centre and near the edge of the pavement

8.2.5 Effect of Water and Loading on Structure Behaviour on Rut Progression

Accelerated load testing of pavements was done with the HVS-NORDIC at VTI in Sweden in 1998 (Wiman, 2001). Figure 8.5 shows the rut depth measurements for a weak pavement comprising a 49mm thick asphalt layer over a bitumen stabilised granular base of thickness 89mm over a sand subgrade 2.5 m thick (mean thicknesses).

After 500 000 passes the increase in rut depth was constant and only 0.88 mm/100 000 passes. Then it was decided to increase the test load from 60–80 kN

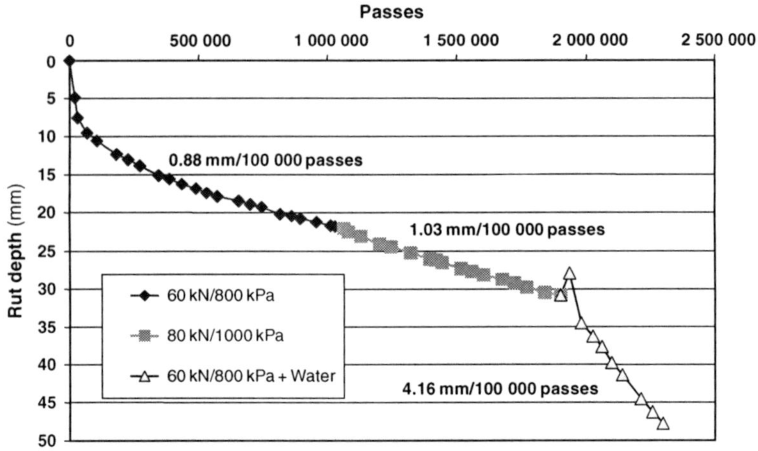

Fig. 8.5 Rut depth propagation rates during test SE01. Figure courtesy L. Wiman, reproduced by permission of VTI

and the tyre pressure from 800 to 1000 kPa. The rut propagation increased but only to 1.03 mm/100 000 passes. The next step was to weakening the sub grade by adding water to the sand to bring the water table to a level 300 mm below the surface of the sub grade – the highest level permitted in the Swedish specifications when constructing new pavements. The test load was at the same time reset to 60 kN and a tyre pressure of 800 kPa. Now the rut propagation increased to 4.16 mm/100 000 passes and the first cracks could be seen at the pavement surface.

8.2.6 Seasonal Variation of Material Parameters

Calculated stiffness values, based on measured deflections under loading of a pavement surface, for a thin pavement structure are given in Fig. 8.6, along with the water content. One can see that the spring-thaw period started in early April as the water content at the three probes increased from 4%–7% to 12%–16% in a very short period of time. When the water content in the lower part of the granular base reached its maximum value (15.2%), the stiffness of that layer reached its minimum value. As the water content during the summer period gradually decreased to 11%, the stiffness increased to its maximum value. The same trend was mainly true for the subgrade as well. The water content of the subgrade though reached its lowest value much later than the granular base and the recovery went on during the whole summer. This is probably due to the subsoil having much higher fines content than the base and the sub-base and, therefore, it takes much longer time for the water to dissipate from the subgrade.

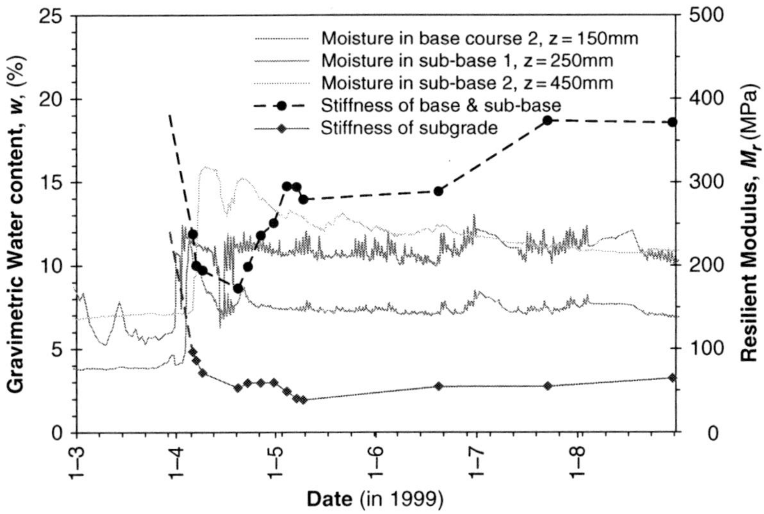

Fig. 8.6 Stiffness and gravimetric water content at one section in SW Iceland

Increased use of dielectric sensors (see Chapter 3, Section 3.2.2) have permitted moisture assessments to be continued during cold-climate winters. By this means it has been observed that complete freezing of all pore moisture doesn't necessarily occur in all the granular pavement layers, even though they are, nominally, within the frost-affected depth.

8.3 Frost and Thawing of Pavements with Frost Susceptible Soils

8.3.1 Frost Heave – Introduction

Frost heave occurs in roads having fine graded, so called frost susceptible material, at a depth to which the freezing front reaches during the winter. The frost heave typically causes an uneven road surface and results in reduced travelling speed and comfort. The main problem though usually arises upon thawing when ice lenses involved with the frost heave melt and result in high water content in the pavement. The increased water content often means reduced bearing capacity and spring-thaw load restrictions are imposed to avoid severe pavement deterioration.

Granular pavement layers normally show a substantial decrease in stiffness with increasing values of moisture. Once thawing commences in the spring season, the granular layers often reach a state of near-saturation that substantially reduces the load carrying capacity. During the winter, short thawing periods have similar effect, especially on granular base courses. Therefore, seasonal changes cause a significant variation in the ability of a pavement to support traffic loads. During the thawing

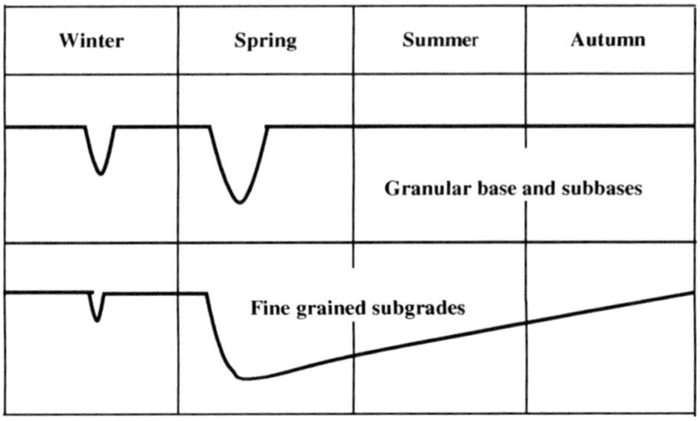

Fig. 8.7 Schematic overview of seasonal variation of stiffness

period, water is melted from the ice lenses and since the layers where the ice lenses are formed have high fines content, the stiffness can drop dramatically. Since the road thaws primarily from the surface downwards, the free water can not drain through the still frozen underlying layers. Another effect of frost heave is that the road layer where the ice lens was formed loses its compaction (density) which is gradually regained under traffic load.

Determination of frost-sensitivity of soils is generally carried out using frost heave tests, such tests are mainly used to classify soils, according to their frost-sensitivity. However, prediction of the mechanical behaviour of such soils in pavements (heave during the frost period or loss of bearing capacity during thawing) is much more complex, because it depends on the climatic and moisture conditions, and on the characteristics of the whole pavement structure.

The principal characteristic of stiffness variation due to environmental effects is given in Fig. 8.7. During winter short thawing periods can cause temporary decreases in the aggregate base and sub-base stiffness. If the thawing penetrates down to the subgrade it also loses its stiffness. As the freezing starts again both layers regain stiffness. During spring thaw, the stiffness of the granular base and sub-base again lowers but the bearing capacity regains soon after the spring thaw period is over. However, if the subgrade has high fines content, it can take longer for the bearing capacity to recover.

8.3.2 Thawing, Field Study, Canada

In cold climates, and with frost susceptible materials, freeze thaw phenomena play a major role in pavement deterioration. Figure 8.8 shows examples of measurements of deflection development, and of water content in the subgrade, performed in a full scale experiment, carried out in Quebec (Savard et al., 2005). The pavement structure consists of 18 cm of bituminous materials over a granular base (40 cm) and a 40 cm thick sand frost-protection layer. The pavement is subjected to frost indices exceeding 1000°C.days. The deep frost penetration (up to 1.5 m during severe winters) leads to large water content variations in the silty, frost susceptible subgrade,

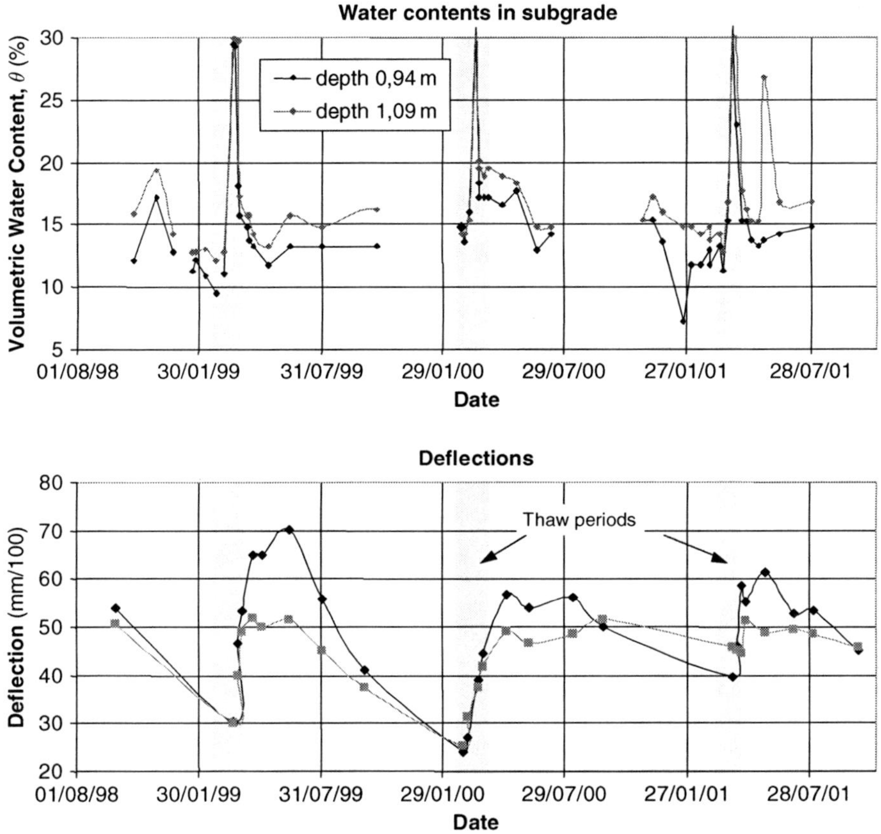

Fig. 8.8 Deflection (caused by loading of pavement) and water content as a function of time – example from Quebec. Grey = Thaw periods

which considerably affect the pavement deflections. The period of reduced bearing capacity lasts about 2 months (thaw period, followed by a recovery period, where the excess moisture dissipates).

8.3.3 Thawing, Field Study, Iceland

Figure 8.9 shows the air temperature together with the volumetric water content at Dyrastadir in Nordurardalur in SW Iceland during spring thaw, monitored through an environmental program run by the Public Roads Administration in Iceland. One can clearly see that as the thawing period starts in early March the water content increases, initially close to the surface and later at greater depth, before it slowly reverts back to normal values. As the water content affects the stiffness of the structure as well as the permanent deformation characteristics, increased deterioration or damage is expected at the high water content if no axle load limitations are applied.

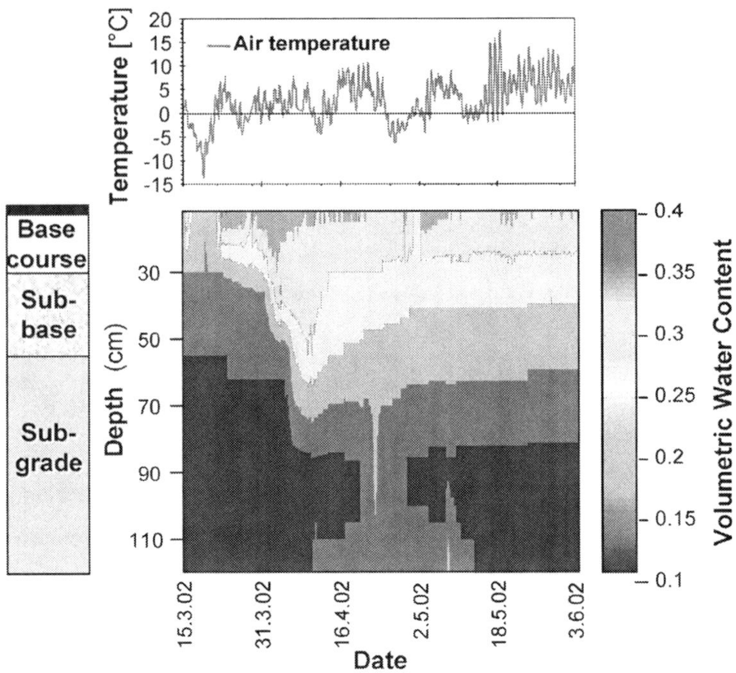

Fig. 8.9 Volumetric water content (*right scale*) during a spring thaw period for a thin pavement structure with an granular base course. The air temperature is also shown, top, as well as a cross section through the low volume pavement structure

8.3.4 Thawing and Bearing Capacity Change, Finnish Example

Figure 8.10 and Table 8.1 show the weakening of a road structure after spring thaw. The stiffness modulus E_2 of the whole pavement structure measured by the FWD is, on average, 13% lower in the spring time than it is before the next freezing period. According to the FWD indices, the reason is the weakening of the upper structure. This is shown since the BCI-indices (which are a representation of deflection in the subsoil) remain about the same, but the upper structure is weakened: the SCI (which is a representation of near-surface deflections) rises by 22%. The reason for this is that there must be more moisture in the structure after the thawing than in late autumn. The temperature of the pavement has been taken into account in the calculation on the indices. The data is from a 5.5 km long old highway section of highway #6 in Finland with an AADT of 6500, before rehabilitation measures.

Figure 8.11 shows an example of a poor quality aggregate base layer in the middle of the above section. In the central length of the road the SCI is very high, indicating that there are a lot of fines in the base layer.

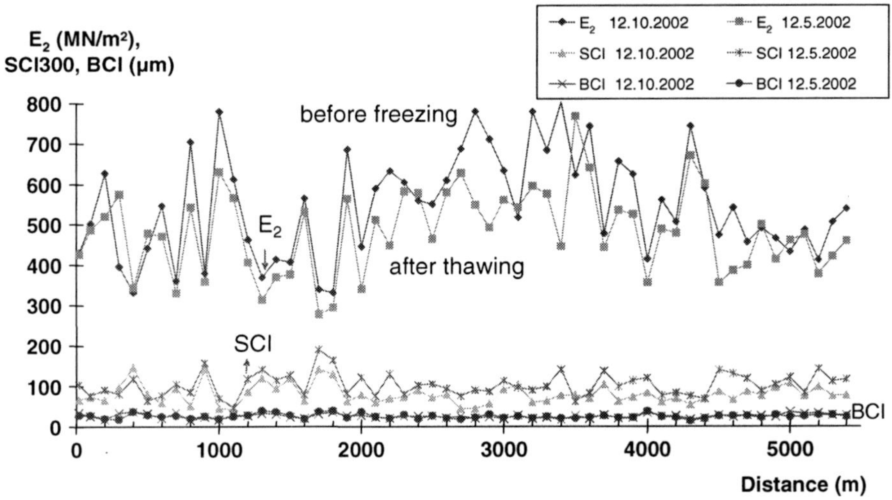

Fig. 8.10 FWD Measurements after thawing and before freezing for a Finnish pavement. SCI = Surface curvature index, BCI = Base Curvature Index, E_2 = stiffness modulus (see Section 10.3)

Table 8.1 Average FWD indices, 0–5500 m

FWD average indices	E_2	SCI	BCI
	(MN/m^2)	(μm)	(μm)
12.05.2002	481	103	25
12.10.2002	545	80	26
Difference	13 %	−22 %	1 %

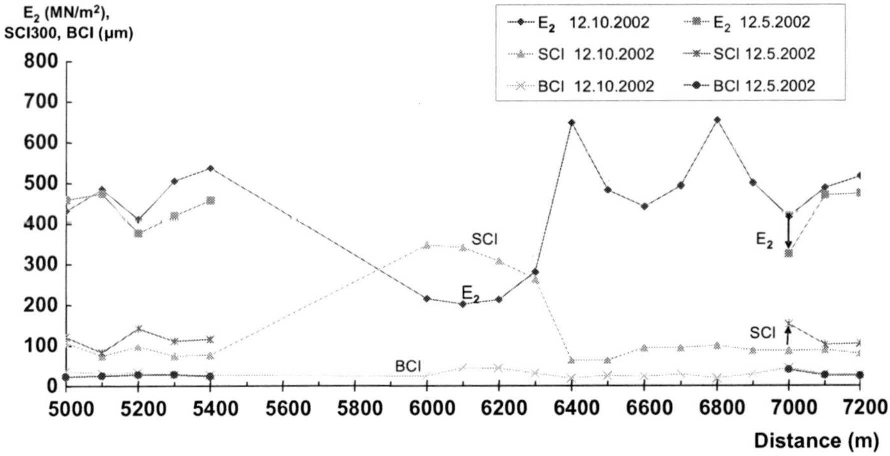

Fig. 8.11 FWD measurement after thawing and before freezing for a poorly performing pavement

8.3.5 Monitoring Frost Depth and Thawing, Finland and Sweden

8.3.5.1 Introduction

Spring-thaw load restrictions are often imposed to avoid severe pavement deterioration during periods of reduced bearing capacity. Equipment that enables monitoring of the pavement strength situation is very important for a road's traffic carrying capacity, as restrictions could be lifted as soon as the pavement regains its capacity. There are projects with this aim being run in both Sweden and Finland. In Finland the Percostation is used to monitor dielectric value, electrical conductivity and temperature with depth. The Swedish approach is to monitor temperature profile only and to distribute this via the Internet enabling direct access from trucks to frost depth readings that are updated twice an hour. Both approaches are now described.

8.3.5.2 The Finnish Percostation

The Percostation monitors dielectric value, electrical conductivity and temperature at different depths through a maximum of eight channels. Both dielectric value and electrical conductivity are sensitive to the amount of unfrozen water, that is, it is clearly indicated when water freezes and when ice melts. Temperature is, of course, also related to freezing and thawing. Figure 8.12 shows dielectric values at depths 0.15 and 0.30 m together with air temperature during the thawing period of the year 2000 at the Koskenkylä Percostation. From the figure it is clear that the dielectric value approximately equals 5 at both depths when the soil is frozen and that it increases at thawing.

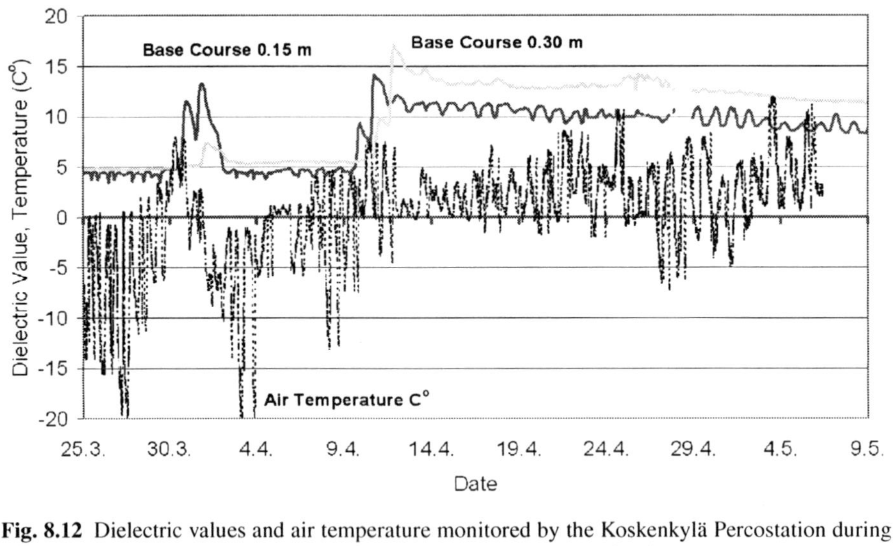

Fig. 8.12 Dielectric values and air temperature monitored by the Koskenkylä Percostation during spring 2000 (Saarenketo et al., 2002). Reproduced by permission of T. Saarenketo

The increase starts at depth 0.15 m and is somewhat delayed at the lower depth as thawing takes place from the surface and downwards. In late spring the dielectric value is highest at the greater depth indicating higher frost susceptibility and higher amount of ice lenses melting. Around April 10–12 the dielectric values show a peak followed by a continuously ongoing decrease when surplus water drains and bearing capacity recovers. The data monitored by the Percostation is of great value when imposing and removing spring-thaw load restrictions.

8.3.5.3 The Swedish Tjäl2004

Tjäl2004, developed by VTI, monitors temperature at every 50 mm down to a depth of 2 m. Temperatures are collected twice an hour and distributed via the Internet. The temperature sensors are calibrated to give highest possible accuracy close to 0°C where freezing starts. Trucks, having mobile Internet, pick up the current freezing situation from the installed Tjäl2004 along the intended roads to travel. Road owners give truckers allowance to use roads as long as the upper part of the pavement is frozen down to a certain depth. This means that in spring, load restrictions are imposed and removed automatically and very frequently. During periods in spring with clear weather the situation might change daily. In the daytime the solar radiation thaws the upper layers, which is followed by re-freezing during the clear and cold night. This means that the heavy loading of trucks is allowed in the early morning but prohibited in the afternoon. Figure 8.13 shows typical repeated freezing and thawing during the period March 4 – April 22 of the year 2008 monitored by one of the Tjäl2004 installed in Sweden. The total frost depth is close to 1.5 m.

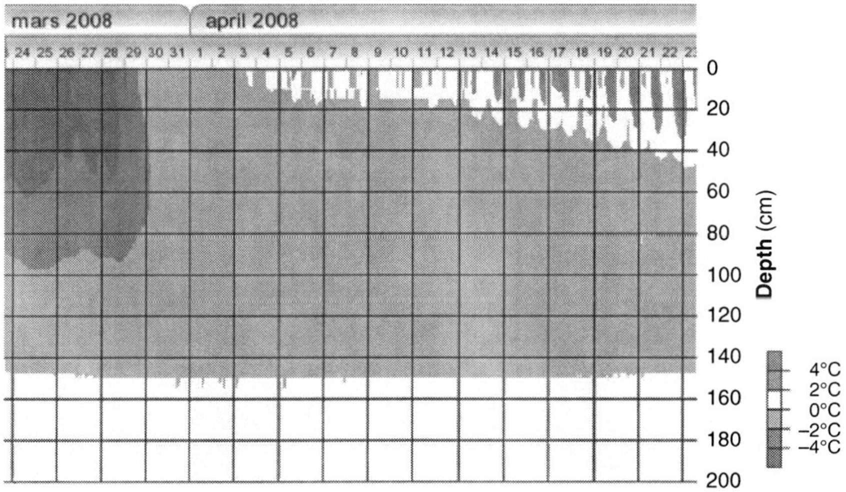

Fig. 8.13 Temperature profile monitored by the Swedish Tjäl2004, distributed via the Internet, see http://www3.vv.se/tjaldjup/

8.4 Conclusions, Implications, Recommendations

Field observations indicate clear and significant variations of moisture in subgrades. This is true both for moderate climates as well as for cold regions where it is related to temperature. In particular, thawing may induce strong increases in moisture levels. The mechanical behaviour as observed in-situ is strongly affected by moisture variations: the wetter the state the lower the stiffness (up to a factor 2 or more), the lower the stiffness then the higher the deflection.

Therefore one may conclude that an efficient drainage system is crucial in order to reduce the road structure's ageing. An analytical assessment of this relationship between moisture and mechanical performance will be undertaken in the following two chapters.

References

Gerke, R.J., 1979, "*Subsurface Drainage: Progress Report*", September, Australian Road Research Board. Internal Report, AIR 317-1.

Hall, K.T. & Crovetti, J.A., 2007, "Effects of Subsurface Drainage on Pavement Performance, Analysis of the SPS-1 and SPS-2 Field Sections", *NCHRP Report* 583, Transportation Research Board, Washington, DC, 92pp. & Annexes.

Lay, M.G., 1986, "*Handbook of Road Technology*, Volume 1: *Planning and Pavements*", Gordon & Breach Science Publishers.

Moris, P.O. & Gray, W.J., 1976, "*Moisture conditions under roads in the Australian environment*". Australian Road Research Board, Research Report, ARR No. 69.

Roy, M. & Johnson, F.D.J., 1979, "M5 Motorway – Blackbrook to Chelston. Report of an Investigation of the Southbound and Northbound Carriageways", Somerset County Council, Taunton, UK, Report 71/79A, (unpublished).

Saarenketo T., Kolisoja P., Luiro K., Maijala P. & Vuorimies N., 2002, "*Percostation for Real Time Monitoring Moisture Variations, Frost Depth and Spring Thaw Weakening*", Paper presented to Transport Research Board Annual Meeting, Washington DC, available on CD Rom.

Savard, Y., de Blois, K., Bontonnet, M., Hornych, P. & Manduit, C., 2005, "Analysis of seasonal bearing capacity correlated to pavement deformation in cold regions", *7th Int. Conf. Bearing Capacity Roads, Railways & Airfields*, Trondheim, Norway, June.

Wiman, L., 2001, "*Accelerated Load Testing of pavements, HVS-Nordic Tests in Sweden 1999*", VTI report 477A-2001; accessible at http://www.vti.se/templates/Page____3250.aspx (*there are 4 underline characters between 'Page' and '3250'*).

Chapter 9
Water Influence on Mechanical Behaviour of Pavements: Constitutive Modelling

Lyesse Laloui*, Robert Charlier, Cyrille Chazallon, Sigurður Erlingsson, Pierre Hornych, Primož Pavšič and Mate Sršen

Abstract This chapter deals with the effects of water on the mechanical behaviour of pavements. The analysis is based on constitutive considerations. Constitutive models devoted to both routine and advanced pavement analysis and design are introduced and both the resilient behaviour as well as the long term elasto-plastic approaches are presented. As soon as the approach considers the material as a two phase (solid matrix and a fluid), the introduction of the effective stress concept is required. In the last section an analysis is made on the extension of the constitutive models to the characterisation of partially saturated materials.

Keywords Constitutive models · resilient behaviour · elasto-plastic models · effective stress · suction effects

9.1 Introduction

Road structures and the underlying soil are subjected to traffic loading. Their mechanical behaviour depends on their initial state, the hydraulic conditions and the temperature. The numerical prediction of the behaviour of such material under such conditions is not a simple matter and the user needs, therefore, to make use of comprehensive constitutive models that include a coupling of the mechanical, hydraulic and thermal aspects. The aim of the present chapter is to present an overview for such constitutive modelling, in particular covering the modelling of effects of water on the mechanical behaviour of pavements.

After first summarising the underlying reasons for the type of mechanical behaviour observed, this chapter presents a consideration of the constitutive relationships of the materials that comprise the pavement and embankment layers. It provides an introduction to the following constitutive models that may be employed in routine and advanced pavement analysis and design:

---

* *Co-ordinating Author:*
✉ L. Laloui
Ecole Polytechnique Fédéral de Lausanne, Swiss Federal Institute of Technology, Switzerland
email: lyesse.laloui@epfl.ch

A.R. Dawson (ed.), *Water in Road Structures,* DOI 10.1007/978-1-4020-8562-8_9,
© Springer Science+Business Media B.V. 2009

- Resilient models: the k-θ model and the Boyce model and their derivatives; and
- Long term elasto-plastic models. These models are split in four categories:
  - Analytical models;
  - Plasticity theory based models;
  - Visco-plastic equivalent models; and
  - Shakedown models.

Routine pavement design is mostly based on an elastic calculation, using a resilient modulus and Poisson's ratio for each layer. The design criterion is usually limited to the maximum vertical strain for a given loading condition. More elaborate models take into account the irreversible behaviour, e.g. the Chazallon-Hornych model, the Suiker model and the Mayoraz elasto-visco-plastic model. References to each of these models is given at the appropriate place in the text.

The final part of the chapter discusses how these models can be adapted to take into account and replicate the effects of variations in suction that occur in partially saturated soils and aggregates. Some suggested research topics are presented towards the end of this chapter.

9.2 Origin of Mechanical Properties in Pavement Materials

The materials that comprise the lower parts of the road and which form the subgrade are all geomaterials – particulate solids with pore spaces occupied by a combination of water and air in varying proportions. The solid particles are, for the most part, crystalline. They are derived, ultimately, from geological sources. Individually the grains have considerable strength which means that the mechanical response (strength, stiffness, resistance to development of rutting) of an assembly of particles is a primarily a consequence of the way the individual grains interact with one another and not of their own properties.

The primary contribution to mechanical property derives from the ease or difficulty with which one particle can be moved adjacent to another particle. This ease or difficulty is controlled by many factors which can, broadly, be grouped into three: physical characteristics of the grains, arrangement of the grains and the fluid conditions in the pores. The list of factors under each heading would be very long, but the following aims to highlight some of the more important:

- Physical characteristics of the grains:
  - Particle shape;
  - Particle mineralogy; and
  - Particle roughness.
- Arrangement of the grains:
  - Size and size distribution of the grains; and
  - Packing of the grains.

- Fluid conditions in the pores:
  - Fluid pressure in the pores;
  - Surface tension effects in the pores between fluids; and
  - Water adsorption to mineral surfaces.

When a stress is applied to a granular or soil material, the stress has to be carried across the assemblage of grains via the inter-particle contact points. These contacts will be subjected to both normal and shear stresses. Both can cause compression that is recoverable and slippage between the particles at the contact or damage and wear to the contact. Recoverable compression of the contacts will contribute to the stiffness behaviour of the whole material while slip and damage will contribute to plastic deformation. In addition the assembly of particles will re-arrange itself by sliding and rolling of particles – also contributing to the stiffness and plastic deformation behaviour of the whole. Changing the shape and nature of the contacts and changing the packing of particles will all, therefore, have an impact on strength, stiffness and resistance to plastic deformation.

As the force carried at an inter-particle contact point increases, the laws of friction dictate that (unless the contact point fails in some way) there will be greater resistance to shear loading. Thus a greater compressive stress applied to an assembly of grains allows the whole material to gain shear strength and resistance to shear deformation which is characterised by the apparent angle of frictional resistance, φ'. The greater compression of the particle contacts also makes further compression more difficult leading to the phenomenon of a non-linear stress-dependent modulus, so often observed in granular materials. Section 9.4 introduces some of the models of mechanical behaviour that are used to replicate these behaviours.

Adding water under pressure to a pore will cause all the particles around the pore to become loaded so that some of the force that previously was carried across the adjacent inter-particle contact points will now be carried by the pore fluid (Fig. 9.1).

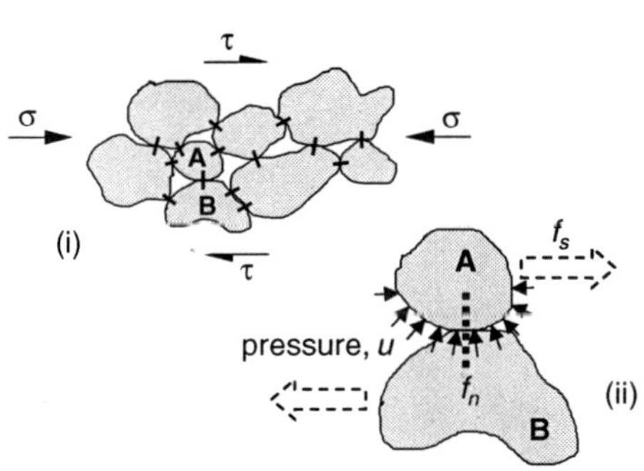

Fig. 9.1 Schematic of inter-particle forces. (i) an assembly of particles is subjected to some external normal, σ, and shear, τ, stresses, which are carried through the assembly at the contact points as shown by the black bars; (ii) when the pore space between particles A and B is pressurised by a fluid at pressure, u, particles A and B experience a pressure on them (illustrated only for A, not illustrated for B) which reduces the inter-particle force, f_n, and makes shear, f_s, more easy to take place because of reduced friction at the contact between the particles

This is the reason behind the effective stress equation, Eq. 1.1, which is further described in this chapter at Eq. 9.19 and following. Because some force is now carried through the pores, the inter-particle forces acting at the contacts are reduced and, therefore, due to the frictional effects, the shear strength, stiffness and resistance to permanent deformation are also reduced.

If water is retained in the pores due to surface tension effects, then the opposite will occur with a suction being applied to the adjacent particles. This causes the inter-particle contact forces to increase and the shear strength, stiffness and resistance to permanent deformation will all rise. These influences of water on mechanical performance are the subject of Section 9.5 and the required models are given in Section 9.6.

9.3 General Objectives, Strength and Deformation

With the use of numerical modelling, engineers aim to obtain the displacement fields as well as the stress values (effective stress, pore pressure and suction) in the road and earthworks sub-structure. The numerical modelling allows the understanding of the behaviour of the geostructure and the analysis of an optimal design. In order to be successful, the computational tool should, then, include the main physical processes of the rheology of the materials that make up the structure.

The behaviour of granular media is mainly dependant on an inter-granular friction as well as on the applied stress which modifies the rigidity and the strength of the material. It is highly non-linear and irreversible. Figure 9.2 summarizes the main stress-strain aspects.

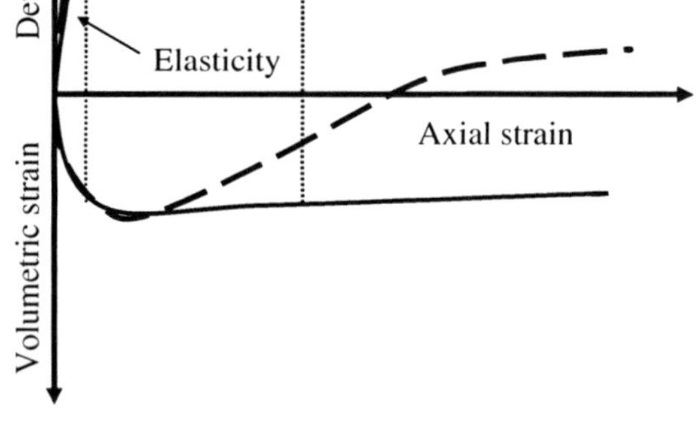

Fig. 9.2 The main stress-strain aspects of a granular medium

9.4 Models for Subgrade Soils and Unbound Granular Materials

The purpose of this section is to introduce some of the constitutive models devoted to routine, as well as advanced, pavement analysis and design. For all these models, moisture or water pressure are not taken into account. They are written, here, in terms of total stresses – i.e. the effects of pore water pressure and/or pore suction are subsumed into the mechanical response of the materials and are not explicitly described.

Nowadays, models for subgrade in pavement engineering have been split in two categories dealing with the main mechanical behaviour which needs to be taken into account:

- resilient behaviour (see Fig. 9.3); and
- long term elasto-plastic behaviour.

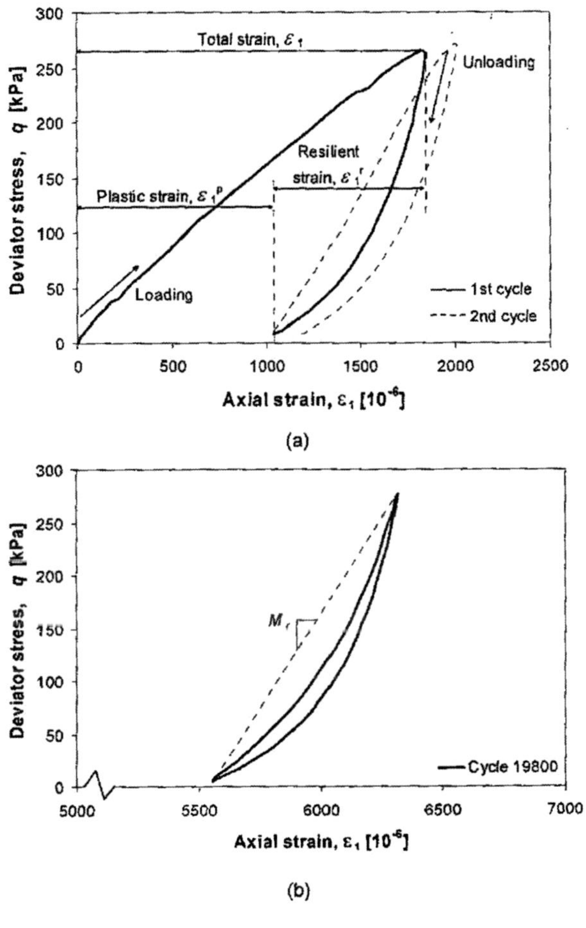

Fig. 9.3 Resilient behaviour shown as deviatoric stress versus axial strain during cyclic testing, (**a**) on initial loading (**b**) after a large number of cycles, after Ekblad (2004). Reproduced by permission of Jonas Ekblad

9.4.1 Resilient Behaviour

9.4.1.1 Routine Pavement Analysis

In practice much routine pavement design is carried out as catalogue-based design. Nevertheless, routine structural analysis and design methods are used as supplemental design methods where the pavement is considered as a multi-layered elastic system (Amadeus Project, 2000). The layers are characterised by Young's modulus, Poisson ratio and thickness. The simplest model for the stress-strain behaviour of isotropic materials is based on linear elasticity, which is described by Hooke's law. In two or three dimensions, the model is written:

$$\sigma_{ij} = E^{e}_{ijkl} \cdot \varepsilon_{kl} \tag{9.1}$$

where σ_{ij}, ε_{kl}, E^{e}_{ijkl} are respectively members of the stress, strain and stiffness tensors $\boldsymbol{\sigma}$, $\boldsymbol{\varepsilon}$ and $\mathbf{E}^{e}$.

A symmetric stiffness matrix is used to describe the constitutive equations in those cases. As a road pavement is a layered structure, the material behaviour might be non-isotropic, with different stiffnesses in horizontal and vertical directions. Thus, the constitutive matrix is described with more independent parameters, which are also difficult to determine in a laboratory test. Hence, materials are usually considered isotropic. If the layers are not too thin, this might be a reasonable simplification.

9.4.1.2 Advanced Pavement Analysis

The behaviour of unbound granular materials in a pavement structure is stress-dependent. For that reason the linear elastic model is not very suitable. A non-linear elastic model, with an elastic modulus varying with the stress and strain level is, therefore, needed.

For isotropic materials, moduli depend only on two stress invariants[1]: the mean stress level, p, and the deviatoric stress, q, which are given in the general, as well as the axi-symmetric case (cylindrical state of stress with $\sigma_1 = \sigma_{axial}$ and $\sigma_2 = \sigma_3 = \sigma_{radial}$), as:

General	Axi-symmetric	
$p = \dfrac{\sigma_{ii}}{3}$	$p = \dfrac{1}{3}(\sigma_1 + 2\sigma_3)$	(9.2)
$q = \sqrt{\dfrac{1}{2}\hat{\sigma}_{ij}\hat{\sigma}_{ij}}$ with $\hat{\sigma}_{ij} = \sigma_{ij} - p\delta_{ij}$	$q = \sigma_1 - \sigma_3$	

In a similar way strain invariants can be introduced. The volumetric strain ε_v and the deviatoric strain ε_q, are defined as:

[1] An invariant has the same value regardless of the orientation at which it is measured.

$$\begin{array}{ll} \text{General} & \text{Axi-symmetric} \\ \varepsilon_\nu = \varepsilon_{ij} & \varepsilon_\nu = \varepsilon_1 + 2\varepsilon_3 \\ \varepsilon_q = \sqrt{\frac{1}{2}\hat{\varepsilon}_{ij}\hat{\varepsilon}_{ij}} \text{ with } \hat{\varepsilon}_{ij} = \varepsilon_{ij} - \frac{\varepsilon_\nu}{3}\delta_{ij} & \varepsilon_q = \frac{2}{3}(\varepsilon_1 - \varepsilon_3) \end{array} \tag{9.3}$$

The stresses and strains are interconnected through the material properties as stated in Eq. 9.1 and the elastic (resilient) response of the material can be expressed according to Hooke's law as a diagonal matrix:

$$\begin{bmatrix} \varepsilon_\nu \\ \varepsilon_q \end{bmatrix} = \frac{1}{E}\begin{bmatrix} 3\,(1-2\nu) & 0 \\ 0 & \frac{2}{3}\,(1+\nu) \end{bmatrix}\begin{bmatrix} p \\ q \end{bmatrix} \tag{9.4}$$

where E and ν are the material stiffness modulus (or the resilient modulus, usually denoted M_r) and Poisson's ratio respectively, defined as:

$$M_r \text{ or } E = \frac{\Delta q}{\Delta\varepsilon_1} \quad \text{and} \quad \nu = \frac{\Delta\varepsilon_3}{\Delta\varepsilon_1} \tag{9.5}$$

where Δ stands for the incremental change during the loading. An alternative formulation is:

$$\Delta\varepsilon_q = \frac{\Delta q}{3G_r} \quad \text{and} \quad \Delta\varepsilon_\nu = \frac{\Delta p}{K_r} \tag{9.6}$$

where K_r and G_r are the bulk and shear moduli of the material. The bulk and the shear moduli are connected to the resilient modulus and the Poisson's ratio through:

$$K_r = \frac{M_r}{3(1-2\nu)} \quad \text{and} \quad G_r = \frac{M_r}{2(1+\nu)} \tag{9.7}$$

for isotropic materials.

The resilient modulus for most unbound pavement materials and soils is stress-dependent but the Poisson's ratio is not, or at least to a much smaller extent. Biarez (1961) described the stress-dependent stress-strain behaviour of granular materials subjected to repeated loading. Independently, similar work was performed in the United States (Hicks and Monismith, 1971). Both results presented the $k-\theta$ model, which is written with dimensionless coefficients like:

$$M_r = k_1 p_a \left(\frac{3p}{p_a}\right)^{k_2} \quad \text{and } \nu = \text{constant} \tag{9.8}$$

where M_r is the resilient modulus, p is the mean stress, p_a is the reference pressure ($p_a = 100\,\text{kPa}$) and k_1, k_2 are coefficients from a regression analyses usually based on repeated load triaxial test results. This model has been very popular for describing non-linear resilient response of unbound granular materials. It assumes a constant Poisson's ratio and that the resilient modulus is independent of the deviatoric stress. To address this latter limitation the Uzan-Witczak model – often called the "Universal" model – has become widely promoted, especially, in recent years, by authors in North America, e.g. Pan et al. (2006). It takes the form:

$$M_r = k_1 p_a \left(\frac{3p}{p_a}\right)^{k_2} \left(1 + \frac{q}{p_a}\right)^{k_3} \quad \text{and } \nu = \text{constant} \tag{9.9}$$

Subgrade soils are also stress-dependent and can also be modelled by one of the k–θ approaches. The principle difference between granular materials and many soils is that the former exhibit a strain-hardening stiffness whereas the latter, typically, exhibit strain-softening behaviour under transient stress loadings. In practice, the incorporation of non-linearity into the stiffness computations for subgrade soils is often less important than for granular materials as the stress pulses due to traffic loading will be a far smaller part of the full stress experienced by the subgrade than is the case for the unbound granular layer. Thus the error introduced by ignoring subgrade non-linearity will be correspondingly smaller.

In 1980, Boyce presented some basis for subsequent work on the stress-dependent modelling of the resilient response of cyclically loaded unbound granular material. The Boyce model takes into account both the mean stress and the deviatoric stress, with the bulk and shear moduli, K and G, of the material calculated as:

$$K_r = \frac{\left(\frac{p}{p_a}\right)^{1-n}}{\frac{1}{K_a} - \frac{\beta}{K_a}\left(\frac{q}{p}\right)^2} \quad \text{and } G_r = \frac{\left(\frac{p}{p_a}\right)^{1-n}}{\left(\frac{1}{G_a}\right)} \quad \text{where } \beta = (1-n)\frac{K_a}{6\,G_a} \tag{9.10}$$

where K_a, G_a, and n are material parameters determined from curve fitting of repeated load triaxial tests results.

Anisotropy of pavement materials is increasingly being recognised as a property that must be modelled if the pavement is to be adequately described (e.g. Seyhan et al., 2005). The Boyce model was modified to include anisotropy in the early 1990's (Elhannani, 1991; Hornych et al., 1998). Hornych and co-workers ntroduced anisotropy by multiplying the principal vertical stress, σ_1 in the expression of the elastic potential by a coefficient of anisotropy γ so that p and q are redefined as follows (c.f. axi-symmetric part of Eq. 9.2):

$$p^* = \frac{\gamma\sigma_1 + 2\sigma_3}{3} \quad \text{and} \quad q^* = \gamma\sigma_1 - \sigma_3 \tag{9.11}$$

and the stress-strain relationships are defined as:

$$\Delta\varepsilon_q{}^* = \frac{2}{3}\left(\Delta\varepsilon_1{}^* - \Delta\varepsilon_3{}^*\right) = \frac{\Delta q^*}{3G_r} \quad \text{and}$$
$$\Delta\varepsilon_v{}^* = \Delta\varepsilon_1{}^* + 2\Delta\varepsilon_3{}^* = \frac{\Delta p^*}{K_r}. \tag{9.12}$$

This yields K_r and G_r, the bulk and shear moduli, respectively as:

$$K_r = \frac{\left(\frac{p^*}{p_a}\right)^{1-n}}{\frac{1}{K_a} - \frac{\beta}{K_a}\left(\frac{q^*}{p^*}\right)^2} \text{ and } G_r = \frac{\left(\frac{p^*}{p_a}\right)^{1-n}}{\left(\frac{1}{G_a}\right)} \text{ with } \beta = (1-n)\frac{K_a}{6G_a} \tag{9.13}$$

The k–θ model, Universal model, Boyce model, and the modified Boyce model must be considered in pavement modelling to ensure a valid stress, strain, and deflection evaluation in pavements. When subjected to repeated loading, two types of deformations are exhibited, linear or non-linear elastic (or resilient) and plastic deformations. Models based on non-linear elasticity deal with resilient deformations only. Their biggest disadvantage is that permanent deformations cannot be modelled.

9.4.1.3 MEPDG

The above methods define stiffness as a function of stress alone. Full incorporation of the effects of moisture (as a pressure or suction) should necessitate use of an effective stress framework (see Section 9.5). However a more simple approach, at least in principle, is to adjust the stiffness value calculated by one of the above relationships using a factor that is dependent on the moisture (and, perhaps, other) condition. The AASHTO 'Mechanistic-Empirical Pavement Design Guide' (MEPDG) takes this approach, though it's attention to many details makes the implementation rather complex (ARA, 2004). In this approach, the reference stiffness value, M_{ropt} (the value of M_r at optimum conditions), is adjusted by a factor, F_{env}, to allow for different environmental effects, with the value of each factor being computed for each of a range of depths, lateral positions and time increments. For moisture the adjustment factor is based on the equation

$$log\left(\frac{M_r}{M_{r\,opt}}\right) = a + \frac{\left(\frac{M_r}{M_{r\,opt}}\right)_{max} - \left(\frac{M_r}{M_{r\,opt}}\right)_{min}}{1 + \exp\left(ln\left(\frac{-\left(\frac{M_r}{M_{r\,opt}}\right)_{max}}{\left(\frac{M_r}{M_{r\,opt}}\right)_{min}}\right) + k_m \cdot \left(S_r - S_{r\,opt}\right)\right)} \tag{9.14}$$

Where k_m is a material parameter and S_r and S_{ropt} = the actual saturation and the saturation ratio at optimum conditions, respectively. The actual saturation value is obtained from the use of the Soil Water Characteristic Curve (SWCC) see Chapter 2, Section 2.7.1. Other adjustments are included in the F_{env} factor to allow for freezing, thawing and temperature. The full approach is too detailed to include here. Interested readers are directed to the relevant report (ARA, 2004).

Long et al. (2006) take another approach, relating modulus to suction and water content rather than to saturation ratio, but still including some stress influence:

$$M_r = \frac{(p - \theta \cdot s)\left(1 + \frac{0.4343}{S_\theta \cdot w}\right)}{\gamma_h(0.435)} \frac{(1+\nu)(1-2\nu)}{(1-\nu)} \tag{9.15}$$

where p is the mean normal stress on the element of soil, θ and w are the volumetric and gravimetric water contents, respectively, s is the matric suction pressure, S_θ is the slope of the soil desorptive curve (the rate of change of the logarithm of s with the logarithm of θ), γ_h is the suction volumetric change index (an indicator of the sensitivity of volume change to change in matric suction) and ν is Poisson's ratio.

9.4.2 Long Term Elasto-Plastic Behaviour

Routine pavement methods are mechanistic-empirical design methods, based on linear elastic calculations. Usually, the only rutting criterion to be used concerns the subgrade soil, and consists in limiting the vertical elastic strains at the top of the subgrade. Rarely is a criterion applied for the unbound granular layers although Dawson and Kolisoja (2004) have shown that in roads with thin bound layers, the road's rutting will largely be the consequence of plastic deformation in the granular layers. Similarly, it is rare for a plastic criterion to be used in design probably because of the much greater difficulty in computing plastic strain fields.

Advanced pavement models are based on the main tests for unbound granular materials: the monotonic triaxial test and the repeated load triaxial test (this test and the models' calibrations are presented in Chapter 10).

These models are split in four categories:

- Analytical models;
- Plasticity theory based models;
- Visco-plastic equivalent models; and
- Shakedown models.

9.4.2.1 Analytical Models

A few material models have been proposed for the development of plastic strains in unbound granular materials in a pavement structure. Lashine et al. (1971), and Barksdale (1972) tested unbound granular material in a repeated load triaxial test for 100 000 cycles. They found that the permanent axial deformation, ε_{p1}, at different

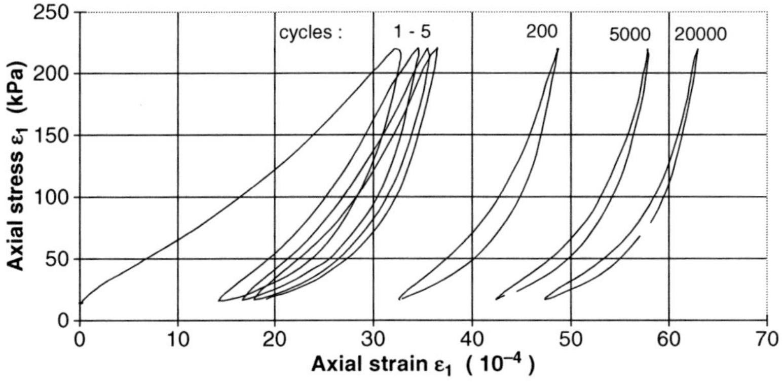

Fig. 9.4 Example of stress-strain cycles obtained in a repeated load triaxial test on a granular material (Hornych et al., 1998)

stress states is proportional to number of load cycles, N (Fig. 9.4). Since 1971, many analytical models have been developed and most of them are listed by Lekarp and Dawson (1998).

However, these models have never been used with finite element (FE) calculations except the Hornych model (Hornych et al., 1993) which has been used with a simplified finite element calculation by de Buhan (Abdelkrim et al., 2003) for a railway track construction and by Hornych et al. (2007) for a full scale flexible pavement.

The mechanical processes that form the basis of a flexible pavement's performance and of a flexible pavement's deterioration can be separated into two categories, namely:

(i) short-term mechanical processes; and
(ii) long-term mechanical processes.

The first category concerns the instantaneous behaviour of a flexible pavement, as activated during the passage of a vehicle, thus the flexible pavement behaviour can be studied by means of (visco)elastic models. de Buhan (Abdelkrim et al., 2003) and Hornych et al. (2007) use, respectively, linear moduli with a Boussinesq stress analysis and the modified Boyce models with a simple FE analysis for the short term behaviour. The second category concerns the mechanical processes, typically characterized by a quasi-static time-dependency, such as long-term settlements under a large number of vehicle axle passages. Together, the approach requires the following parameters:

- Elastic behaviour: E, ν or K_a, G_a, γ and n.
- Plastic behaviour:
  - Rupture (i.e. shear failure) parameters: m, s.
  - Plasticity parameters: $\varepsilon_1^{p^0}$ B and n.

The Hornych model is another formulation of the Paute model (Paute et al., 1988) which was adopted also in an European norm – EN 13268-7 (CEN, 2000). Some other alternatives were suggested in Lekarp and Dawson (1998), but no overall framework has been established yet to explain completely the behaviour of unbound granular materials under complex repeated loading.

9.4.2.2 Plasticity Theory Based Models

The plasticity theory based models require the definition of yield surface, plastic potential, isotropic hardening laws, and simplified accumulation rules (Bonaquist & Witczak, 1997 and Desai, 2002), or kinematic hardening laws (Chazallon et al., 2006). Some of these models have been used for finite element modelling of pavement. Now, the main concepts of these models are presented.

The model developed by Bonaquist is based on the plasticity model developed by Desai et al. (1986). These two models differ from each other by the simplified calculation method for large number of cycles. Consequently, the basis of the Desai model (Desai, 2002) is first presented and the different accelerated analysis procedures of each model are introduced.

The Desai formulation is based on the disturbed state concept which provides a unified model that includes various responses such as elastic, plastic, creep and micro-cracking. The idea is that the behaviour of a deforming material can be expressed in terms of the behaviour of the relatively intact or continuum part and of the micro-cracked part. During the deformation, the initial material transforms continuously into the micro-cracked state and, at the limiting load, the entire material element approaches the fully micro-cracked state. The transformation of the material from one state to another occurs due to the micro-structural changes caused by relative motion such as translation, rotation and interpenetration of the particles and softening or healing at the microscopic scale. The disturbance expresses such micro-structural motions. Under repetitive loading, an accelerated procedure exists. From experimental cyclic tests, the relation between the deviatoric plastic strain trajectory and the number of loading cycles can be expressed as a power function of the number of cycle. Pavement finite element modelling feasibility has been carried out with this model (Desai, 2002). The $k-\theta$ model was used for the elastic part and the model requires the following parameters:

- Elastic behaviour: $k_1,\ k_2,\ \nu$.
- Plastic behaviour:
  - Rupture and characteristic parameters: $3R,\ n,\ \gamma$.
  - Plasticity parameters: $\beta,\ a_1,\ n_1,\ N_r,\ b$.

Bonaquist's approximate accelerated analysis is based on the total plastic strain at the end of each cycle and defined by a power function of the cycle number which depends on the ratio: maximum deviatoric stress and the corresponding deviatoric stress at rupture (for the same q/p ratio). Instead of the N_r and b parameters a parameter ξ_b is required.

The model developed by Chazallon and Hornych (Chazallon et al., 2006) is based on the model of Hujeux (1985) in its simplest formulation. This formulation is a non-associated elasto-plastic model and reproduces the saturated monotonic behaviour of sand and clay. A kinematic hardening has been added to reproduce the accumulation of plastic strains under repeated loadings. Each cycle is calculated, nevertheless, by a simplified approach based on the decoupling of the calculation of the elastic strains and the plastic strains. A pavement finite element modelling feasibility has been carried out with this model, see Hornych et al. (2007).

The elastic part is computed with the anisotropic Boyce model and the model requires the following parameters:

- Elastic behaviour: K_a, G_a, γ and n.
- Plastic behaviour:
  - Rupture and dilatancy parameters: C_0, M, and M_c.
  - Monotonic plasticity parameters: β, P_{C_0}, a, b.
  - Cyclic plasticity parameters: r_c^{el}, P_{uc}, P_{lc}.(where the subscript l and u indicate loading and unloading, respectively – see Figs. 9.5 and 9.6).

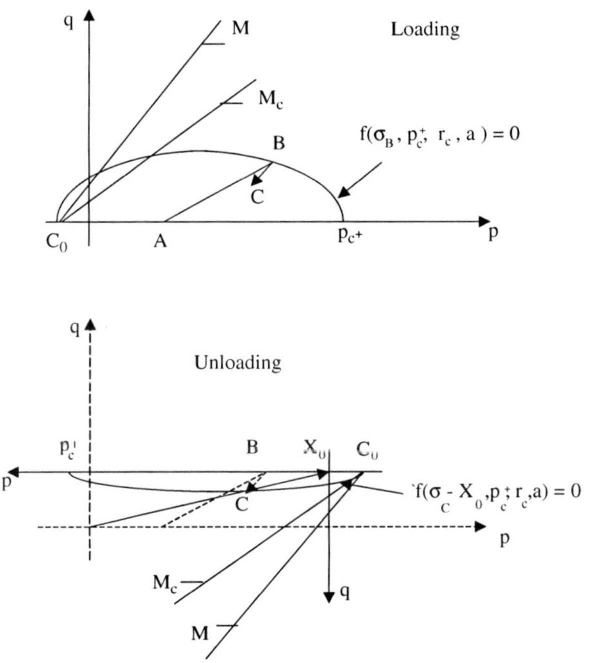

Fig. 9.5 Yield surfaces during loading and unloading in the $p - q$ space (Chazallon et al., 2006). With permission from ASCE

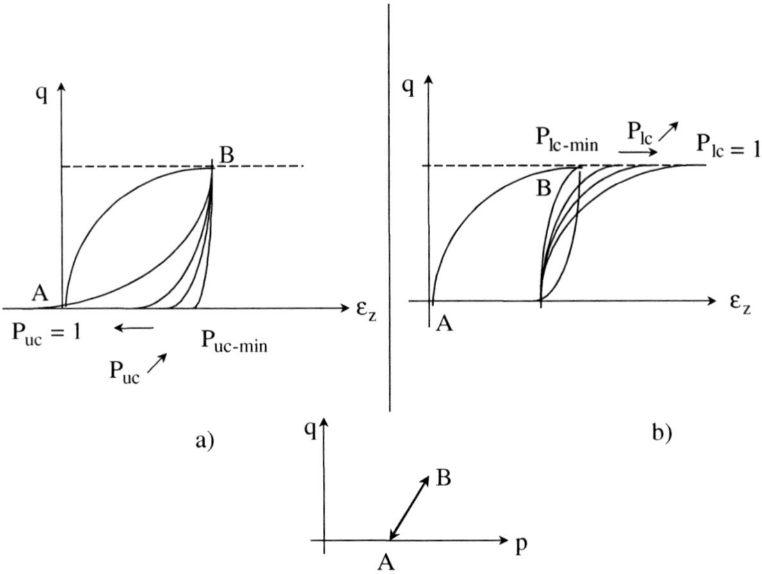

Fig. 9.6 Representation of the influence of the $P_{uc/lc}$ parameters on plastic strains when an unloading and a reloading occur (Chazallon et al., 2006) With permission from ASCE

9.4.2.3 Visco-Plastic Equivalent Models

Visco-plastic equivalent models based on an equivalent time: number-of-cycles relationship, have been developed by Suiker and de Borst (2003) for the finite element modelling of a railway track structure and by Mayoraz (2002) for a laboratory study of sand.

Suiker has developed a cyclic densification model. It is based on repeated load triaxial tests carried out on two ballast materials. The idea was to develop a model that captures only the envelope of the maximum plastic strain generated during the cyclic loading process. The unloading is considered as elastic. The plastic deformation behaviour is composed of two different mechanisms namely: frictional sliding and volumetric compaction. In general, both mechanisms densify the material. This model is based on the Drucker-Prager yield surface and Cap surface. The stress space is divided into four regimes (see Fig. 9.7):

(i) the shakedown regime, in which the cyclic response of the granular media is fully elastic.
(ii) the cyclic densification regime, in which the cyclic loading submits the granular material to progressive plastic deformations.
(iii) the frictional regime, in which frictional collapse occurs, since the cyclic load level exceeds the static peak strength of the granular material.
(iv) the tensile failure regime, in which the non cohesive granular material instantaneously disintegrates, as it can not sustain tensile stresses.

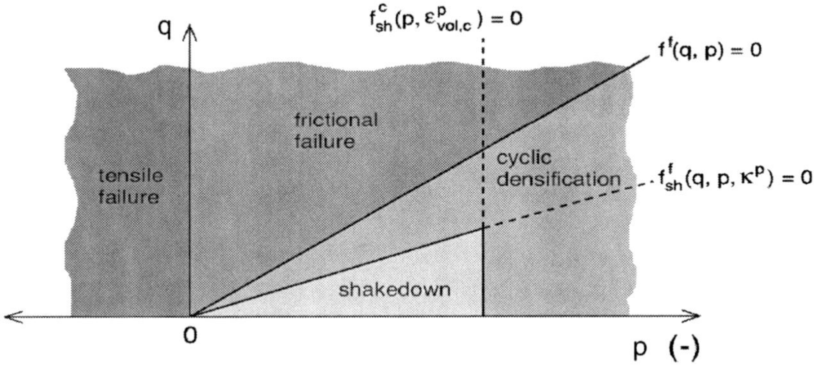

Fig. 9.7 Map of various response regimes in $(p,\ q)$ plane during cyclic loading (Suiker and de Borst, 2003). Copyright John Wiley & Sons Limited. Reproduced with permission

The model requires the following parameters:

- Elastic behaviour: $k_1,\ k_2,\ \nu$; and
- Plastic behaviour:
  - Monotonic parameters: $p_t,\ h_m,\ h_0,\ p_0,\ d_0,\ d_m,\ \zeta_f,\ \zeta_c$.
  - Cyclic plasticity parameters: $p_t,\ h_0,\ h_m,\ \zeta_f,\ \alpha_f,\ \alpha_c,\ \gamma_c,\ p_0,\ \eta_f,\ \eta_c,\ d_0,\ d_m$.

The monotonic parameters initialise the state of stress and strains in the railway track structure which are required for the cyclic model.

Mayoraz has developed a visco-plastic equivalent model based on the associated modified Cam-Clay model with no elastic part. Permanent deformations comparisons with the results of repeated load triaxial tests performed on a sand have been carried out. This model requires only 5 parameters and is based on the Perzyna concept (Perzyna, 1966) developed for visco-plastic creep of clay.

Parameters needed for the model are:

- Plastic behaviour:
  - Rupture parameters: M; and
  - Plasticity parameters: $n_2,\ \Gamma,\ J_{init}^0$ and β^*.

9.4.2.4 Shakedown Models

New concepts have been developed to determine the long term mechanical behaviour of unbound materials under repeated loadings. All these concepts are presented in a special issue (Yu, 2005) of the International Journal of Road Materials and Pavements Design.

The shakedown concept applied to pavements was introduced first by Sharp and Booker (1984). The various possible responses of an elastic-plastic structure to a cyclical load history are indicated schematically in Fig. 9.8. If the load level is sufficiently small, the response is purely elastic, no permanent strains are induced

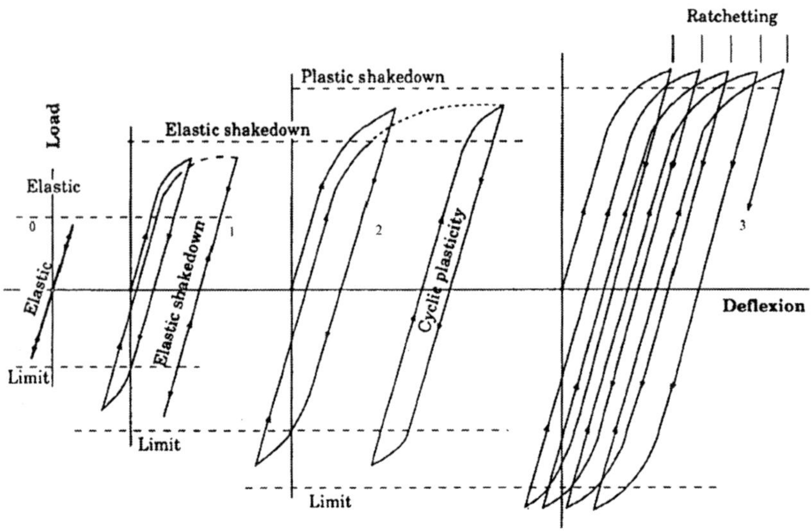

Fig. 9.8 Classical elastic/plastic shakedown behaviour under repeated cyclic tension and compression. Reprinted from Wong et al. (1997), © 1997, with permission from Elsevier

and the structure returns to its original configuration after each load application. However, if the load level exceeds the elastic limit load, permanent plastic strains occur and the response of the structure to a second and subsequent loading cycle is different from the first. When the load exceeds the elastic limit the structure can exhibit three long term responses depending on the load level (Fig. 9.8). After a finite number of load applications, the build up of residual stresses and changing of material properties can be such that the structure's response is purely elastic, so that no further permanent strain occurs. When this happens, the structure is said to have shaken down: it is in the elastic shakedown region. In a pavement this could mean that some rutting, subsurface deterioration, or cracking occurs but that, after a certain time, this deterioration ceases and no further structural damage occurs.

At still higher load levels however, shakedown does not occur, and either the permanent strains settle into a closed cycle (plastic shakedown behaviour) or they go on, increasing indefinitely (ratcheting behaviour). Contributions of Yu & Hossain (1998), Collins and Boulbibane (2000), Arnold et al. (2003) and Maier et al. (2003) are based on the fact that if either of these latter situations occurs, the structure will fail. The critical load level below which the structure shakes down and above which it fails is called the shakedown load and it is this parameter that is the key design load. The essence of shakedown analysis is to determine the critical shakedown load for a given pavement. Pavements operating above this load are predicted to exhibit increased accumulation of plastic strains under long-term repeated loading conditions that eventually lead to incremental collapse. Those pavements operating at loads below this critical level may exhibit some initial distress, but will eventually settle down to a steady state in which no further mechanical deterioration occurs.

The direct calculation of the shakedown load is difficult. Indeed lower and upper bounds are usually calculated using Melan's static or Koiter's kinematic theorems, respectively. These procedures are similar to the familiar limit analysis techniques for failure under monotonic loading, except that now the elastic stress field needs to be known and included in the calculation. Finally, the material is assumed to be perfectly plastic with an associated flow rule.

Shakedown models require an elasticity framework and parameters, for example as provided by the Universal model, and the knowledge of the rupture parameters of the Drucker-Prager or Mohr-Coulomb surfaces. 2D finite element plane strain calculations of pavements have been carried out in this way.

Parameters needed are:

- Elastic behaviour: $k_1,\ k_2,\ k_3,\ \nu$; and
- Plastic behaviour: $c,\ \varphi$.

Contributions of Habiballah and Chazallon (2005) and Allou et al. (2007) are based on the theory developed by Zarka and Casier (1979) for metallic structures submitted to cyclic loadings. Zarka defines the plastic strains at elastic shakedown with the Melan's static theorem extended to kinematic hardening materials. The evaluation of the plastic strains when plastic shakedown occurs is based on his simplified method. Habiballah has extended the previous results to unbound granular materials with a non-associated elasto-plastic model. The Drucker-Prager yield surface is used with a von Mise plastic potential.

This approach requires the elasticity parameters of the $k-\theta$ or other model, the rupture line and the law describing the development of the kinematic hardening modulus. This model requires a "multi-stage" procedure, developed by Gidel et al., (2001) which consists, in each permanent deformation test, of performing, successively, several cyclic load sequences, following the same stress path, with the same q/p ratio, but with increasing stress amplitudes. Finite element calculations have been carried out under axi-symmetric condition and 3D. The initial state of stress is determined with the $k-\theta$ or other model, then the plastic strains are calculated.

Parameters required are:

- Elastic behaviour: $k_1,\ k_2,\ \nu$ (assuming that the $k-\theta$ model was selected); and
- Plastic behaviour: $c,\ \varphi,\ \underset{\cdot}{\mathrm{H}}(p,\ q)$.

9.5 Effective Stress Approach

The constitutive models introduced in the previous sections express the constitutive stress-strain relation of the material. As soon as the water is involved, the material has to be considered as a multi-phase porous media with two phases: the solid matrix (for which we introduced the stress-strain constitutive relations) and the water phase. The two phases are coupled since the pressure acting in the water may affect the mechanical behaviour of the material. Also, the material deformation may modify the pressure in the water. Such hydro-mechanical coupling is well represented by the Terzaghi effective stress concept for saturated conditions (when water is filling all the pores) (Terzaghi, 1943). It shows the importance of the consideration of the water phase in the analysis of the mechanical behaviour of the material.

In the case of non saturation (water is no longer filling all the pores) the effective stress, expressed as a function of the externally applied stresses and the internal fluid pressures, converts a multi–phase porous media to a mechanically equivalent, single-phase, single-stress state continuum (Khalili et al., 2005). It enters the elastic as well as elasto-plastic constitutive equations of the solid phase, linking a change in stress to strain or any other relevant quantity of the soil skeleton; e.g. see Laloui et al. (2003). As a first approximation, let us consider a force F_n applied on a porous medium (constituted by a solid matrix and pores) through an area A. In this case, we can define a total stress:

$$\sigma = \frac{F_n}{A} \tag{9.16}$$

If we consider only the part of the load acting on the solid matrix (and deforming it), we may define an effective stress as the part of the load acting on the solid area (ΣS_i) (Fig. 9.9):

$$\sigma' = \frac{F_n}{\sum S_i} \tag{9.17}$$

The effective stress may be simply defined as that emanating from the elastic (mechanical) straining of the solid skeleton:

$$\boldsymbol{\varepsilon}^e = \mathbf{C}^e \boldsymbol{\sigma}' \tag{9.18}$$

in which $\boldsymbol{\varepsilon}^e$ is the elastic strain of the solid skeleton, $\mathbf{C}^e$ is the drained compliance matrix, and $\boldsymbol{\sigma}'$ is the effective stress tensor.

In a saturated medium, the effective stress is expressed as the difference between total stress, σ, and pore water pressure, u (Terzaghi, 1943):

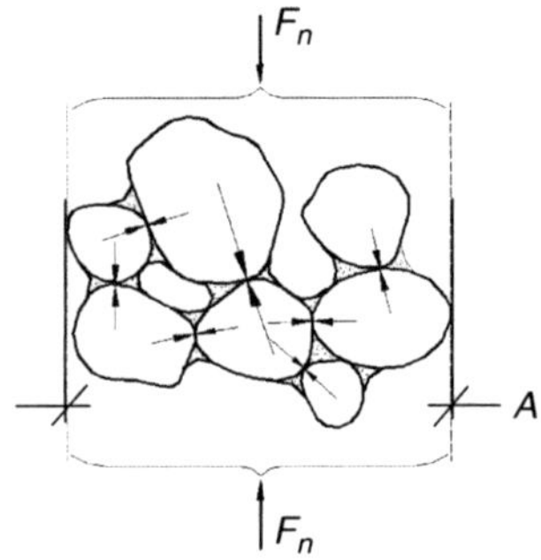

Fig. 9.9 An illustration of inter-granular stresses

$$\sigma' = \sigma - u \tag{9.19}$$

In an unsaturated granular material with several pressures of different fluid constituents, the effective stress is expressed as:

$$\boldsymbol{\sigma}' = \boldsymbol{\sigma} - \sum_{m=1}^{n} \alpha_m u_m \mathbf{I} \tag{9.20}$$

in which α_m is the effective stress parameter, u_m is the phase pressure, and $m = 1, 2, \ldots n$ represents the number of fluid phases within the system. **I** is the second order identity tensor. This equation is close to the one of Bishop (1959) for a three-phase material (solid, water and air):

$$\sigma' = (\sigma - u_a) + \chi (u_a - u) \tag{9.21}$$

where u is the pore water pressure, u_a is the pore air pressure, χ is an empirical parameter, which has a value of 1 for saturated soils and 0 for dry soils. It represents the proportion of soil suction that contributes to the effective stress. Several attempts have been made to correlate this parameter to the degree of saturation and the suction (Bishop, 1959; Khalili & Khabbaz 1998). As the parameter χ seems path-dependent, several authors, starting from Bishop and Blight (1963), proposed the use of two sets of independent "effective" stress fields combining the total stress σ, and the pore-air and pore-water pressures, u_a and u (Fredlund & Morgenstern, 1977). In the literature the net stress $\bar{\sigma} = \sigma - u_a$ and the suction $s = u_a - u$ are commonly chosen (Alonso et al, 1990). In general, this net stress concept will be defined in invariant terms using the independent stress variables $\bar{p}(= (\bar{\sigma}_1 + \bar{\sigma}_2 + \bar{\sigma}_3)/3)$, q and s.

Another way to describe the behaviour is to use the "saturated effective stress" $\sigma' = \sigma - u_w$ and the suction, s (Laloui et al., 2001). This combination has the advantage – among others – of permitting a smooth transition from fully saturated to unsaturated condition.

Continuing with the approach having two sets of independent stresses, the strain rate obtained for the elasto-plastic behaviour may be decomposed into elastic and plastic parts:

$$\dot{\varepsilon}_{ij} = \dot{\varepsilon}^e_{ij} + \dot{\varepsilon}^p_{ij} \tag{9.22}$$

each of which results from mechanical and suction variations, as follows. The elastic increment $\dot{\varepsilon}^e_{ij}$ is composed of a mechanical and a hydraulic strain increment:

$$\dot{\varepsilon}^e_{ij} = \dot{\varepsilon}^{e^m}_{ij} + \frac{1}{3}\dot{\varepsilon}^{e^h}_v \delta_{ij} = E^{e-1}_{ijkl}\dot{\sigma}'_{kl} + \frac{1}{3}\kappa^{-1}\dot{s}\delta_{ij} \tag{9.23}$$

where $\dot{\varepsilon}_{ij}^{em}$ is the elastic mechanical strain increment induced by the variation of the effective stress σ', $\frac{1}{3}\dot{\varepsilon}_{\nu}^{eh}$ is the reversible hydraulic strain increment, $\mathbf{E}^{e}$ is the classical elastic tensor and κ is a proportionality coefficient which describes the hydraulic behaviour.

Similarly the plastic strain increment is also deduced from mechanical and hydraulic loads by considering two plastic mechanisms derived from two yield limits:

$$\dot{\varepsilon}_{ij}^{p} = \dot{\varepsilon}_{ij}^{p^m} + \frac{1}{3}\dot{\varepsilon}_{\nu}^{p^h}\delta_{ij} \tag{9.24}$$

Where $\dot{\varepsilon}_{ij}^{p^m}$ is the mechanical plastic strain increment, associated with the mechanical yield surface and $\frac{1}{3}\dot{\varepsilon}_{\nu}^{p^h}$ is the hydraulic plastic strain increment, associated with the hydraulic yield surface.

Using an effective stress approach together with the non-linear models presented in the preceding section allows users to partly take into account the moisture variation effects on mechanical behaviour. However, more fundamental modifications are probably needed. The next section indicates some tools to advance in that direction.

9.6 Constitutive Modelling and Partial Saturation, Suction Coupling, Water Interaction on Mechanical Behaviour

Routine pavement design is based on an elastic calculation, with a resilient modulus. The design criterion is, typically, a limit placed on the maximum vertical strain.

More elaborated models take into account the irreversible behaviour, e.g.:

- The Chazallon-Hornych model is based on the Hujeux multi-mechanism yield surface improved by a kinematical hardening; and
- The Suiker and Mayoraz elasto-visco-plastic models evaluate the irreversible strains on the basis of an overstress (Perzyna theory) which is the distance between the stress level and a visco-plastic potential.

Each of these elaborated models is then based on a yield surface, a potential surface, a limit surface, in all cases a surface typical of the granular soil mechanics, with a frictional mechanism, possibly a cap contractive mechanism, a dependency not only on the shear/von Mise's stress but also on the mean stress ($p - q$ plane), and on the Lode angle.

How can we adapt these models to take into account the suction variation effects? For routine pavement design, only the elastic moduli need to be adapted. For the higher-level models, the yield surface and hardening mechanism also need to be adapted.

During the two last decades a number of models for partly saturated soils have been proposed (for a review, see e.g. Laloui et al., 2001). Most of them are based on the suction as an additional variable, with the same status as the stress tensor.

The so-called ***Barcelona Basic Model – BBM***, proposed by Alonso et al (1990) is probably one of the best known. It is now the reference for most new developments in mechanics of geomaterials under partial saturation.

The BBM is based on the well-known CamClay model. It is written within the framework of the independent stresses state variables $\bar{p} - q - s$ defined in Section 9.5.

The BBM yield surface depends not only on $p - q$ stresses but also on the independent stresses state variables $\bar{p} - q - s$. Two lines are added with respect with the modified Cam-Clay model. On a wetting path (a loading path along which the suction decreases), the Loading-Collapse, LC, line allows a normally consolidated material to support irreversible plastic strains and hardening, and the plastic slope to depend on the suction level (as there will be an increase of stiffness with suction). For low stress level, the cohesion only depends on the suction level. For the case illustrated in Fig. 9.10, a capillary cohesion is postulated, which depends linearly on the suction. Eventually, under very high suction (a consequence of the drying process) irreversible strains may also occur. This is at the plane SI in the figure – the suction increase surface.

However, neither do the BBM, nor the other published models, introduce any suction dependency into the elastic moduli formulations.

From this illustrative model, it appears that building a coupled model for repeated loading and suction variation on granular soil material needs the following developments:

- An elastic (resilient) model with a modulus depending on the stress and suction level. A rigorous development should lead to a hyper-elastic model. Such a model would be sufficient for routine pavement design. It seems not to exist at present.
- Improving the available models, such as the Chazallon-Hornych, the Suiker or the Mayoraz models, implies the addition of yield/potential surfaces and a dependency on the suction. The elastic stress space, lying inside the yield surface would be higher for high suction; wetting would reduce the elastic domain and then increases the irreversible strains that occur under each load cycle.

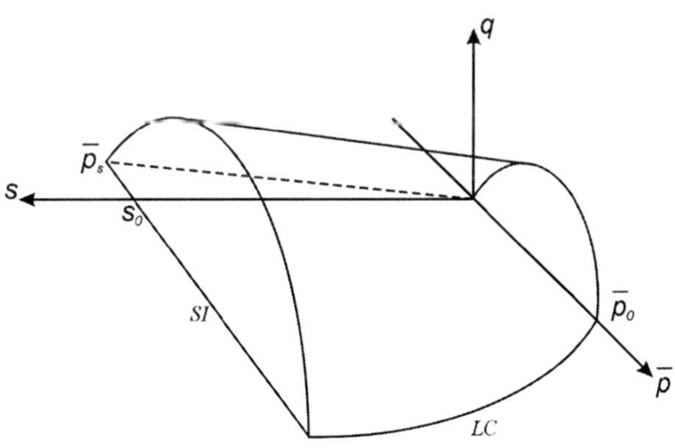

Fig. 9.10 BBM yield surface – coupling $\bar{p} - q - s$ state variables

- Using the generalised effective stress or the net stress approach allows developing coupled moisture – mechanics models to be developed.
- For any development an experimental basis will be needed to calibrate and validate the models.

9.7 Conclusions

This chapter deals with the constitutive modelling of the effects of water on the mechanical behaviour of pavements. It has been shown that routine pavement design is based on an elastic calculation, with a resilient modulus. The design criterion is, typically, a limitation of the maximum resilient vertical strain. Such design approaches do not model in a realistic manner the observed irreversible behaviour seen as rutting and as other forms of distress. To achieve a design approach that can replicate more closely the observed behaviour is likely to require use of the concepts of elasto plasticity and visco-plasticity. More elaborate models of soil and granular material will be needed to take into account these concepts and several approaches have been reviewed that attempt to do this.

At present few of these newer approaches explicitly include the effect of water pressures and suctions within the soil or aggregate pores, so the chapter has discussed how the available constitutive models could be improved to take into account suction and suction variation effects. Some research topics have, also, been suggested to enable further development.

References

Abdelkrim, M., De Buhan, P. & Bonnet G., 2003, "A computational procedure for predicting the long term residual settlement of a platform induced by repeated traffic loading", *Comput. Geotech.*, **30** pp. 463–476.

Allou, F., Chazallon, C., Hornych, P., 2007, "A numerical model for flexible pavements rut depth evolution with time", *Int. J. Numer. Anal. Methods Geomech.*, **31**, pp. 1–22.

Alonso, E. E., Gens, A. & Josa, A. 1990, "A constitutive model for partially saturated soils", *Géotechnique*, **40**(3), pp. 405–430.

Amadeus Project, 2000, "*Advanced Models for Analytical Design of European Pavement Structures*", Final report, European Commission.

ARA Inc, 2004, "*Guide for the Mechanistic-Empirical Design of New and Rehabilitated pavement Structures*", Final Report No 1-37A, Part 2, "Design Inputs", Chapter 3 "Environmental Effects", March, 50pp.

Arnold, G., Hughes, D.A.B., Dawson, A.R. & Robinson, D., 2003, "Design of granular pavements", *Transport. Res. Board*, 1819, **2**, pp. 194–200.

Barksdale, RD., 1972, "Laboratory Evaluation of Rutting in base Course Materials", *Proc. 3rd Int. Conf. Structural Design of Asphalt Pavements*, London, pp. 161–174.

Biarez, J., 1961, "*Contribution à l'étude des propriétés mécanique des sols et des matériaux pulvérulents*", PhD Thesis, Faculty of Sciences, University of Grenoble, France.

Bishop, A.W., 1959, "*The principle of effective stress*", Tecknisk Ukelbad, **39**, pp. 859–863.

Bishop, A.W. & Blight, G.E., 1963, "Some aspects on effective stress in saturated and partly saturated soils", *Géotechnique*, **13**(3), pp. 177–197.

Bonaquist, R.F. & Witczak, M.W., 1997, "A comprehensive constitutive model for granular materials in flexible pavements structures", *Proc. 8th Int. Conf. Asphalt Pavements*, Seattle, U.S.A., pp. 783–802.

Boyce, JR., 1980, "A non-linear model for elastic behaviour of granular materials under repeated loading", *Proc. Int. Symp. Soils under Cyclic and Transient Loading*, Swansea, pp. 280–294

CEN, 2000, "*Unbound and hydraulically bound mixtures – Part 7: Cyclic load triaxial test for unbound mixtures*", European standard EN 13286-7, Comité Européen de Normalisation.

Chazallon C., Hornych P. & Mouhoubi S., 2006, "An elasto-plastic model for the long term behaviour modelling of unbound granular materials in flexible pavements", *Int. J. of Geomech.*, Am. Soc. Civil Eng., **6**(4), pp. 279–289.

Collins, I.F. & Boulbibane, M., 2000, "Geomechanical analysis of unbound pavements based on shakedown theory", *J. Geotech. Geoenviron. Eng*, Am. Soc. Civil Eng'rs., **126**(1) pp. 50–59.

Dawson, A.R. & Kolisoja, P., 2004, "*Permanent deformation*", Report to ROADEX II project, Highland Regional Council, 47pp. Available at www.roadex.org.

Desai, C.S., 2002, "Mechanistic Pavement Analysis and Design using Unified Material and Computer Models", *Proc., 3rd Int. Symp. 3D Finite Elements for Pavement Analysis*, Amsterdam, The Netherlands, pp. 1–63.

Desai, C.S., and Somasundaram, S. & Frantziskonis, G., 1986, "A hierarchical approach for constitutive modelling of geologic materials", *Int J. Numer. Anal. Methods Geomech.*, **10**(3), pp. 225–257.

Ekblad, J., 2004, "*Influence of water on resilient properties of coarse granular materials", Licentiate thesis, Civil and Architectural Engineering*", Royal Inst. Technology (KTH), Stockholm, 107pp.

Elhannani, M., 1991, "*Modélisation et Simulation Numérique des Chaussées Souples*", PhD Thesis, University of Nantes, France.

Fredlund D.G. & Morgenstern, N.R., 1977, "Stress state variables for unsaturated soils". *J. Geotech. Eng.*, Am. Soc. Civil Eng., **5**, pp 447–465.

Gidel, G., Hornych, P., Chauvin, J-J., Breysse, D. & Denis A., 2001, "Nouvelle approche pour l'étude des déformations permanentes des graves non traitées à l'appareil triaxial à chargement répétés", *Bulletin de liaison des Laboratoire des Ponts et Chaussées*, pp. 5–22.

Habiballah, T.M. & Chazallon, C., 2005, "Cyclic plasticity based model for the unbound granular materials permanent strains modelling of flexible pavements", *Int. J. Numer. Anal. Methods Geomech.*, **29**(6), pp. 577–596.

Hicks, R.G. & Monismith, C.L., 1971, "Factors influencing the resilient response of granular materials", *Highways Research Record*, No. 345, Highways Research Board, pp. 15–31.

Hornych, P., Chazallon, C., Allou, F. & El Abd, A., 2007, "Prediction of Permanent Deformations of Unbound Granular Materials in Low Traffic Pavements", *Int. J. Road Mater. Pavement Des.*, **8**(4), pp. 643–666.

Hornych, P., Corte J.F. & Paute, J.L., 1993, "Etude des déformations permanentes sous chargements répétés de trois graves non traitées." *Bulletin de liaison des Laboratoires des Ponts et Chaussées*, Presse de l'ENPC, Paris, France, No 184, pp. 45–55.

Hornych, P., Kazai, A. & Piau, JM., 1998, "Study of the resilient behaviour of unbound granular materials", *Proc. Bearing Capacity Roads, Railways & Airfields*, 1998, Trondheim, **3**, pp. 1277–1287

Hujeux J.C., 1985, "Une loi de comportement pour le chargement cyclique des sols", in *Génie Parasismique*, Presses des Ponts et Chaussées, Paris, pp. 316–331.

Khalili, N. & Khabbaz, M.H., 1998, "A unique relationship for the determination of the shear strength of unsaturated soils", *Géotechnique*, **48**(5), pp. 681–687.

Khalili N., Witt, R., Laloui, L., Vulliet, L. & Koliji, A., 2005, "Effective stress in double porous media with two immiscible fluids", *Geophy. Res. Lett.*, **32**(15), L15309.

Laloui L., Geiser, F. & Vulliet, L., 2001, "Constitutive modeling of unsaturated soils", *Revue française de génie civil*, **5**, pp. 797–807.

Laloui, L., Klubertanz, G. & Vulliet, L., 2003, "Solid-liquid-air coupling in multiphase porous media", *Int. J. Numer. Anal. Methods Geomech.*, **27**(3), 183–206.

Lashine, AK., Brown, SF., Pell, PS., 1971, "*Dynamic properties of soils*", Dept. Civil Eng'g., University of Nottingham.

Lekarp, F. & Dawson, A., 1998, "Modelling permanent deformation behaviour of UGM.", *Construction and Building Materials*, **12**(1), pp. 9–18.

Long, X., Aubeny, C.P., Rifat, B. & Lytton, R.L., 2006, "Two-dimensional moisture flow – soil deformation model for application to pavement design", *Transport. Res. Rec.*, **1967**, pp. 121–131.

Maier, G., Pastor, J., Ponter, A.R.S. & Weichert, D., 2003, "Direct methods of limit and shakedown analysis", *Numerical & Computational Methods*, Chapter 12 in volume **3**, R. de Borst and H.A. Mang, (eds.), Elsevier-Pergamon, Amsterdam.

Mayoraz F., 2002, "*Comportement mécanique des milieux granulaires sous sollicitations cycliques: Application aux fondations des chaussées souples*", Ph. D. Thesis, Ecole Polytechnique Fédérale de Lausanne, Switzerland.

Pan, T., Tutumluer, E. & Anochie-Boateng, J., 2006, "Aggregate Morphology Affecting Resilient Behavior of Unbound Granular Materials", *Transport. Res. Rec.*, **1952**, Transportation Research Board, Washington, DC, pp. 12–20.

Paute, J.L., Jouve, P., Martinez, P. & Regneau, E., 1988, "*Modèle de calcul pour le dimensionnement des chaussées souples*", Bulletin de liaison des Laboratoire des Ponts et Chaussées, **156**, pp. 21–36.

Perzyna, P., 1966, "Fundamental problems in viscoplasticity", *Rec. Adv. Appl. Mech.*, **9**, pp. 243–377.

Seyhan, U., Tutumluer, E., & Yesilyurt, H., 2005, "Anisotropic aggregate base inputs for mechanistic pavement analysis considering effects of moving wheel loads," *J. Mater. Civil Eng*, Am. Soc. Civil Eng'rs., **17**(5), October, pp. 1–8.

Sharp, RW. & Booker, JR., 1984, "Shakedown of Pavements under Moving Surface Loads", *J. Transport. Eng.*, Am. Soc. Civil Eng'rs., **110**(1), pp. 1–14.

Suiker, A.S.J. & de Borst R., 2003, "A numerical model for cyclic deterioration of railways tracks", *Int. J. Numer. Methods Eng.*, **57**(4), pp. 441–470.

Terzaghi K., 1943. "*Theoretical soil mechanics*", Chapman and Hall, London.

Wong, S.K., Kapoor, A. & Williams, J.A., 1997, "Shakedown limits on coated and engineered surfaces", *Wear*, **203–204**, pp. 162–170.

Yu H.S., 2005. Special issue on the Shakedown theory, *Int. J. Road Mater. Pavement Des.*, **6**(1), pp. 1–134.

Yu, H.S., & Hossain, M.Z., 1998, "Lower bound shakedown analysis of layered pavements discontinuous stress fields", *Comput. Methods Appl. Mech. Eng.*, **167**, pp. 209–222.

Zarka, J. & Casier, J., 1979, "Elastic plastic response of structure to cyclic loading: practical rules", *Mechanics Today*, **6**, Ed Nemat-Nasser, Pergamon Press, pp. 93–198.

Chapter 10
Water Influence on Mechanical Behaviour of Pavements: Experimental Investigation

Cane Cekerevac*, **Susanne Baltzer, Robert Charlier, Cyrille Chazallon, Sigurður Erlingsson, Beata Gajewska, Pierre Hornych, Cezary Kraszewski and Primož Pavšič**

Abstract This chapter presents laboratory and *in-situ* experimental techniques used to describe the mechanical behaviour of pavement material at different saturation stages. The use of repeated triaxial load testing to obtain stiffness characteristics as well as the ability of the material to withstand accumulation of permanent deformation during cyclic loading is considered. For unsaturated soils, in addition to mechanical variables, it is shown that a moisture/suction control should be added. Several techniques are described to assist in this. A brief presentation of model parameters and tests needed for model calibration are introduced. Evaluation of pavement structural capacity based on deflection measurements with non-destructive testing equipment are presented. Finally, some examples of laboratory and *in-situ* measurement are shown.

Keywords Laboratory testing · suction control · repeated triaxial test · CBR test · parameter calibration · field testing · laboratory and in-situ experimental results

10.1 Introduction

Besides proper construction design, the materials used in the construction of pavements should be such that the effects of excess moisture are minimized. To help achieve this purpose, several laboratory and in-situ methods can be used. For aggregates used within the unbound layer, performance, when subjected to traffic and moisture is mainly characterized by:

- the compaction properties;
- the amount of degradation;
- the composition (as determined by sieving analysis); and
- the evaluation of the quality of fines (in some cases).

---

* *Co-ordinating Author:*
✉ C. Cekerevac
Stucky SA, Switzerland
e-mail: ccekerevac@stucky.ch

A.R. Dawson (ed.), *Water in Road Structures*, DOI 10.1007/978-1-4020-8562-8_10,
© Springer Science+Business Media B.V. 2009

All of these are index tests, which do not give us proper insight into the behaviour of the material under traffic loads and in different moisture conditions. A better understanding of actual in-situ behaviour can be obtained with the cyclic load triaxial test and with the Falling Weight Deflectometer (FWD) and other in-situ tests. From June 2004 the manufacturers of aggregates in the EU are required to undertake the initial type test and Factory Production Control to ensure that their products conform with European Standard EN 13242 (CEN, 2002).

10.2 Laboratory Investigation: Cycling, Suction/Saturation Control

Although there are a wide variety of tests for assessing soils and road materials, any serious investigation will need to know the mechanical behaviour of these materials when subjected to repeated loading that simulates the effects of trafficking and under moisture conditions (water content and suction) that simulate that found in the layers of the road construction and embankment. Various devices have been developed including the k-mould (Semmelink et al., 1997) and the Springbox (Edwards et al., 2004), but the cyclic triaxial test has secured the greatest following for material assessment over many years and is now the subject of European, US and Australian standards (CEN 2004, AASHTO 2000, Standards Australia 1995). It is the use of this test that is described in this section.

In laboratory testing procedures it is well known that the size of the sample may have a very important influence on the results. If the size of the sample is not appropriate for a test procedure, the results obtained may be corrupted and not valuable. In order to have a continuum condition in the sample, it is necessary to satisfy some conditions with regard to its microstructure and sample size. For fine soils, it can be assumed that the test sample should be some centimetres in diameter, roughly from 2 to 6 cm. However, for granular materials, the sample diameter should be much higher, from about 5 to 10–15 cm, depending on maximum grain size.

10.2.1 Repeated Load Triaxial Testing of Unbound Granular Materials

The repeated load triaxial testing (RLT) (also known as the cyclic triaxial test) method is commonly used to establish the mechanical characteristics of granular materials. During the testing, a cylindrical specimen is compacted to a desired level and then tested by applying confining and vertical stresses. Two variants exist:

- a constant confining pressure (CCP) method; and
- a variable confining pressure (VCP) method.

In the CCP-method the sample is initially subjected to a hydrostatic confining pressure σ_c, which induces an initial strain ε_c (unmeasured in the test, but it is the same in

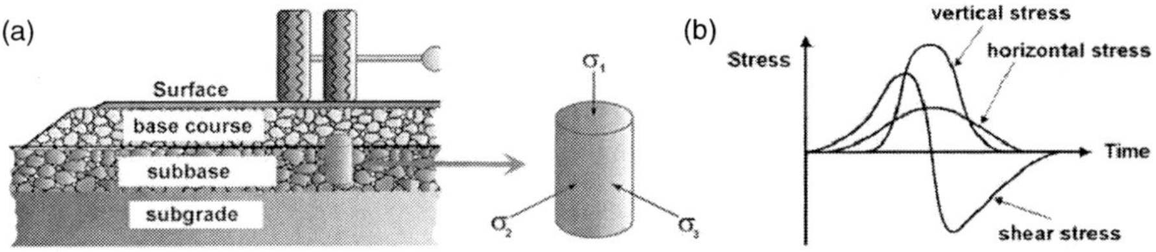

Fig. 10.1 Stresses in an unbound granular material layer. (**a**) Typical pavement structure and stresses, (**b**) induced stresses in a pavement element due to moving wheel load (Erlingsson, 2007)

all directions for isotropic material behaviour and, thus, can be estimated). The axial stress is then cycled at a constant magnitude q, which induces the cyclic resilient axial strain $\Delta\varepsilon$. In the VCP-method both the axial and the radial stresses are cycled. The RLT can be used to obtain both the stiffness characteristics as well as the ability of the material to withstand accumulation of permanent deformation during pulsating loading (Gomes-Correia et al., 1999; Erlingsson, 2000). Figure 10.1 illustrates the general stress regime experienced in an unbound granular layer in a pavement structure as a result of a moving wheel load within the plane of the wheel track. Due to the wheel load, pulses of vertical and horizontal stress, accompanied by a double pulse of shear stress with a sign reversal, affect the element (Brown, 1996).

This stress regime associated with the vertical as well as the horizontal stress pulses can be established using the VCP-method in the RLT. Using the CCP-method the variation of the horizontal stresses is neglected, as the confining pressure is kept constant. For further details please refer to CEN standard EN 13286-7 (2004).

Resilient testing of granular material is usually divided into two phases: (i) a conditioning phase and (ii) a testing phase. During the conditioning phase 20,000 symmetric haversine load pulses are applied with the frequency of 5 Hz to stabilize the response from the specimen. Thereafter, different stress paths are applied to estimate the specimen's response. Since the unbound granular materials show stress dependency behaviour, it is very important to apply a number of stress paths in order to observe such behaviour. During each stress path 100 symmetric haversine load cycles are applied with a rise time of 50 ms (total length of pulse 0.1 s) followed by a 0.9 s rest time. During the last ten load cycles data from the transducers as well as the axial load are collected to evaluate the specimen response, see Fig. 10.2.

For the permanent deformation estimation, constant amplitude symmetric haversine load pulses are applied in the axial direction usually with a frequency of 5 Hz, without any rest time between the pulses (Fig. 10.3). The stresses, cyclic numbers as well as the axial and sometimes the radial deformations are recorded during the test at appropriate intervals.

10.2.2 Control or Measurement of Suction/Moisture

As shown in Chapter 9, Section 9.5, a complete description of a material's behaviour necessitates an effective stress approach with the pore pressures (or pore

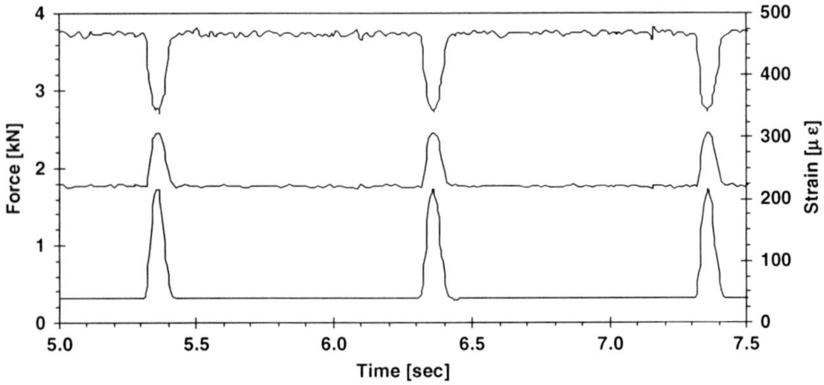

Fig. 10.2 Measured response of an unbound granular specimen (Erlingsson & Magnusdottir, 2002) Note: Three symmetric haversine load pulses (the lowest curve) are shown and the strain response from one of the vertical transducers (the top curve) as well as the horizontal transducer (the middle curve). The vertical transducer shows compression during the load pulses and the horizontal one shows extension.

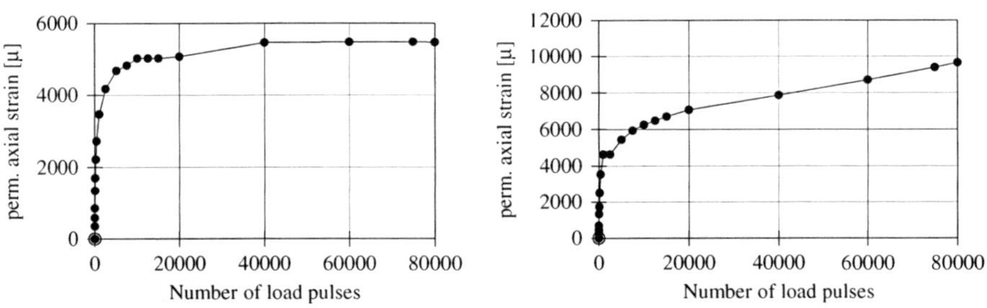

Fig. 10.3 Typical results from permanent deformation testing where the accumulated axial strain is shown as a function of the number of load pulses in the permanent deformation tests (Erlingsson, 2000)

suctions) being separately controlled, or monitored, from the applied pressures. Because most road materials are coarse grained and partially saturated and/or above the ground water table, it usually proves impossible – and certainly it is impractical in most situations – to monitor the pore suctions during each transient pulse. For this reason, almost all testing programs determine parameter values for resilient and incrementally-developed plastic strain models in terms of total, not effective, stresses. Instead, test procedures typically seek to control the moisture or suction conditions so that data is collected at representative conditions.

As described above, unsaturated soils exert an attraction on water, either by capillarity in the pores, between grains of the soil, or by the physico-chemical effects. The energy necessary to extract water from the unsaturated soil, for example from a dense compacted swelling clay used for engineering barriers in nuclear waste, can be higher than 10 MPa. However, a granular soil will have a suction that is several tens of kPa (Delage 2001). The following techniques are commonly used for suction/moisture control:

i) Axis translation system
This system is based on ceramic low porosity stone, called High Air Entry Value (HAEV) porous stone. The principle of the system is that the pores of the ceramic are too small to de-saturate, even when a high air pressure is applied. In other words, the capillary air-water menisci located at the surface of the ceramic can resist an air pressure applied to it, thereby maintaining a continuous water column from stone to specimen. The system has been employed for determination of the water retention curve (i.e. the soil water characteristic curve) (as in the Richards cell) as well as for suction control in oedometer and triaxial testing (Péron et al., 2007; Cuisinier & Laloui, 2004). The retention curve can be determined as follows. A saturated sample should be placed in the cell and the air pressure increased in a step-by-step progression. For a given air pressure, suction equilibrium should be achieved in a time period of a few days (typically 3–10). At the end of equilibrium phase, the sample should be quickly withdrawn and weighed to determine the water content. The sample is then placed into the cell and the air pressure increased to bring about a new equilibrium with a higher suction and lower water content and lower degree of saturation. The process is then repeated as desired.

ii) Osmotic technique
This was initially used by Kassif and Ben Shalom (1971) in an oedometer test to study expansive soils. Several authors extended this technique to a standard triaxial apparatus (Laloui et al., 2006; Komorik et al., 1980; Delage et al., 1987). The technique is based on drainage of the specimen caused by the process of osmosis. The soil sample is placed in a tube-shape cellular semi-permeable membrane, which is immersed in an aqueous solution of PolyEthyleneGlycol (PEG). Since the PEG molecules are too big to cross the semi-permeable membrane, the difference between concentration of the PEG solution and that of pore water results in an osmotic pressure. Thus, the suction value depends on the concentration of the solution: the higher the concentration of the polyethylene glycol (PEG) solution, the higher the suction of soil specimen in the semi-membrane after the equilibrium state is reached (Dineen & Burland, 1995; Delage et al., 1998).

iii) Vapour phase method
Usually, this method is used to determine the water retention curve of soils with high suctions, though it can be used to bring a triaxial or other test specimen to a particular moisture-suction condition prior to mechanical testing. The method is based on the theoretical standpoint that water potential (i.e. suction) is related to a particular relative vapour pressure of the water in the soil-water system (Qian et al., 2006). The relative vapour pressure of water in equilibrium with the system is characterised by its relative humidity. Therefore, suction can be established by creating the relative humidity that is related to the concentration of a solution identical with the composition of the soil water. The soil specimen should be placed in a desiccator containing an aqueous solution of

a given chemical compound. Depending on the physico-chemical properties of the compound, a relative humidity is imposed within the desiccator. Water exchange occurs by vapour transfer between the solution and the sample via the vapour and a given suction is imposed when vapour equilibrium is reached. It should be noted that the technique generally requires a very long equilibrium time that can be a few months (1–2 and up to 6). Marcial et al. (2002) introduced an air circulation technique to reduce the equilibrium time from 6 months down to 2–4 weeks. Also, it should be noted that either the same products at various concentrations or various saturated saline solutions may be used.

In conventional practice, osmotic and vapour phase methods have quite often been combined to bring soil specimens to any desired point on the soil-water characteristic curve. The osmotic technique should be used to control low matrix suction (less than 2 MPa) and the vapour phase method for measurements of higher suctions (> 2MPa).

10.2.3 Identification and Estimation of Model Parameters

A general overview of conventional and some advanced numerical models used in practice has been given in Chapter 9, Section 9.4. Therefore, tests needed for the parameters of these models will now be presented.

- Resilient behaviour models
  - Routine pavement design model: in practice much routine pavement design is carried out as catalogue – based design. The pavement is considered as a multi-layered elastic system with constant stiffness parameters in each layer.
  - Advanced pavement models: RLT tests are required with variable confining pressure (CEN standard EN 13286-7 (2004)), they correspond to strain stabilization. Parameters are determined with curve fitting by the least squares method applied to the equations of the model and to RLT tests results in the stress – elastic strain planes $(p,\ \varepsilon_v)$ and $(q,\ \varepsilon_q)$. Both phenomenological models which describe behaviour from an observational standpoint (such as the k–θ and "Universal" Models) as well as theoretically-derived models (such as the Boyce and modified Boyce model) require this kind of testing. The more complex models require VCP test procedures (see Section 10.2.1, above).

- Permanent deformations models
  - Analytical models: they require 3 monotonic triaxial tests for the rupture parameters and 3 VCP tests ($q/p = 1$, 2 and 3 for example) for the plasticity parameters. Curve fitting can be used with the least squares method applied to VCP tests results and to the analytical equation of the model ($\varepsilon^p_{\text{vertical}} = \text{f}(N)$) in the plane (N, $\varepsilon^p_{\text{vertical}}$).

- Plasticity theory based models
  - Bonaquist and Desai models
    - Elastic behaviour: the parameters require RLT tests ($q/p = 3$) at various confining pressures (AASHTO T307-99 (2000)).
    - Monotonic plasticity parameters: 3 monotonic triaxial tests till rupture for the rupture characteristics and hardening parameters.
    - Approximate accelerated analysis: 1 RLT test with one stage at ($q/p = 3$) is required.
  - Chazallon and Hornych model
    - Elastic behaviour: RLT tests are required with variable confining pressure (CEN standard EN 13286-7 (2004)), they correspond to strain stabilization.
    - Monotonic plasticity parameters: 1 oedometric test, and 3 triaxial tests till rupture are required.
    - Cyclic plasticity parameters: 1 RLT test with one stage is required ($q/p = 2$).
- Elasto-visco-plastic equivalent models
  - Suiker elasto-visco-plastic model
    - Elastic behaviour: these parameters are required for the initial state of the material. They require at least 2 monotonic triaxial tests at two different confining pressures when 100 cycles have been performed ($q/p = 3$).
    - Monotonic plasticity parameters: they are required for the initial state of the material: 3 monotonic triaxial tests are required.
    - Cyclic plasticity parameters: RLT tests ($q/p = 3$) are required.
- Mayoraz visco-plastic model
  - Plasticity parameters require 3 triaxial tests till rupture and a RLT test ($q/p = 3$).
- Shakedown models
  - Perfectly plastic models
    - Elasticity parameters: RLT tests (CEN standard EN 13286-7 (2004)) are required with variable confining pressure till stabilization.
    - Plasticity parameters: 3 triaxial monotonic tests till rupture are required.
  - Kinematic hardening models
    - Elasticity parameters: RLT tests (CEN standard EN 13286-7 (2004)) are required with variable confining pressure till stabilization.

– Plasticity parameters: 3 triaxial monotonic tests till rupture are required for rupture parameters and 3 VCP tests ($q/p = 1, 2$ and 3 for example) with 3 stages for each stress path.

10.3 Bearing Capacity Measurements In-Situ

There are many ways of evaluating pavement structural capacity or adequacy and it is very common to perform deflection measurements with non-destructive testing equipment (COST Action 325, 1997). Once again, there is usually no knowledge of the pore pressure or pore suctions in the soil or pavement layer being assessed, so a total stress interpretative framework is necessary even though an effective stress framework would be more desirable.

There are several reasons for deflection measurements to be carried out: for quality assurance, to evaluate the bearing capacity of the unbound material of the granular base, sub-base and subgrade layers, to identify weak parts of the road, to investigate reinforcement requirements, to establish priorities for road strengthening and for research purposes.

Deflection measurements can be performed in various ways, using:

i) static deflection measurement equipment;
ii) automated beam deflection measurement equipment;
iii) dynamic deflection measurement equipment;
iv) deflection instruments with a harmonic load; and
v) deflection measurement equipment with an impulse load.

*i) **Static devices*** include: static and dynamic plate loading (bearing) tests and Benkelman beam.

Plate bearing test. This involves measurement of the deflection caused by a known static load applied through a hydraulic jack on the pavement layer surface by a circular plate. Circular plates are of specified diameters −300, 450 and 600 mm with loads up to 60 kN. It is a slow and laborious test and needs a heavy vehicle acting as stationary reaction frame against which the hydraulic jack reacts. The plate test device may be mounted inside a van or at the rear of a lorry.

Dynamic plate bearing test. This is performed by lifting and dropping a known load from a known height onto the circular plate and measuring deflections. In this way, the dynamic effect of vehicles to the pavement is simulated.

The two preceding plate bearing tests are generally only used for measurement of bearing capacity of unbound layers.

Benkelman beam. This device consists of two main parts: a stand and a beam (Fig. 10.4). One end of the beam rests on the road surface but the other one is connected to a dial test indicator. The beam is suspended on

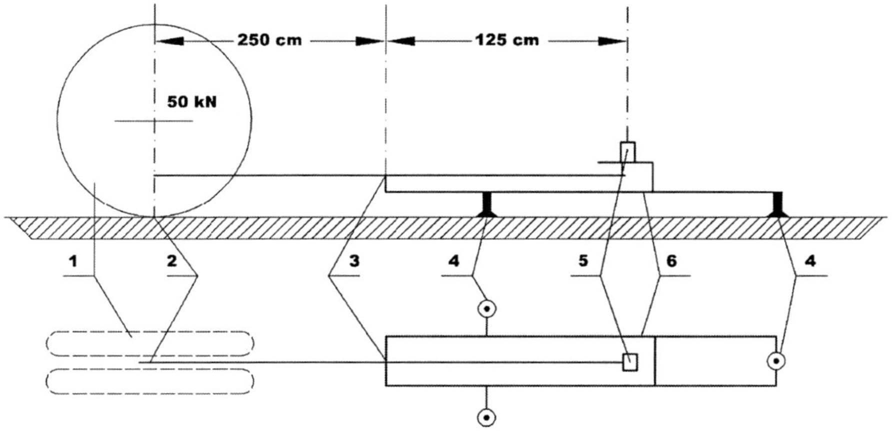

Fig. 10.4 Benkelman beam: (1) dual tyres of a loaded vehicle, (2) tip of the beam, (3) ball bearing, (4) adjustable support legs, (5) dial test indicator, (6) stand (Tehnicne specifikacije za ceste, TSC 06.630, 2002). Reproduced by permission of Direkcija Republike Slovenije za Ceste

the stand at two thirds of its length from the end being in contact with the road surface. In this way, when this end moves in one direction, the other end moves in the opposite direction half of that movement and it will be measured by dial gauge. The tip of the beam is placed between the dual tyres of a loaded vehicle. The vertical movement of the surface is then recorded as the truck moves slowly away from the loading area as this results in the rebound of the deflection that was caused by the application of the loaded vehicle. The rebound deflection/deflection ratio depends on the condition of the underlying road layers. For example, high water content caused by thawing can result in a relatively low ratio. If the deflection bowl is large, as in the case of weak subgrades, and the supporting legs are placed in the deflection area, this may produce inaccurate measurements.

ii) ***Automated beam deflection measurement equipment*** is represented by the Lacroix Deflectograph. It was developed in order to carry out measurements more quickly than by the time-consuming Benkelman beam. The measurement beam is automatically displaced along the measurement direction, while a measurement vehicle proceeds at slow speed. Equipment includes a T-shaped frame towed between the axles of lorry, enabling deflection to be measured in both wheel paths simultaneously (Fig. 10.5).

The measurement starts when the rear wheels of lorry are at a certain distance behind the tip of measuring beam and ends when the wheels reach a certain distance in front of it. Then the entire T-frame is moved forward a specified distance in the direction of movement. The axial load can vary from 80 kN to 130 kN.

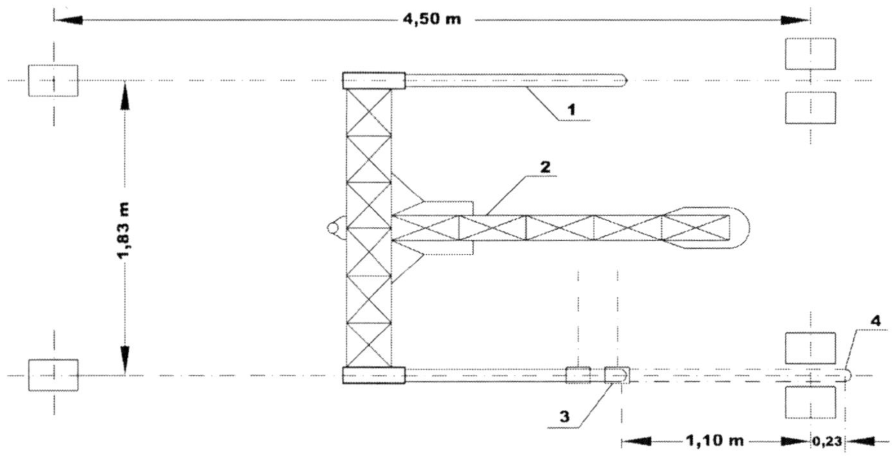

Fig. 10.5 Deflectograph Lacroix equipment: (1) measuring beam, (2) T frame, (3) starting position of tip of measuring beam, (4) end position of tip of measuring beam (Tehnicne specifikacije za ceste, TSC 06.630, 2002). Reproduced by permission of Direkcija Republike Slovenije za Ceste

iii) ***Curviameter*** records the surface deflection under a ***dynamic load***. The vertical displacement is determined in the right-hand wheel path by means of velocity sensors when the loaded twin-wheel rear axle of the measuring lorry passes over. Like the Lacroix deflectograph, loads can vary from 80 kN to 130 kN. The High Speed Deflectograph also belongs to this category. At a driving speed of 80–90 km/h, laser Doppler sensors measure the movement of the pavement surface under a 10-tonne axle.

iv) ***Deflection measurement by harmonic load*** can be performed by equipment that can exert a sinusoidal vibration in a road pavement. A well-known example of this type of equipment is the Dynaflect, in which the power is transferred to the road by means of crank at a specified frequency via two steel wheels. The deflections are measured by five geophones spaced at various offsets from the centre of load.

v) ***Bearing capacity assessment***. The most common device to measure bearing capacity of roads is the Falling Weight Deflectometer (FWD). In a FWD test an impulse loading is applied to the pavement which is similar in magnitude and duration to that of a single heavy moving wheel load. The vertical response of the pavement system is measured on the surface with deflection sensors at different distances from the loading. Usually only the peak values are registered, see Fig. 10.6.

Frequently six or seven sensors are used but the spacing between the sensors can vary. Typical values are given in Table 10.1.

The deflections can be used to back-calculate the layer moduli of the structure. This requires information on the number of layers in the pavement structure and

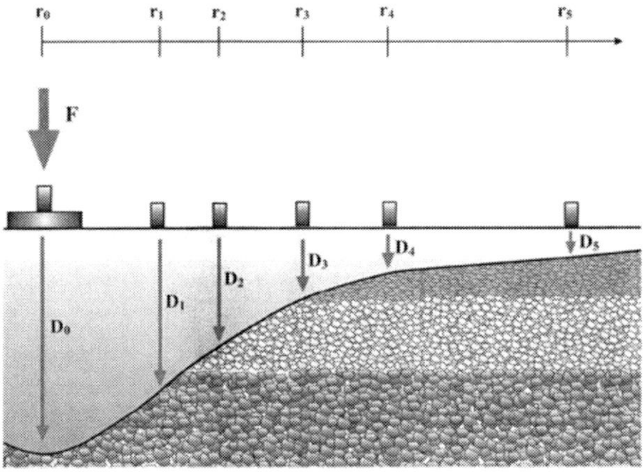

Fig. 10.6 Schematic overview of Falling Weight Deflectometer testing

their thicknesses. This is an inverse process where a single, correct solution does not exist. A number of software solutions exist to perform such back-analysis.

The deflections can also be used to estimate some simple parameters which are related to the condition of the pavement, some of the more common ones are:

$$SCI = D_0 - D_2 \tag{10.1}$$

$$BDI = D_2 - D_4 \tag{10.2}$$

$$BCI = D_4 - D_5 \tag{10.3}$$

$$AREA = \frac{1}{D_0}\sum_{i=0}^{N-1}\left[(D_{i-1} + D_i)(r_i - r_{i-1})\right] \tag{10.4}$$

where SCI = the Surface Curvature Index, BDI = the Base Damage Index, BCI = the Base Curvature Index and AREA = the area of the deflection basin (all assuming radii as per the light pavement in Table 10.1).

A falling weight deflectometer is mounted on a trailer or in a van, so it needs a stable surface. On soft ground or in trenches the smaller portable light weight FWDs (Fig. 10.7) can be used. These equipments measure the central deflection, resulting in a surface modulus.

Table 10.1 Some typical set-ups for FWD testing

Radial position of falling weight deflectometer sensors								
Sensor no.	i	0	1	2	3	4	5	6
Radius, light pavement [cm]	r_i	0	20	30	45	60	90	–
Radius, heavy pavement [cm]	r_i	0	30	60	90	120	150	210

Data in bottom row taken from Highways Agency (1999).
Deflection D_i is measured at distance r_i from centre of the impact of the load.

Fig. 10.7 Portable FWD. Photo courtesy of Grontmij/Carl Bro A/S. Reproduced by permission of Carl Bro Pavement Consultants A/S

10.4 Examples of Test Results

10.4.1 Laboratory Results

Wetting of unsaturated soil reduces the suction in the soil, the pore pressure approaches the pore air pressure and the effective stress is reduced. Because of this, increasing moisture is associated with decreases in shear strength, stiffness and resistance to plastic deformation in all soils and aggregates and we can observe a decrease in bearing capacity and lower moduli of elasticity and increases in deformability under the same applied loading.

10.4.1.1 CBR Tests

This influence of moisture on the bearing capacity of soils can be easily observed in the simple CBR test. This test measures the resistance of the compacted soil to the penetration of a piston, to evaluate its bearing capacity. CBR values for three typical soils related to water content are presented in Fig. 10.8. The sensitivity to moisture variations is particularly important for the silty sand (a).

Decrease of bearing capacity and of unconfined compression strength of clay with increasing water content can also be observed in Fig. 10.9.

The CBR test gives a good indication of the moisture sensitivity of subgrade soils, but its results are only qualitative. The influence of moisture on the resilient modulus and resistance to permanent deformations of soils and unbound granular materials can be studied using repeated load triaxial tests.

10.4.1.2 Repeated Load Triaxial Tests – Unbound Granular Aggregates

Unbound granular materials, which are continuously graded materials containing fines, are also sensitive to moisture. Examples of influence of moisture on the

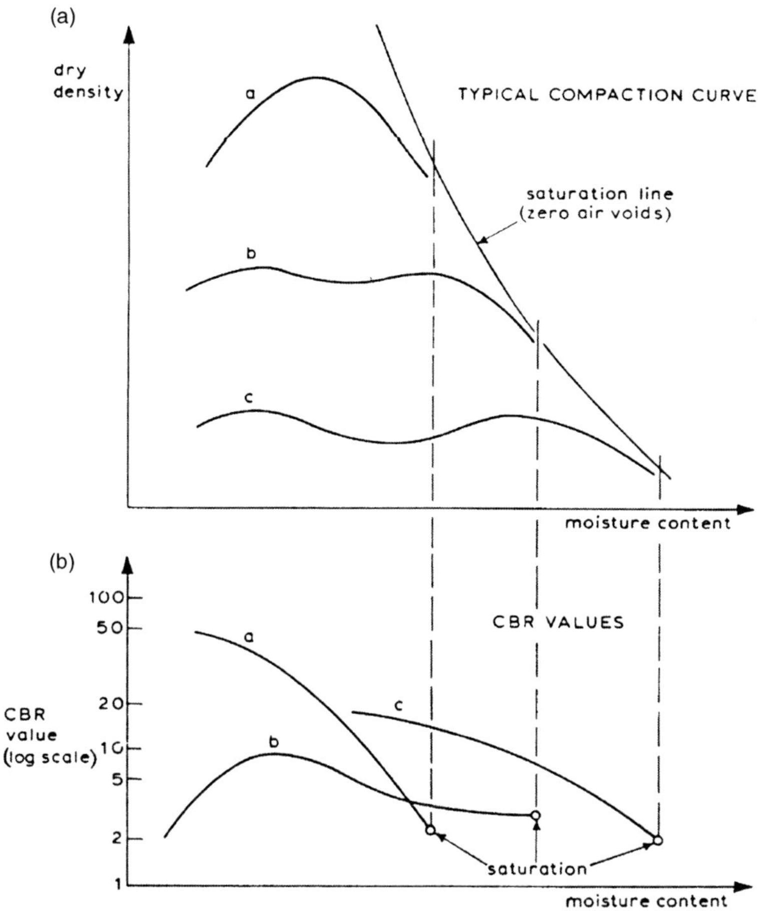

Fig. 10.8 CBR values related to moisture (water) content and compaction curves for typical soils: (**a**) well-graded silty sand with clay, (**b**) uniform fine sand, (**c**) heavy clay (Head 1994). © 1996, copyright John Wiley & Sons Limited. Reproduced with permission

resilient modulus and on the permanent deformation of 3 French unbound granular materials, of different mineralogy (hard and soft limestone, micro-granite) are shown in Fig. 10.10. All 3 materials present a decrease of their resilient modulus as the water content approaches the modified Proctor Optimum (w_{OPM}), but the sensitivity to moisture is much more important for the limestone than for the igneous rock material (micro-granite). The permanent axial strains become very large for all 3 materials when the water content approaches w_{OPM}.

Other examples, showing the influence of water content on the permanent deformation and modulus of elasticity of several unbound granular materials from Slovenia, are presented in Fig. 10.11. Again, the permanent strains appear to be more sensitive to moisture than the resilient modulus.

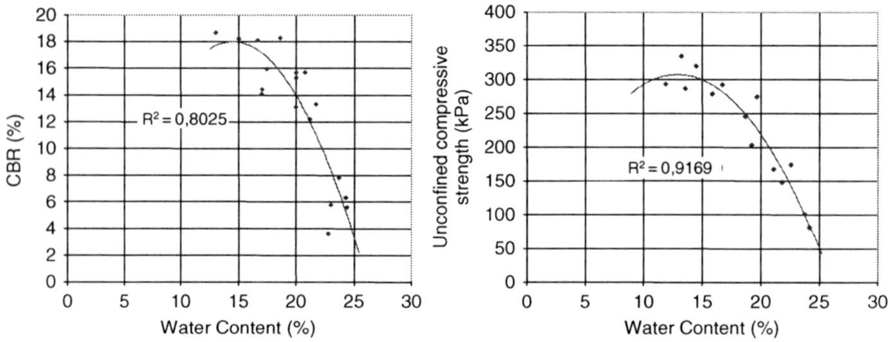

Fig. 10.9 CBR values and unconfined compression strength related to water content for a clay from the Lenart area in Slovenia (adapted from Petkovšek et al., 2003). Reproduced by permission of Direkcija Republike Slovenije za Ceste

Several studies carried out in France on a large number of different unbound granular materials have shown that their sensitivity to moisture is strongly related to their mineralogical nature, and is particularly important for soft limestone materials. This is illustrated in Fig. 10.12 which presents values of resilient modulus obtained for different natures of granular materials and different water contents.

The igneous materials present relatively low resilient moduli (generally between 300 and 500 MPa), but are not very sensitive to moisture. The soft limestone materials present significantly higher moduli at low water contents (up to 1000 MPa), but these moduli drop when the water content approaches the optimum (w_{OPM}).

Ekblad (2004) investigated the influence of water on the resilient properties of coarse unbound granular materials in the saturated as well as the unsaturated state. This study was limited to one type of aggregate of different gradings (with

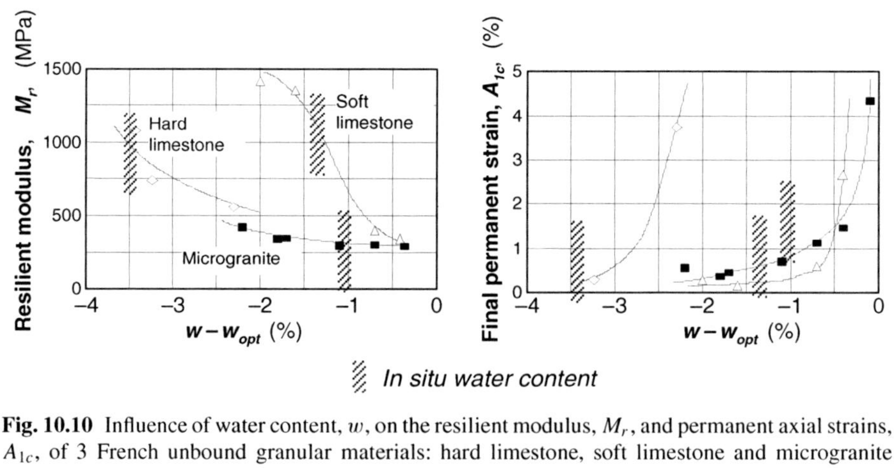

Fig. 10.10 Influence of water content, w, on the resilient modulus, M_r, and permanent axial strains, A_{1c}, of 3 French unbound granular materials: hard limestone, soft limestone and microgranite (Hornych et al., 1998)

Note: A_{1c} is a level of strain anticipated once plastic strain has stabilised.

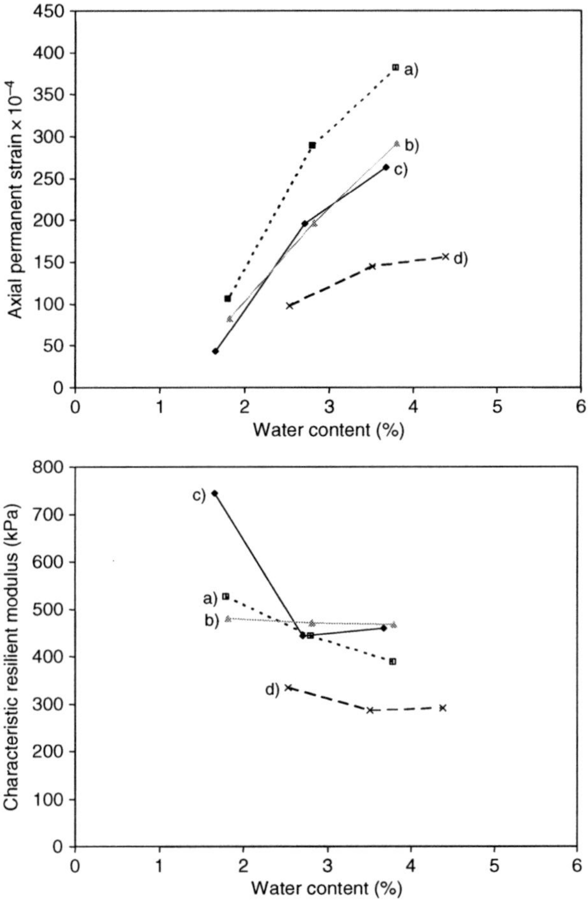

Fig. 10.11 Influence of water content on permanent deformations and modulus of elasticity for some unbound granular materials. (**a**) Crushed gravel from gravel pit Hrušica, (**b**) dolomite from quarry Lukovica, (**c**) dolomite from quarry Kamna Gorica, (**d**) gravel from Hoče (Pavšič, 2006)

maximum particle size 90 mm). The aggregate comes from Skärlunda in Östergötland in Sweden. Ekblad's tests on unbound granular materials of different granulometric curves showed that the influence of water content on resilient properties depends on the material grading.

First, the dependency of resilient modulus, M_r, on confining stress was established (Fig. 10.13). Increased confining pressure leads to a substantial increase in resilient modulus. Confining pressures of 100 kPa were reached by Ekblad. Triaxial tests at different water contents were also performed. The water content was successively increased from an initially low water content to a soaked condition (representing full saturation) and then the sample was allowed to drain freely. All these triaxial tests were performed at a confining pressure of 40 kPa (Fig. 10.13).

Finally, to achieve a summary comparison, the resilient response for a mean normal stress of 100 kPa at a confining pressure of 40 kPa was calculated as a function of the degree of saturation. From Fig. 10.14 it can be observed that the relative reduction in modulus seems to depend on the grading coefficient, with a lower

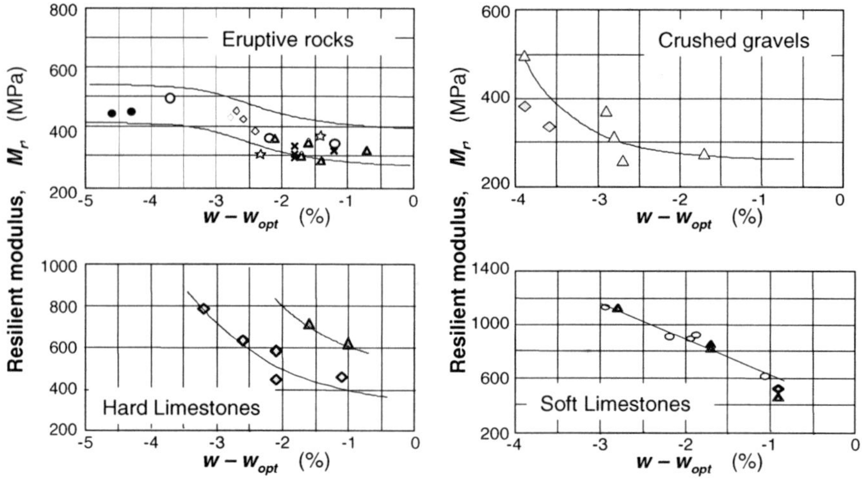

Fig. 10.12 Sensitivity to moisture of unbound granular materials of different origin (Hornych et al., 1998)

grading parameter (i.e. a higher proportion of fine particles) yielding a larger modulus reduction upon saturation.

10.4.1.3 Triaxial Tests – Soils

Examples of variation of resilient modulus and permanent deformations of subgrade soils with moisture content are presented in Figs. 10.15 and 10.16. Figure 10.15 presents results obtained on a clayey sand (14% fines, optimum moisture content $w_{\text{OPM}} = 8\%$).

Figure 10.16 presents results obtained for a silt (85% fines, optimum moisture content $w_{\text{OPM}} = 14\%$). For the 2 soils, the resilient modulus (determined for two different levels of stress) decreases by a factor of 3–4 when the water content increases from $w_{\text{OPM}} - 2\%$ to $w_{\text{OPM}} + 2\%$ (typical *in-situ* moisture contents). For the same change, the permanent axial strains (determined after 200 000 load cycles with cyclic stresses $p = 26\,\text{kPa}$ and $q = 80\,\text{kPa}$) increase considerably.

Brüll (1983) performed triaxial tests on 2 different soils, a loam (from Sterrebeek in Belgium) and a slightly clayey sand (from Noucelles in Belgium). The tests were carried out for different dry volumetric masses, for suctions between 0 and 600 kPa, for different confining pressures (between 10 and 50 kPa) and for different deviatoric stress (between 5 and 25 kPa). The soil water characteristic curves (also known as retention curves) as a function of the volumetric mass of the samples are presented in Fig. 10.17.

The Resilient Modulus, M_r, of the different samples were determined from the triaxial tests. The influence of the volumetric mass and stresses have been analysed. For the tested materials, Brüll attempted to establish a linear correlation between the Resilient Modulus, M_r, and the suction, s (Fig. 10.18).

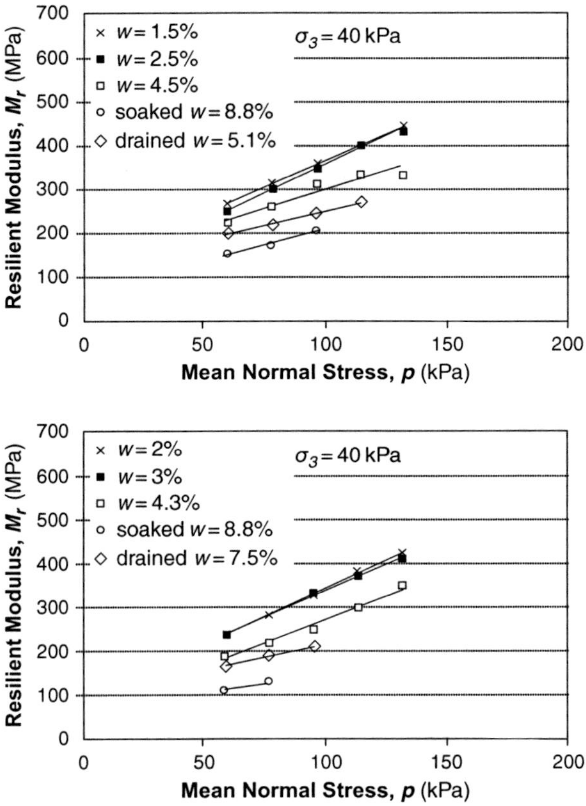

Fig. 10.13 Resilient response at different water contents for grading 0.4 or 0.3 (Ekblad and Isacsson, 2006). Reproduced with permission of Lavoisier

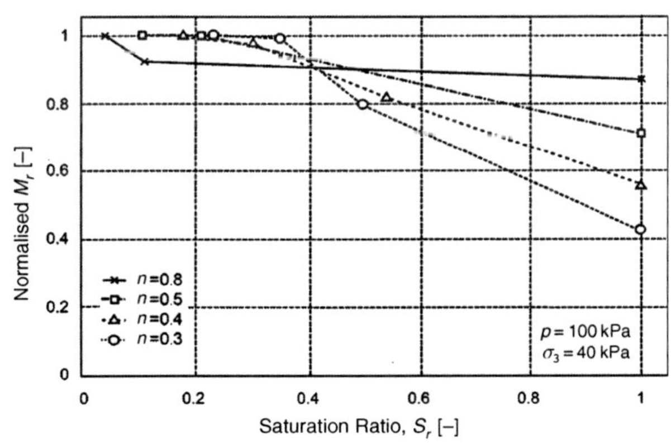

Fig. 10.14 Normalized resilient modulus as a function of degree of saturation ($p = 100$ kPa and $\sigma_3 = 40$ kPa) (Ekblad and Isacsson, 2006). Reproduced with permission of Lavoisier

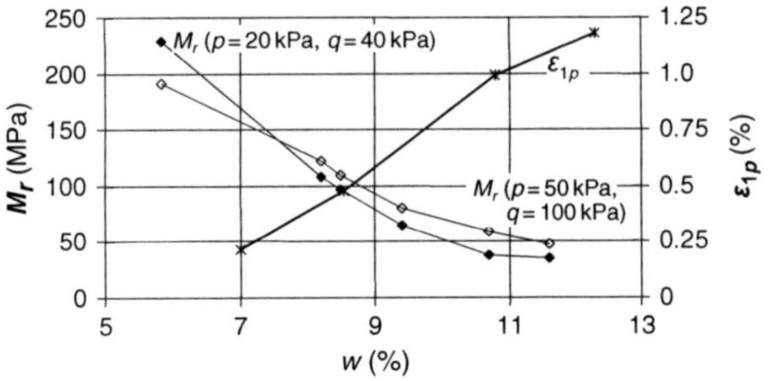

Fig. 10.15 Variations of resilient modulus M_r and permanent axial strains with moisture content for a clayey sand (Hornych et al., 1998)

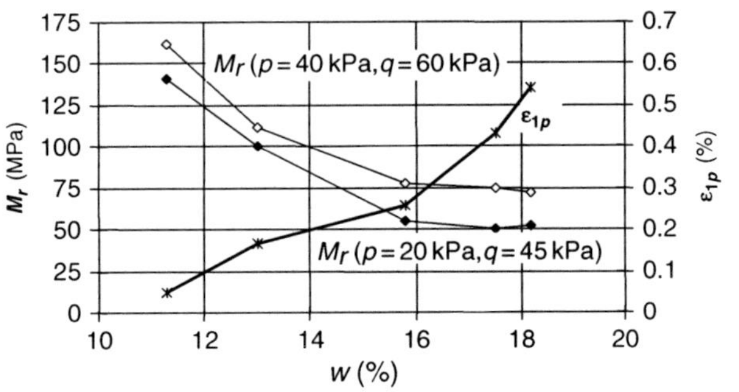

Fig. 10.16 Variations of resilient modulus, M_r, and permanent axial strains with moisture content for a silt (Hornych et al., 1998)

Cui (1993) tested a remoulded Jossigny silt in an unsaturated state and gave an evaluation of Resilient Modulus, M_r, and shear modulus, G_r, as a function of confining pressure and of suction level (Fig. 10.19).

10.4.2 In-Situ Results

The deterioration of strength, stiffness and resistance to the development of permanent deformation, or the reduction in pavement life, with increasing moisture levels is a common observation. Trial pavement studies in which the water content of the construction has been changed and reduced performance observed are quite numerous. In recent years work in Finland has been reported by Korkiala-Tanttu and Dawson (2007) showing the much faster rutting of a test pavement with a high water table than one in which it was lower. In an earlier study by accelerated trafficking, Vuong et al. (1994) found that the life of a crushed-rock base was very dependent on the degree of saturation in the aggregate base course. Assuming a water content for optimal behaviour, then a 5% change increase in relative water content could lead to a 400% reduction in pavement life. Sharp et al. (1999) reported significant deterioration of in-situ moduli values at an accelerated pavement testing site in lateritic gravel bases and sub-bases upon wetting (or improvement on drying) by more than a factor of 2. The silty sand subgrade at the test site also changed stiffness to a similar degree.

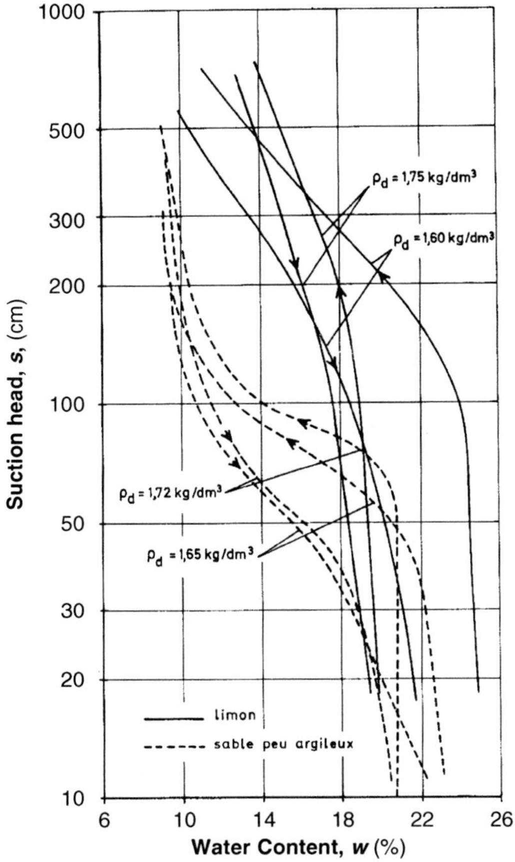

Fig. 10.17 Retention curves (SWCCs) of two different types of materials (Brüll, 1983). Figure courtesy of the Belgian Road Research Centre
Silt = *solid lines*;
slightly clayey sand = *dashed lines*.

Thus site studies broadly support the laboratory and theoretical work reported elsewhere in this book. As an example of a particular study of the effects of moisture change on bearing capacity, the following case record is instructive.

10.4.2.1 Moisture Content Effect on Bearing Capacity – a Danish Study

Relationships between bearing capacity and moisture content were studied in the Danish Road Testing Machine (ALT-facility) (Krarup, 1995). A typical Danish pavement was built in the facility, and for almost one year the only condition that was changed was the level of the water table. No load was applied except for FWD-measurements. The layer stiffness moduli (E-values) of the pavement layers were calculated from FWD-measurements with the program ELMOD (Dynatest, 1989). Based on the test results, a relationship between E-value and suction, pore pressure and degree of saturation was established. Simple linear regression describes the

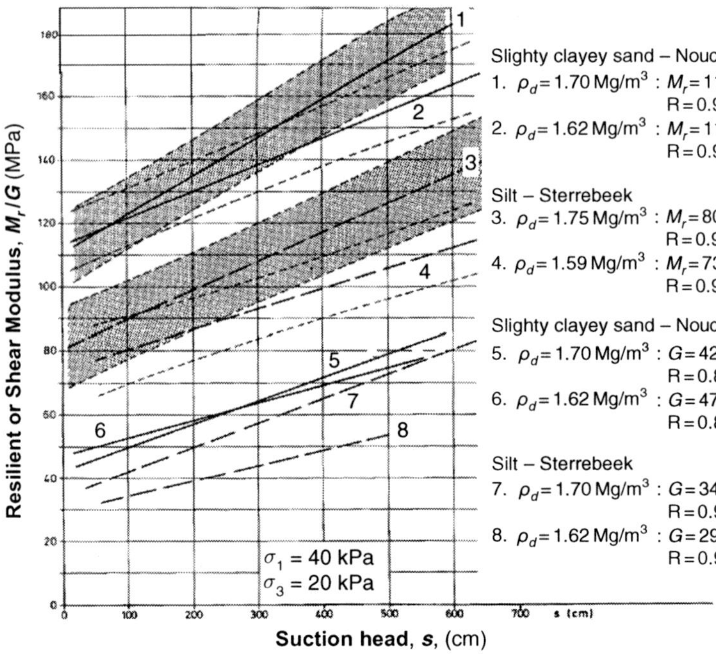

Fig. 10.18 Correlations between the resilient modulus and the suction for the two soils analyzed for a confining pressure of 20 kPa and a deviatoric stress of 20 kPa (Brüll, 1983). Figure courtesy of the Belgian Road Research Centre
Note: M_r = Resilient Modulus (MPa), G = Shear Modulus (MPa), R = coefficient of correlation, N = number of data points, s = suction (cm), light dotted lines show 95% confidence limits for M_r.

relation between the value of the E-value and suction measured by the tensiometer in the form: $E = B_1 + B_2 \times s$, where

- E is the layer E-value in MPa from the ELMOD calculation;
- B_1 and B_2 are constants; and
- s is the measured suction in kPa.

The regressions reveal high levels of R-squared within the range of the measured suction values for all unbound layers. The measured values and the linear regression lines from three of the tensiometer depths are plotted in Fig. 10.20.

When calculating the stress in a pavement (a continuous body), the stress according to Boussinesq is independent of the E-modulus of the material. Combining this statement with Terzaghi's principle of effective stress, it becomes unpredictable to what extent E-values from FWD-tests relate to positive pore pressure measured in standpipes. Positive pore water pressure only appears in what is considered to be a saturated condition. In the Krarup study, the upper pavement layers only became saturated for very short periods and water/suction versus time series exist for this granular base course layer data.

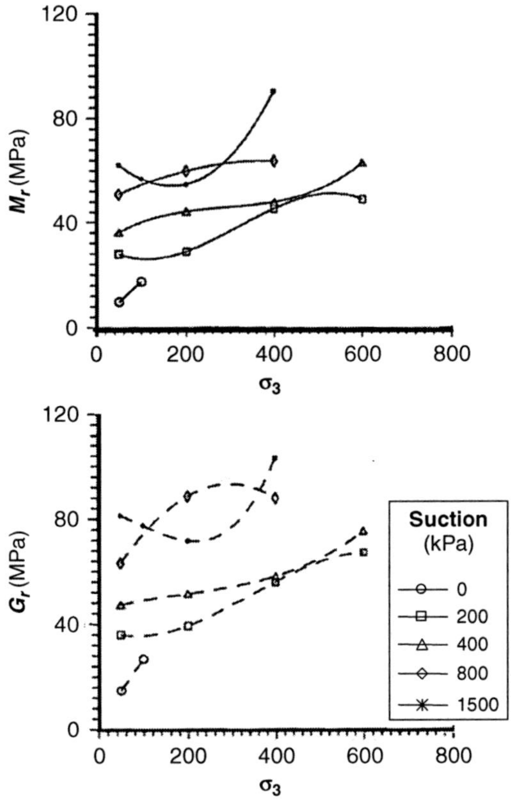

Fig. 10.19 Resilient modulus, M_r, and shear modulus, G_r, as functions of suction level and of lateral stress for a compacted Jossigny silt (Cui, 1993). Figures courtesy of Y-J. Cui

The available measured data, linear and linear-exponential regressions were tested on the data to investigate any dependency. The linear regression came up with a R-squared value ranging from 0.56 to 0.88, whereas the linear-exponential regression had the best fit at 0.72. The data and the linear regressions were plotted in Fig. 10.21. Note the much lower values of stiffness were determined *in-situ* than those typically measured in laboratory triaxial assessments (cf. Fig. 10.12 for example).

The reason for the decrease of the lower unbound layer material stiffness over the monitoring period is expected to be the water that adheres to the surface of the mineral's granular materials, more than any pore water pressure phenomenon.

Saturation: Saturation and E-values are not expected to reveal a linear relationship, as the change in E-value dependent on degree of saturation occurs at a certain level of saturation determined by the void size distribution. The results plotted in Fig. 10.22 are derived from time series and are not scattered data points. During the monitoring period the saturation increased and decreased, so the curves should be read from left to right and then back again so as to follow time.

The two time series of saturation data from 20 to 40 cm below surface are plotted with the E-values assigned to the sub-base layer, and therefore plotted with the same

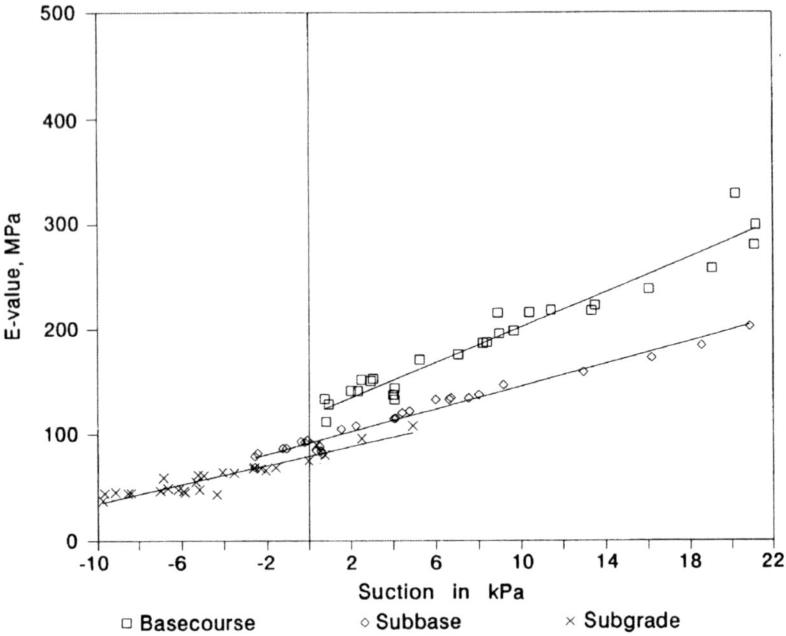

Fig. 10.20 E-values versus granular material suction measured at three depths (Krarup, 1995). Reproduced by permission of the Danish Road Institute

set of E-values. Figure 10.23 shows that granular base course and sub-base layers tended to have threshold E-values in the test pavement. Keeping the low E-values of the subgrade at the construction time in mind, the E-value of the natural till subgrade might continue to decrease as long as the water table is above the material.

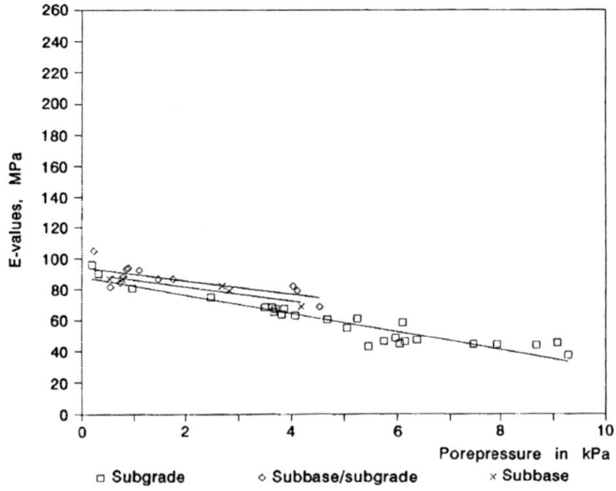

Fig. 10.21 E-values versus positive pore water pressure (Krarup, 1995). Reproduced by permission of the Danish Road Institute

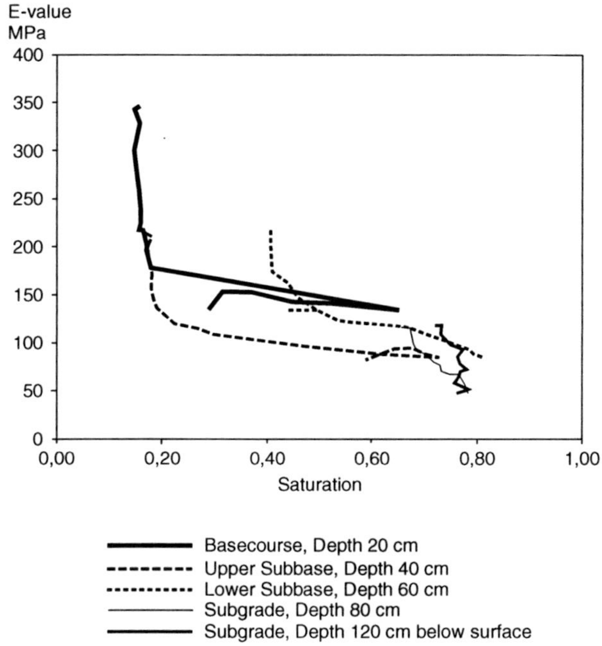

Fig. 10.22 Measured saturation (nuclear probe) and E-values (ELMOD) (Krarup, 1995). Reproduced by permission of the Danish Road Institute

As a rather strong relationship between E-value and suction was confirmed from the measurements, the relation between suction and saturation becomes interesting. In soil water research the relationship is the so-called soil-water characteristic curve or the retention curve. From experiments in the laboratory the relationship can be found for small soil samples. Data measured in the test pavement were plotted as Fig. 10.23.

10.5 Concluding Remarks

This Chapter presents in-situ and laboratory experimental techniques used to describe mechanical behaviour of pavement materials (soils and aggregates) at different saturation stages. Repeated triaxial load testing can be applied to obtain both stiffness characteristics and assessments of the ability of the material to withstand accumulation of permanent deformation during cyclic loading. For unsaturated soils, in addition to mechanical variables, a moisture/suction control should be added, which can be imposed by several techniques as explained in the chapter. A brief presentation of the model parameters and tests needed for model calibration was introduced with particular reference to the modelling approaches described in Chapter 9. Evaluations of pavement structural capacity based on deflection measurements with non-destructive testing equipment have also been presented. Finally, some examples of laboratory and *in-situ* measurement are shown.

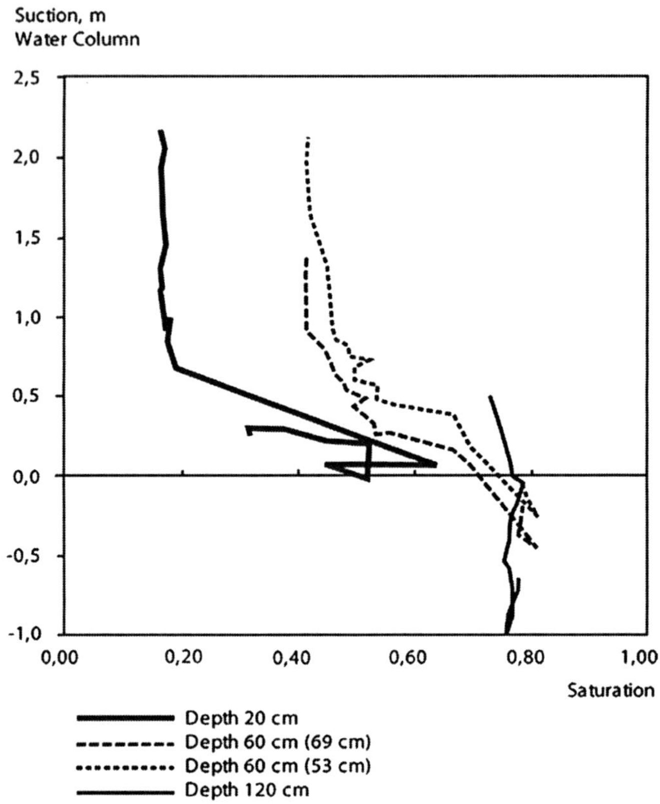

Fig. 10.23 Soil suction measured with tensiometers and saturation measured with the moisture/density probe (Krarup, 1995). Reproduced by permission of the Danish Road Institute
Note: Where measurements were carried out at different depths, the depths of the tensiometer are given in brackets (69 cm). The curves are time series beginning from the left (high suction).

Based on experimental results presented above it can be concluded as follows:

i) Bearing capacity and unconfined compressive strength decrease with increase in moisture content.
ii) The permanent axial strain increases when water content approaches to w_{OPM}.
iii) Resilient modulus decreases as the water content approaches to w_{OPM}. The resilient modulus of soils decreases by a factor 4–5, for realistic (temperate) seasonal variation of moisture contents.
iv) Reduction in resilient modulus with suction depends also on the grading coefficient: lower grading parameters (i.e. more fine particles) yields larger modulus reductions as saturation is approached.
v) In-situ experimental data confirms that resilient moduli decrease with decrease in suction. The soil-water characteristic curve depends on the grading of tested material meaning that the modulus-suction relationship is likely to be very soil-specific.

References

AASHTO, 2000, "Determining the resilient modulus of soils and aggregate materials", standard: T 307-99 in *Standard Specifications for Transportation Materials and Methods of Sampling and Testing*, 20th ed'n., Am. Assoc. State Highway & Transportation Eng'rs., Washington, DC.

Brown, S.F., 1996, "Soil mechanics in pavement engineering", *Geotechnique* **46**(3), pp. 382–426.

Brüll, A., 1983, "*Effets mécaniques de l'eau interstitielle – Détermination des modules complexes de déformation longitudinale et de cisaillement des sols en fonction de la succion*", Centre de Recherches Routières, Bruxelles, Belgique, 180pp.

CEN, 2002, "*Aggregates for unbound and hydraulically bound materials for use in civil engineering work and road construction*", Comité Européen de Normalisation, European standard EN 13242.

CEN, 2004, "*Unbound and hydraulically bound mixtures – Part 7: Cyclic load triaxial test for unbound mixtures*", Comité Européen de Normalisation, European standard EN 13286-7.

COST Action 325, 1997, "*New Road Monitoring equipment and Methods*", Final Report of the Action, Reference EUR 17547, European Commission, Directorate General Transport.

Cui, Y.-J., 1993, "*Etude du comportement d'un limon compacté non saturé et de sa modélisation dans sun cadre élastoplastique*". Thèse de doctorat de l'Ecole nationale des Ponts et Chaussées, Paris.

Cuisinier, O., & Laloui, L., 2004, "Fabric evolution during hydro-mechanical loading of compacted silt", *Int'l J. Numerical & Analytical Methods in Geomech.*, **28**, pp. 483–499.

Delage, P., 2001, "Experimental techniques used in investigation of coupled THMC behaviour of geomaterials", in *Environmental Geomechanics*, ed. Charlier, R. & Gens, A., *Revue française de genie civil*, **5**(6), pp. 777–796.

Delage, P., Suraj de Silva, G.P.R. & De Laure, E., 1987, "Un novel appareil triaxial pour les sols non saturés", *Proc. 9th Europ. Conf. Soil Mech. & Foundation Eng'g.*, **1**, pp. 26–28.

Delage, P., Howat, M.D. & Cui, Y.J., 1998, "The relationship between suction and swelling properties in a heavily compacted unsaturated clay", *Engineering Geology*, **50**, pp. 31–48.

Dineen, K. & Burland, J., 1995, "A new approach to osmotically controlled oedometer testing", in *Proc. 1st Int'l Conf. Unsaturated Soils*, Eds. Alonso, E.E. & Delage, P., Balkema/Press des Ponts et Chaussées, Paris, **2**, pp. 459–465.

Direkcija Republike Slovenije za ceste, 2002, "*Technical specification for roads*": TSC 06.630:2002, Lastnost voznih povrsin Podajnost, Direkcija Republike Slovenije za ceste, Ljubljana.

Dynatest, 1989, "*ELMOD/ELCON, Evaluation of Layer Moduli and Overlay Design*", User's Manual, Dynatest Engineering, Denmark.

Edwards, J.P., Thom, N.H., & Fleming, P.R., 2004, "Development of a simplified test for unbound aggregates and weak hydraulically bound materials utilising the NAT", in *Pavements Unbound* (proc. 6th Int'l. Symp. Unbound Aggregates in Roads, UNBAR6), pp. 3–12, Balkema, Leiden.

Ekblad, J., 2004, "*Influence of Water on Resilient Properties of Coarse Granular Materials*", Licenciate Thesis, KTH Civil and Architectural Engineering, Stockholm, Sweden, 107pp.

Ekblad, J. & Isacsson, U., 2006, "Influence of water on resilient properties of coarse granular materials", *Road Materials and Pavement Design*, **7**(3), pp. 369–404.

Erlingsson, S., 2007, "Numerical Modelling of Thin Pavements Behaviour in Accelerated HVS Tests", *Int'l. J. Road Mat'ls & Pavement Design*, **8**(4), pp. 719–744.

Erlingsson, S. & Magnusdottir, B., 2002, "Dynamic Triaxial Testing of Unbound Granular Base Course Materials," *Proc. 6th Int'l Conf. Bearing Capacity of Roads, Railways & Airfields*, Lisbon, Portugal, pp. 989–1000.

Erlingsson, S., 2000, "Dynamic triaxial testing of unbound base course materials". *Proc. XIII Nordic Geotech.l Conf., NGM 2000*, Finnish Geotechnical Society, Helsinki, pp. 69–76.

Gomes-Correia, G., Hornych, P & Akou, Y., 1999, "Review of models and modelling of unbound granular materials". In G. Gomes-Correia (ed.), *Proceedings of an international workshop on modelling and advanced testing for unbound granular materials*, Lisbon, Balkema, pp. 3–15.

Head, K.H., 1994, "*Manual of Soil Laboratory Testing; Volume 2: Permeability, Shear Strength and Compressibility Tests*", John Wiley & Sons, Inc., 454pp.

Highways Agency, 1999, "Structural Assessment Methods", HD 29/94, *Design Manual for Roads & Bridges*, HMSO, London.

Hornych P., Hameury, O. & Paute, J.L., 1998, "Influence de l'eau sur la comportement mécanique des graves non traitées et des sols supports de chaussées", *Symposium Int'l AIPCR sur le Drainage des Chaussées*, PIARC/AIPCR, Grenada, Spain, November, pp. 249–257.

Kassif, G. & Ben Shalom, A., 1971, "Experimental relationship between swell pressure and suction", *Géotechnique*, **21**(3), pp. 245–255.

Komorik, A., Livneh, M. & Smucha, S., 1980, "Shear strength and swelling of clays under suction", *Proc. 4th Int'l Conf. Expansive Soils*, Denver, **1**, pp. 206–226.

Korkiala-Tanttu, L. & Dawson, A.R., 2007, "Relating Full-Scale Pavement Rutting to Laboratory Permanent Deformation Testing", *Int'l. Jnl. Pavement Eng.*, **8**(1), pp. 19–28.

Krarup, J., 1995, "Bearing Capacity and Water, Part III: *Measured Pavement Performance*", Danish Road Institute, note 253.

Laloui, L., Péron, H., Geiser, F., Rifa'i, A. & Vulliet, L., 2006, "Advances in volume measurement in unsaturated triaxial tests", *Soils and Foundations*, **46**(3), pp. 341–350.

Marcial, D., Delage, P. & Cui, Y.J., 2002, "On the high stress compression of bentonites, *Canadian Geotech. J.*, **39**, pp. 812–820.

Pavšič, P., 2006, "Influence of water on mechanical behaviour of soils and aggregates", in *Active geotechnical design in infrastructure development*, edited by Logar, J., Gaberc, A., and Majes, B., Slovensko geotehniško društvo (Slovenian Geotechnical Society), Ljubljana, **2**, pp. 109–111.

Petkovšek, A., Kokot, D., Leben, B. & Pavšič, P., 2003, "*Influence of seasonal moisture and temperature changes in unbound layers on bearing capacity and deformability of pavements: research report (2000–2003)*", Final Report, (Sezonski vplivi spreminjanja vlage in temperature v nevezanih plasteh na nosilnost in deformabilnost voziščne konstrukcije: razvojno-raziskovalna naloga (2000–2003), končno poročilo razvojno raziskovalnega projekta), Gradbeni inštitut ZRMK, Zavod za Gradbeništvo Slovenije, Ljubljana.

Péron, H., Hueckel, T. & Laloui, L., 2007, "An improved volume measurement in determining soil water retention curve", *Geotech. Testing J.*, **30**(1), pp. 1–8.

Qian, L, Ye, W. & Chen, B., 2006, "Soil-Water Characteristic Curves and Permeability of Shanghai Soft Soils", in *Unsaturated Soils 2006*, Proc. 4th Int'l. Conf. Unsaturated Soils, Eds. Miller, G.A., Zapata, C.E., Houston, S.L. & Fredlund, D.G., Am. Soc. Civil Eng'rs., Geotech. Special Publ'n. 147, pp. 1571–1582.

Semmelink, CJ, Jooste, FJ & de Beer, M, 1997, "Use of the K-mould in determination of the elastic and shear properties of road materials for flexible pavements", *8th Int. Conf. Asphalt Pavements*, **II**, August, Seattle, Washington, USA, pp. 1643–58.

Sharp, K.G., Vuong, B.Y., Rollings, R.S., Baran, E. & Metcalf, J., 1999, "The Performance of Lateritic Gravel Pavements", *Proc. 1st Acclerated Pavement Testing Conf.*, Reno, Nevada, Paper CS11-02, 21pp.

Standards Australia, 1995, "*Determination of the Resilient Modulus and Permanent Deformation of Granular Unbound Materials*", Australian Standard, AS 1289.6.8.1.

Vuong, B., Sharp, K.G., Baran, E. & Vertessy, N.J., 1994, "The Performance of a Fine-Grained Marginal Material Under Accelerated Loading," *Proc. 17th Australian Road Research Conf.*, Part 2, Australian Road Research Board, Victoria, Australia, pp. 145–164.

Chapter 11
Modelling Coupled Mechanics, Moisture and Heat in Pavement Structures

Robert Charlier*, **Lyesse Laloui, Mihael Brenčič, Sigurður Erlingsson, Klas Hansson and Pierre Hornych**

Abstract Different physical problems have been analysed in the preceding chapters: they relate to water transfer, to heat transfer, to pollutant transfer and to mechanical equilibrium. All these problems are governed by differential equations and boundary conditions but analytical solutions are, in general, unobtainable because of the complex interaction of the various aspects which are always present in real-world situations. In such circumstances, numerical modelling can give a valuable alternative methodology for solving such highly coupled problems. The first part of this chapter is dedicated to a brief statement of the finite element method for highly coupled phenomena. In the second part, a number of numerical simulations are summarised as an illustration of what could be done with modern tools. The chapter shows that it is possible to achieve realistic results although, at present, some simplification is often required to do so.

Keywords Finite element method · multi-physics coupling · partial differential equation · examples of applications

11.1 Introduction – Problems to be Treated

When trying to replicate *in-situ* behaviour by computational techniques, a number of different physical phenomena (Gens, 2001) need to be considered, including:

- The non-linear solid mechanics and especially granular unbound or bound material mechanics: we consider the relations between displacements, strains, stresses and forces within solids. The material behaviour is described by a constitutive model, which can take into account elasto-plasticity or elasto-visco-plasticity;
- The fluid flow within porous media: fluid can be a single phase of various natures (water, air,...) or it can be an association of two fluids, leading to unsat-

* *Co-ordinating Author:*
✉ R. Charlier
University of Liège, Belgium
e-mail: robert.charlier@ulg.ac.be

A.R. Dawson (ed.), *Water in Road Structures*, DOI 10.1007/978-1-4020-8562-8_11,
© Springer Science+Business Media B.V. 2009

urated media (water and air,...). In the second case, partial saturation leads to permeability and storage terms depending on the saturation degree or on the suction level, involving non-linear aspects;

- The thermal transfers within porous media: conduction is the leading process in a solid (in the geomaterial matrix), but convection can also occur in the porous volume, as a consequence of the fluid flow. Radiation transfer could also occur inside the pores, but it will be neglected here. Conduction coefficients and latent heat may depend on the temperature; and
- The pollutant transport or any spatial transfer of substance due to the fluid flow: the pollutant concentration may be high enough to modify the densities, involving non-linear effects.

All these problems are non-linear ones, and can be formulated with sets of partial differential equations. However, only three types of differential equations have to be considered, concerning respectively:

i) solid mechanics;
ii) diffusion; and
iii) advection-diffusion problems.

11.1.1 Solid Mechanics

On the one hand, solid mechanics can be modelled on the following basis. The equilibrium equation is:

$$\partial_i \sigma_{ij} + P_j = 0 \tag{11.1}$$

Where P_j is a member of $\underline{P}$, the vector of volume forces, σ_{ij} is a member of σ, the Cauchy stress tensor, and ∂ represents the spatial partial derivative operator:

$$\partial_i \equiv \frac{\partial}{\partial x_i} \tag{11.2}$$

The stress tensor is obtained by the time integration of an (elastic, elasto-plastic or elasto-visco-plastic) constitutive equation (see Chapter 9, Section 9.4.2; Laloui, 2001; Coussy & Ulm, 2001):

$$\dot{\sigma}_{ij} = fn(\sigma, \dot{\varepsilon}, Z) \tag{11.3}$$

where $\dot{\sigma}_{ij}$ is the stress rate, $\dot{\varepsilon}$ is the strain rate and Z is a set of history parameters (state variables, like e.g. the preconsolidation stress). In the most classical case of elasto-plasticity, this equation reduces to:

$$\dot{\sigma}_{ij} = E^{ep}_{ijkl}\, \dot{\varepsilon}_{kl} \tag{11.4}$$

where E^{ep}_{ijkl} is a member of the elasto-plastic constitutive (stress-strain) tensor, $\mathbf{E}^{ep}$. Most constitutive relationships for geomaterials are non-linear ones and not as previously introduced in Eq. 9.1 in a linear version.

When modelling a solid mechanics problem with the finite element method, the most commonly used formulation is based on displacements that make up the vector $\underline{l}$ or on actual coordinates that make up the vector $\underline{x}$. If one considers only small strains and small displacements, the strain rate reduces to the well-known Cauchy's strain rate:

$$\dot{\varepsilon}_{ij} = \frac{1}{2}\left(\partial_i l_j + \partial_j l_i\right) \tag{11.5}$$

The time dimension is not addressed for solid mechanics problems, except when a viscous term is considered in the constitutive model. Generally, the time that appears in the time derivatives in Eqs. 11.3, 11.4 and 11.5 is only a formal one.

11.1.2 Diffusion

Thermal conduction exchanges (Chapter 4) in solids and diffusion of contaminants (Chapter 6) are modelled by similar diffusion equations.

The balance equation is written:

$$\partial_i f_i + Q = \dot{S} \tag{11.6}$$

where f_i represents a flux of fluid or heat, Q represents a sink term and S represents the storage of fluid or of heat. When modelling a diffusion problem with the finite element method, the most often used formulation is based on fluid pore pressure, u, or on temperature, T.

Then the Darcy's law for **fluid flow** in porous media gives the fluid flux (this equation has been presented in a slightly different form in Chapter 1 (Eq. 1.2) and in Chapter 2, Eqs. 2.15 and 2.16):

$$f_i = -\frac{K}{\mu}(\partial_i u + \partial_i \rho g z) \tag{11.7}$$

with the intrinsic permeability K (possibly depending on the saturation degree), the dynamic viscosity, μ, the density, ρ, the fluid pressure, u, the altitude, z, and the gravitational acceleration, g. The fluid storage term, S, depends on the saturation degree, S_r, and on the fluid pressure (see Chapter 2, Section 2.7):

$$\dot{S} = fn(u, S_r) \tag{11.8}$$

For **thermal conduction** one obtains Fourier's law – see Eq. 4.1, rewritten here as:

$$f_i = -\lambda \partial_i T \tag{11.9}$$

with the conductivity coefficient, λ. The heat storage (enthalpy) term depends on the temperature, T (Chapter 4, Section 4.4):

$$\dot{S} = fn(T) \tag{11.10}$$

Diffusion of contaminant follows a similar law (Chapter 6, Section 6.3.1). The diffusion problem is non-linear when:

- the permeability depends (directly or indirectly) on the fluid pore pressure;
- the fluid storage is a non-linear function of the pore pressure;
- partial saturation occurs;
- the conductivity coefficient depends on the temperature; and
- the enthalpy is a non-linear function of the temperature.

When the storage term is considered, the time dimension of the problem has to be addressed.

11.1.3 Advection – Diffusion

Transport of pollutant or of heat in porous media is governed by a combination of advection and diffusion (Chapter 6, Section 6.3.1). The advection phenomenon is related to the transport (noted as a flow $\underline{f}_{adv}$) of any substance by a fluid flow, described by the fluid's velocity, $\underline{f}_{diff}^{fluid}$:

$$\underline{f}_{adv} = C \underline{f}_{diff}^{fluid} \tag{11.11}$$

The substance concentration, C, is generally supposed to be small enough not to influence the fluid flow. In porous media, due to the tortuosity of the pore network, and due to the friction, advection is always associated with a diffusion characterised by the diffusion-dispersion tensor, $\mathbf{D}$. Therefore, the total flux of substance is:

$$\underline{f}_{i,adv} = C \underline{f}_{diff}^{fluid} - \lambda \mathbf{D} \partial_i C \tag{11.12}$$

Balance equations and storage equations may be written in a similar way to the one for diffusion problems Eqs. 11.6, 11.8 and 11.10.

Compared to the diffusion constitutive law, Eqs. 11.7 and 11.9, here an advection term appears which doesn't depend on the concentration gradient, but directly on the concentration. This is modifying completely the nature of the equations to be solved. Problems dominated by advection are very difficult to solve numerically (Charlier & Radu, 2001). In order to evaluate the relative advection effect, it is useful to evaluate the Peclet's number, *Pe*, which is the ratio between the diffusive and advective effects:

$$Pe = \frac{f_{diff}^{fluid} L}{2D_h} \tag{11.13}$$

where L is an element dimension and D_h is the hydrodynamic dispersion coefficient (see Chapter 6, Section 6.3.1).

11.1.4 Boundary Conditions

In the preceding section, differential equations were given for three types of problems. In order to solve these equations, we need to define boundary and initial conditions. Classical boundary conditions may be considered: imposed displacements or forces for solid mechanics problems and imposed fluid pressures, temperatures, concentrations or fluxes for diffusion and advection-diffusion problems.

However, it may be useful to consider much more complex boundary conditions. For example, in solid mechanics, unilateral contact with friction or interface behaviour is often to be considered.

When coupling phenomena, the question of boundary conditions increases in complexity and has to be discussed.

11.2 Numerical Tools: The Finite Element Method

11.2.1 Introduction

An approximated solution of most problems described by a set of partial differential equations may be obtained by numerical methods like the finite element method (FEM), the discrete element method (DEM), the finite difference method (FDM), the finite volume method (FVM), or the boundary element method (BEM). For the problems concerned here, the most commonly used methods are the finite element and the finite difference procedures. Commonly, non-linear solid mechanics is better solved using the finite element method. Boundary element methods have strong limitation in the non-linear field. Finite difference methods are not easy to apply to tensorial equations (with the exception of the FLAC code, developed by Itasca).

Diffusion and advection-diffusion problems are often solved by finite difference or finite element methods. Some finite difference (or finite volume) codes are very popular for fluid flow, like e.g. MODFLOW, TOUGH2 (Pruess et al., 1999) for aquifer modelling or ECLIPSE (Schlumberger 2000) for oil reservoir modelling. These codes have been developed over a number of years and possess a number of specific features allowing users to take numerous effects into account. However, they suffer from some drawbacks, which limit their potential for modelling coupled phenomena. Therefore only a little information will be included here concerning finite difference approaches.

11.2.2 Finite Element Method

The basic idea of the finite element method is to divide the field to be analysed into sub-domains, the so-called *finite elements*, of simple shape: e.g. triangles, quadrilaterals with linear, parabolic or cubic sides for two-dimensional analysis. In each finite element, an analytically simple equation is postulated for the variable to be determined, i.e. the coordinate or displacement for solid mechanics, and the fluid pressure, temperature or concentration for diffusion problems. In order to obtain continuity, the unknown variable field has to be continuous at the limit between finite elements. This requirement is obtained thanks to common values of the field at specific points, the so-called *nodes*, which are *linking* the finite elements together. The field values at nodal points are the discretised problem unknowns.

For most solid mechanics and diffusion problems, isoparametric finite elements seem to be optimal (Zienkiewicz et al., 1988). The unknown field $\underline{l}$ (here representing a set of displacements in all directions) may then be written, for solid mechanics cases[1] as:

$$\underline{l} = N_L(\xi, \eta)\,\underline{x}_L \quad L = 1, m \tag{11.14}$$

where m is the number of nodes in the model. This unknown field, $\underline{l}$, then depends on the nodal unknowns $\underline{x}_L$ (not only referring to the x-direction) and on shape functions N_L, which, themselves, depend on the isoparametric coordinates, ξ, η, defined on a reference, normalised, space. The strain rate and the spin may then be derived thanks to Eq. 11.5, the stress rate is obtained by Eqs. 11.3 and 11.4 and is time integrated. Eventually, equilibrium (Eq. 11.1) has to be checked.

For scalar diffusion or advection-diffusion problems, the unknown field, p, representing a general pressure (which could be pore pressure, u (the use here), temperature, T, or concentration, C, by appropriately changing the notation) may then be written:

$$p = N_L(\xi, \eta)\,p_L \quad L = 1, m \tag{11.15}$$

Where p depends on the nodal unknowns, p_L, and on the shape functions, N_L. Then Darcy's fluid velocity and the storage changes may be derived thanks to Eqs. 11.7 and 11.8 (or, respectively, Eqs. 11.9 and 11.10). No time integration is required here. Finally, the balance equation (Eq. 11.6) has to be checked.

The finite element method allows an accurate modelling of the boundary condition, thanks to an easily adapted finite element shape. Internal boundaries of any shape between different geological layers or different solids can be modelled. Specific finite elements for interfacial behaviour or for unilateral boundaries have also been developed (e.g. Charlier & Habraken, 1990). Variations of the finite element

---

[1] For the sake of simplicity, we limit ourselves here to two-dimensional cases.

size and density over the mesh are also easy to manage, with the help of modern mesh generators.

11.2.3 Finite Difference Method

The finite difference method doesn't postulate explicitly any specific shape of the unknown field. As we are concerned with partial differential equations, exact derivatives are replaced by an approximation based on neighbouring values of the unknown (still denoted as p):

$$\left(\frac{\partial p}{\partial x}\right)_i = \frac{p_{i+1} - p_{i-1}}{2L} \tag{11.16}$$

where the subscript i denotes the cell number and L denotes the cell size. For an orthogonal mesh, such derivatives are easily generalised to variable cell dimensions. However, non-orthogonal meshes pose problems that are highly difficult to solve and are generally not used. Boundary conditions have then to be modelled by the juxtaposition of orthogonal cells, giving a kind of stepped edge for oblique or curved boundaries. Similarly, local refinement of the mesh induces irreducible global refinement. These aspects are the most prominent drawbacks of the finite difference method compared to the finite element one. On the other hand computing time is generally much lower with finite differences then with finite elements.

11.2.4 Solving the Non-Linear Problem – The Newton-Raphson Method

Let us now concentrate on the finite element method. The fundamental equation to be solved is the equilibrium Eq. 11.1 (or the balance Eq. 11.6 for diffusion phenomena). As the numerical methods give an approximate solution, the equilibrium/balance equation has to be solved with the best compromise. This is obtained by a global weak form of the local equation. Using weighted residuals, for solid mechanics, one obtains:

$$\int_V \left[\sigma_{ij}\, \delta\varepsilon_{ij}\right] dV = \int_V P_i \delta l_i dV + \int_A \bar{p} \delta l_i dA \tag{11.17}$$

And for diffusion phenomena:

$$\int_V \left[\dot{S} \delta p - f_i \partial_i \left(\delta p\right)\right] dV = \int_V Q \delta p dV + \int_A q \delta p dA \tag{11.18}$$

where $\bar{p}$ and q are surface terms of imposed loads/fluxes. The weighting functions are denoted $\underline{\delta l}$ and δp, and $\underline{\delta \varepsilon}$ represents a derivative of the weighting function based on the Cauchy's strain derivate operator. An equivalent equation could be obtained based on the virtual power principle. The $\underline{\delta l}$ and δp terms would then be interpreted as virtual arbitrary displacements and pressures. Within the finite element method, the global equilibrium/balance equation will be verified for a number of fundamental cases equivalent to the degrees of freedom (d.o.f.) of the problem, i.e. the number of nodes times the number of degrees of freedom per node, minus the number of imposed values. The corresponding weighting functions will have simple forms based on the element shape functions.[2]

Giving a field of stress or of flux, using the weighting functions, one will obtain a value for each d.o.f., which is equivalent to a nodal expression of the equilibrium/balance equation.

More precisely, for solid mechanics problems, one will obtain internal forces equivalent to stresses at each node, L:

$$F_{Li}^{int} = \int_V \sigma_{ij} B_{Lj} dV \tag{11.19}$$

where B_{Lj} is a member of the matrix, $\mathbf{B}$, of derivatives of the shape functions, $\mathbf{N}$. If equilibrium is maintained from the discretised point of view, these internal forces are equal to external forces (if external forces are distributed, a weighting is necessary):

$$F_{Li}^{\text{int}} = F_{Li}^{ext} \tag{11.20}$$

Similarly, for diffusion phenomena the nodal internal fluxes are equivalent to the local fluxes:

$$F_L^{\text{int}} = \int_V \left[\dot{S} N_L - f_i \partial_i N_L\right] dV \tag{11.21}$$

If the balance equation is respected from the discretised point of view, these internal fluxes are equal to external ones:

$$F_L^{int} = F_L^{ext} \tag{11.22}$$

However, as we are considering non linear-problems, equilibrium/balance cannot be obtained immediately, but requires iteration. This means that the equations (Eqs. 11.20 and 11.22) are not fulfilled until the last iteration of each step.

Non-linear problems have been solved for some decades, and different methods have been used. From the present point of view, the Newton-Raphson method is the

[2] This concerns Galerkin's approximation. For advection dominated problems, other weighting functions have to be used.

reference method and probably the best one for a large number of problems. Let us describe the method. In Eq. 11.20 the internal forces F_{Li}^{int} are dependant on the basic unknown of the problem, i.e. the displacement field. Similarly in Eq. 11.22 the internal fluxes are dependant on the pressure (temperature, concentration...) field.

If the external forces/fluxes don't equilibrate, the question to be treated can be formulated in the following manner. Following the Newton-Raphson method, one develops the internal force as a first order Taylor's series around the last approximation of the displacement field:

$$F_{Li}^{int} = F_{Li}^{int}\left(l_{(i)}\right) + \frac{\partial F_{Li}^{int}}{\partial l_{Kj}} dl_{Kj} + O^2 = F_{Li}^{ext}, \tag{11.23}$$

where the subscript (i) indicates the iteration number and O^2 represents second order, infinitely small terms. This is a linearization of the non-linear equilibrium equation. It allows one to obtain a correction of the displacement field:

$$\Delta l_{Kj} = \left(\frac{\partial F_{Li}^{int}}{\partial l_{Kj}}\right)^{-1} \left(F_{Li}^{int}(l_{(i)}) - F_{Li}^{ext}\right) = E_{Li,Kj}\left(F_{Li}^{int}(l_{(i)}) - F_{Li}^{ext}\right) \tag{11.24}$$

Here, the matrix, **E**, represented here by its member term $E_{Li,Kj}$, is the so-called stiffness matrix. With the corrected displacement field, one may evaluate new strain rates, new stress rates, and new improved internal forces. Equilibrium should then be improved.

The same meaning may be developed for diffusion problems using Taylor's development of the internal fluxes with respect to the pressure/temperature/concentration nodal unknowns.

The iterative process may be summarised as shown in Fig. 11.1 for a one-d.o.f. solid mechanics problem. Starting from a first approximation of the displacement field $l_{(1)}$ the internal forces $F^{int}{}_{(1)}$ (point $A^{(1)}$ in the figure) are computed to be lower then the imposed external forces F^{ext}. Equilibrium is then not achieved and a new approximation of the displacement field is sought. The tangent stiffness matrix is

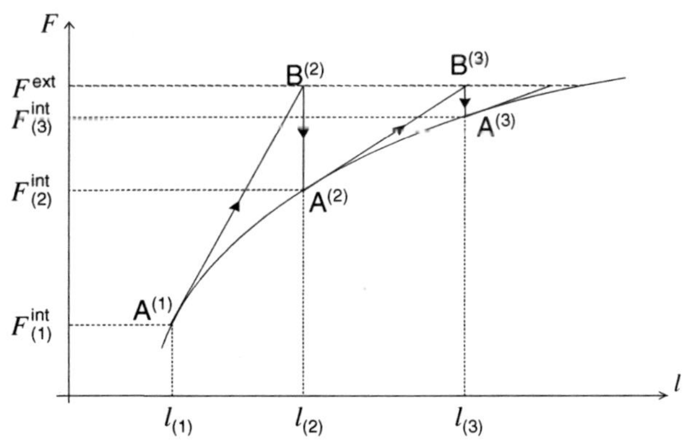

Fig. 11.1 Illustration of the Newton-Raphson process

evaluated and an improved displacement is obtained $l_{(2)}$ (point $B^{(1)}$) (the target being as in Eq. 11.22. One computes again the internal forces ${F^{int}}_{(2)}$ (point $A^{(2)}$) that are again lower then the external forces F^{ext}. As equilibrium is not yet fulfilled, a new approximation of the displacement field is sought, $l_{(3)}$ (point $B^{(2)}$). The procedure has to be repeated until the equilibrium/balance equation is fulfilled with a given accuracy (numerical convergence norm). The process has a quadratic convergence, which is generally considered as the optimum numerical solution.

However the Newton-Raphson method has an important drawback: it needs a large amount of work to be performed as well as to be run on a computer. The stiffness matrix, **E**, is especially time-consuming for analytical development and for numerical inversion. Therefore other methods have been proposed:

- An approximate stiffness matrix, in which some non-linear terms are neglected.
- Successive use of the same stiffness matrix avoiding new computation and inversion at each iteration.

It should be noted that each alternative is reducing the numerical convergence rate. For some highly non-linear problems, the convergence may be lost, and then no numerical solution will be obtained.

Some other authors, considering the properties and the efficiency of explicit time schemes for rapid dynamic problems (e.g. for shock modelling) add an artificial mass to the problem in order to solve it as a quick dynamic one. It should be clear that such a technique might degrade the accuracy of the solution, as artificial inertial effects are added and the static equilibrium Eq. 11.1 is not checked.

11.2.5 The Stiffness Matrix

From Eq. 11.24, it appears that the stiffness matrix is a derivative of the internal forces:

$$E_{Li,Kj} = \frac{\partial F_{Li}^{int}}{\partial l_{Kj}} = \frac{\partial}{\partial l_{Kj}} \left(\int_V \sigma_{ij} B_{Lj} dV \right) \tag{11.25}$$

	1	2
1	Derivative of problem **1** nodal forces with respect to problem **1** nodal unknowns	Derivative of problem **1** nodal forces with respect to problem **2** nodal unknowns
2	Derivative of problem **2** nodal forces with respect to problem **1** nodal unknowns	Derivative of problem **2** nodal forces with respect to problem **2** nodal unknowns

Fig. 11.2 Illustrative layout of stiffness matrix

Two contributions will be obtained (Fig. 11.2). On the one hand, one has to derive the stress state with respect to the strain field, itself depending on the displacement field. On the other hand, the integral is performed on the volume, and the B matrix depends on the geometry. If we are concerned with large strains and if we are using the Cauchy's stresses, geometry is defined in the current configuration, which is changing from step to step, and even from one iteration to the other. These two contributions, the material one, issued from the constitutive model, and the geometric one, have to be accurately computed in order to guarantee the quadratic convergence rate.

A similar discussion may be given for diffusive problems. However, the geometry is not modified for pure diffuse problems, so only the material term is to be considered.

11.2.6 Transient Effects: The Time Dimension

The time dimension appears in the form of a first order time derivative in the constitutive mechanical model (Eqs. 11.3, 11.4) and in the diffusion problems though the storage term (Eq. 11.6). We will here discuss the time integration procedure and the accuracy and stability problems that are involved.

11.2.6.1 Time Integration – Diffusion Problems

The period to be considered is divided into time steps. Linear development of the basic variable with respect to the time is generally considered within a time step:

$$p = \frac{t - t_A}{t_B - t_A} p_B + \frac{t_B - t}{t_B - t_A} p_A \qquad (11.26)$$

where the subscripts A, B denote, respectively, the beginning and the end of a time step. Then the pressure rate is:

$$\dot{p} = \frac{dp}{dt} = \frac{p_B - p_A}{t_B - t_A} = \frac{\Delta p}{\Delta t} \qquad (11.27)$$

This time discretisation is equivalent to a finite difference scheme. It allows the evaluation of any variable at any time within a time step.

The balance equation should ideally be satisfied at any time during any time step. Of course this is not possible for a discretised problem. Only a mean assessment of the balance equation can be obtained. Weighted residual formulations have been proposed in a similar way as for finite elements (Zienkiewicz et al., 1988). However, the implementation complexity is too high with respect to the accuracy. Then the easiest solution is to assess only the balance equation at a given time, denoted t_τ, inside the time step t_A to t_B, such that a time variable, τ, is defined:

$$\tau = \frac{t_\tau - t_A}{t_B - t_A} \tag{11.28}$$

All variables have then to be evaluated at the reference time, t_τ. Different classical schemes have been discussed for some decades:

- Fully explicit scheme – $\tau = 0$: all variables and the balance are expressed at the time step beginning, where everything is known (from the solution of the preceding time step). The solution is, therefore, very easily obtained.
- Crank-Nicholson scheme or mid-point scheme – $\tau = 1/2$
- Galerkin's scheme – $\tau = 2/3$
- Fully implicit scheme – $\tau = 1$

The last three schemes are functions of the pore pressure/temperature/concentration at the end of the time step, and may need to be solved iteratively if non-linear problems are considered.

For some problems, phase changes, or similar large variations of properties, may occur abruptly. For example, icing or vaporising of water is associated with latent heat consumption and abrupt change of specific heat and thermal conductivity. Such rapid change is not easy to model. The change in specific heat may be smoothed using an enthalpy formulation, because enthalpy, H, is an integral of the specific heat, c. Then the finite difference of the enthalpy evaluated over the whole time step gives a mean value, $\bar{c}$, and so allows an accurate balance equation:

$$H = \int_T c dT \tag{11.29}$$

$$\bar{c} = \frac{H_B - H_A}{t_B - t_A} \tag{11.30}$$

11.2.6.2 Time Integration – Solid Mechanics

For solid mechanics problems, the constitutive law form (Eqs. 11.3 and 11.4) is an incremental one and differs from the ones for diffusion problems (Eq. 11.7). The knowledge of the stress tensor at any time implies that a time-integrated constitutive law is required. The stress tensor is a state variable that is stored and transmitted from step to step based on its final/initial value, and this value plays a key role in the numerical algorithm.

Then, in nearly all finite element codes devoted to modelling, equilibrium is expressed at the end of the time steps, following a fully implicit scheme ($\tau = 1$), and using the end of step stress tensor value.

However, integrating the stress history with enough accuracy is crucial for the numerical process stability and global accuracy. Integration of the first order differential equation (Eq. 11.4):

$$\underline{\sigma}^B = \underline{\sigma}^A + \int_{t_A}^{t_B} \mathbf{E}^{ep} \dot{\underline{\varepsilon}}\, dt \tag{11.31}$$

can be based on similar concepts as the one described in the preceding paragraph (the superscripts of σ here indicating the time at which σ is evaluated). Various time schemes based on different τ values may be used for which similar reflections on stability and accuracy can be made.

When performing large time steps, obtaining enough accuracy can require the use of sub-stepping: within each global time step (as regulated by the global numerical convergence and accuracy problem) the stress integration is performed at each finite element integration point after division of the step into a number a sub-steps allowing higher accuracy and stability.

11.2.6.3 Scheme Accuracy

The theoretical analysis of a time integration scheme accuracy and stability is generally based on a simplified problem (Zienkiewicz et al., 1988). Let us consider diffusion phenomena restricted to the linear case. Introducing the discretised field (Eq. 11.15) into the constitutive equations gives Darcy law (Eq. 11.7) (neglecting here the gravity term for the sake of simplicity and using the more general pressure p in place of u) in the following form:

$$f_i = -\frac{K}{\mu}\partial_i p = -\frac{K}{\mu}(\partial_i N_L) p_L \tag{11.32}$$

Similarly the storage law (linear case) can be re-written:

$$\dot{S} = r\dot{p} = rN_L \dot{p}_L \tag{11.33}$$

where r is a storage parameter (cf. Eq. 11.8). Neglecting source terms, the weak form of the balance equation, Eq. 11.21, then produces:

$$\begin{aligned} &\int_V \left[\dot{S}\delta p - f_i \partial_i(\delta p)\right] dV = \\ &\int_V rN_L \dot{p}_L N_K \delta p_K dV - \int_V -\frac{K}{\mu}\partial_i N_L p_L \partial_i N_K \delta p_K dV = 0 \end{aligned} \tag{11.34}$$

Considering that nodal values are not affected by the integration, this becomes:

$$\begin{aligned} &\left(\int_V rN_L N_K dV\right)\dot{p}_L \delta p_K + \left(\int_V \frac{K}{\mu}\partial_i N_L \partial_i N_K dV\right) p_L \delta p_K = \\ &R_{KL}\dot{p}_L \delta p_K + K_{KL} p_L \delta p_K = 0 \end{aligned} \tag{11.35}$$

which is valid for any arbitrary perturbation $\underline{\delta p}$. Thus:

$$R_{KL}\dot{p}_L + K_{KL}p_L = 0 \tag{11.36}$$

which is a simple system of linear equations with a time derivative, a storage matrix **R** (of which R_{KL} is an element) and a permeability matrix **K** (of which K_{KL} is an element). One can extract the eigenvalues of this system and so arrive at a series of scalar independent equations of similar form:

$$\dot{p}_L + \alpha_L^2 p_L = 0 \text{ (no summation)} \tag{11.37}$$

where L represents now the number of the eigenmode with the eigenvalue α_L although it will be omitted in the following. The exact solution for Eq. 11.35 is a decreasing exponential function of time, t:

$$p(t) = p(t_0)e^{-\alpha^2 t} \tag{11.38}$$

This problem represents the damping of a perturbation for a given eigenmode. Numerically, the modelling is approximated and numerical errors always appear. If Eq. 11.36 is well modelled, any numerical error will be rapidly damped, if the error source is not maintained. Following this analysis, the whole accuracy and stability discussion may be based on these last scalar Eqs. 11.37 and 11.38.

Introducing the time discretisation (Eqs. 11.26 and 11.27) into Eq. 11.37 gives:

$$\frac{p_B - p_A}{\Delta t} + \alpha^2 \left[(1 - \tau)p_A + \tau p_B \right] = 0 \tag{11.39}$$

which allows evaluation of the end-of-step pressure as a function of the pressure at the beginning of the step:

$$p_B = U p_A \tag{11.40}$$

with the amplification factor, U:

$$U = \frac{1 - (1 - \tau)\alpha^2 \Delta t}{1 + \tau \alpha^2 \Delta t} \tag{11.41}$$

To ensure the damping process of the numerical algorithm, which is the *stability condition*, it is strictly necessary that the amplification factor remains lower than unity:

$$-1 < U < 1 \tag{11.42}$$

This condition is always verified if $\tau \geq 1/2$, and conditionally satisfied otherwise:

$$\Delta t \frac{2}{(1 - 2\tau)\alpha^2} \quad \text{if } \tau < 1/2 \tag{11.43}$$

This last equation is not easy to verify, as it depends on the eigenvalues, which are generally not computed. Therefore, for classical diffusion processes considered in geomaterials, the condition $\tau \geq 1/2$ is generally used.

It should be noted that the amplification factor becomes negative for large time steps, except for the fully implicit scheme. Then, the perturbed pressure decreases monotonically in amplitude but with changes of sign. This may be questionable for some coupled phenomena, as it could induce oscillation of the coupled problem.

Let us now consider the accuracy of the numerical schemes. Developing in Taylor's series the exact and numerical solution allows a comparison:

$$\begin{aligned} U_{\text{exact}} &= 1 - x + \frac{1}{2}x^2 - \frac{1}{6}x^3 + \ldots \\ U_{\text{numerical}} &= 1 - x + \theta x^2 - \theta^2 x^3 + \ldots x = \alpha^2 \Delta t \end{aligned} \tag{11.44}$$

It appears that only the Crank-Nicholson scheme $\tau = 1/2$ has second order accuracy properties. However this conclusion is limited to infinitesimal time steps. For larger time steps, as in most numerical models, the Galerkin's scheme $\tau = 2/3$ gives the optimal compromise and should generally be used.

The whole discussion related to the stability and accuracy of the proposed time numerical schemes was based on eigenmodes of a linear problem. Can we extrapolate them to general problems? The eigenvalue passage is only a mathematical tool allowing consideration of scalar problems, and has no influence on our conclusions. Conversely, the non-linear aspects could sometimes modify our conclusions. However, it is impossible to develop the analysis for a general non-linear problem, and the preceding conclusions should be adopted as guidelines, as they appear to be fruitful in most cases.

11.2.7 Advection Diffusion Processes

Let us first consider a purely advective process. In this case, the transport is governed by the advection Eq. 11.11 and by the balance Eq. 11.6. Associating these two equations, one obtains:

$$(\underline{\nabla}^T C).\underline{f}_{diff}^{fluid} + \dot{C} = 0 \tag{11.45}$$

which is a hyperbolic differential equation. It cannot be solved by the finite element or finite difference problem, but by characteristic methods. The idea is to follow the movement of a pollutant particle by simply integrating step by step the fluid velocity field. This integration has to be accurate enough, as errors are cumulated from one step to the next.

On the other hand, if advection is very small compared to diffusion, then the finite element and finite difference methods are really efficient.

For most practical cases, an intermediate situation holds. It can be checked by Peclet's number, Eq. 11.13, which is high for mainly advective processes and low for mainly diffusive ones. As diffusion has to be taken into account, the numerical solution must be based on the finite element method (the finite difference one may also be used but will not be discussed here). However, numerical experiments show that the classical Galerkin's formulation gives very poor results with high spatial oscillations and artificial dispersion. Thus, new solutions have been proposed (Zienkiewicz & Taylor, 1989, Charlier & Radu, 2001). A first solution is based on the use in the weighted residual method of a weighting function that differs from the shape one by an upwind term, i.e. a term depending in amplitude and direction on the fluid velocity field. The main advantage of this method is the maintenance of the finite element code formalism. However, it is never possible to obtain a highly accurate procedure. Numerical dispersion will always occur.

Other solutions are based on the association of the characteristic method for the advection part of the process and of the finite element method for the diffusive part (Li et al., 1997). The characteristic method may be embedded in the finite element code, which has a strong influence on the finite element code structure. It is also possible to manage the two methods in separated codes, as in a staggered procedure (cf. Section 11.3.4).

11.3 Coupling Various Problems

11.3.1 Finite Element Modelling: Monolithical Approach

Modelling the coupling between different phenomena should imply the need to model each of them and, simultaneously, all the interactions between them. A first approach consists in developing new finite element and constitutive laws especially dedicated to the physical coupled problem to be modelled. This approach allows taking accurately all the coupling terms into account. However there are some drawbacks that will be discussed in Section 11.3.4. Constitutive equations for coupled phenomena will be discussed in the following sections.

The number of basic unknowns and, consequently, the number of degrees of freedom – d.o.f. – per node are increased. This has a direct effect on the computer time used for solving the equation system (up to the third power of the total d.o.f. number). Coupled problems are highly time consuming.

Isoparametric finite elements will often be considered. However, some specific difficulties may be encountered for specific problems. Nodal forces or fluxes are computed in the same way as for decoupled problems. However, the stiffness matrix evaluation is much more complex, as interactions between the different phenomena are to be taken into account. Remember that the stiffness or iteration matrix, Eq. 11.25, is the derivative of internal nodal forces/fluxes with respect to the nodal unknowns (displacements/pressures/etc...). The complexity is illustrated by the following scheme of the stiffness matrix, restricted to the coupling between two problems.

The part of the stiffness matrix in cells 1-1 and 2-2 are similar, or simpler, than the ones involved in uncoupled problems. The two other cells, 1-2 and 2-1, are new and may be of a greater complexity. Remember also that the derivative considers internal nodal forces/fluxes as obtained numerically, i.e. taking into account all numerical integration/derivation procedures. On the other hand, the large difference of orders of magnitude between different terms may cause troubles in solving the problem and so needs to be checked.

Numerical convergence of the Newton-Raphson process has to be evaluated carefully. It is generally based on some norms of the out-of-balance forces/fluxes. However, coupling often implies the mixing of different kinds of d.o.f., which may not be compared without precaution. Convergence has to be obtained for each basic problem modelled, not only for one, which would then predominate in the computed indicator.

11.3.2 Physical Aspects: Various Terms of Coupling

A large number of different phenomena may be coupled. It is impossible to discuss here all potential terms of coupling, and we will restrict ourselves to some basic cases often implied in environmental geomaterial mechanics. In the following paragraphs, some fundamental aspects of potential coupling are briefly described.

11.3.2.1 Hydro-Mechanical Coupling

In the case of hydro-mechanical coupling, the number of d.o.f. per node will be 3 (2 displacements +1 pore pressure) for 2D analysis and 4 (3 displacements +1 pore pressure) for 3D analysis.

Coupling mechanical deformation of soils or rock mass and water flow in pores is a frequent problem in geomechanics. The first coupling terms are related to the influence of pore pressure on mechanical equilibrium through Terzaghi's postulate (or through any other effective stress concept or net stress use, cf. Eq. 9.20):

$$\boldsymbol{\sigma} = \boldsymbol{\sigma}' + u\mathbf{I} \tag{11.46}$$

with the effective stress tensor $\boldsymbol{\sigma}'$ related to the strain rate tensor thanks to the constitutive Eq. 11.3, and the identity tensor **I**.

The second type of coupling concerns the influence of the solid mechanics behaviour on the flow process, which comes first through the storage term. Storage of water in saturated media is mainly due to pores strains, i.e. to volumetric changes in the soil/rock matrix:

$$\dot{S} = \dot{\varepsilon}_v \tag{11.47}$$

Another effect, which may be considered, is the permeability change related to the pore volume change, which may, for example, be modelled by the Kozeny-Carman law as a function of the porosity $K = K(n)$.

Biot proposed an alternative formulation for rocks where contacts between grains are much more important than in soils. Following Biot, the coupling between flow and solid mechanics are much more important (Detournay & Cheng, 1991; Thimus et al., 1998).

The time dimension may cause some problems. First, an implicit scheme is used for the solid mechanics equilibrium and various solutions are possible for the pore pressure diffusion process. Consistency would imply use of fully explicit schemes for the two problems. Moreover, it has been shown that time oscillations of the pore pressure may occur for other time schemes. Associated to Terzaghi's postulate, oscillations could appear also on the stress tensor, which can degrade the numerical convergence rate for elasto-plastic constitutive laws.

When using isoparametric finite elements, the shape functions for geometry and for pore pressure are identical. Let us consider, for example, a second order finite element. As the displacement field is of second order, the strain rate field is linear. For an elastic material, the effective stress tensor rate is then also linear. However, the pore pressure field is quadratic. Then Terzaghi's postulate mixes the linear and quadratic fields, which is not very consistent. Some authors have then proposed to mix in one element quadratic shape functions for the geometry and linear shape functions for pore pressure. But then problems arrive with the choice of spatial integration points (e.g. should there be 1 or 4 Gauss points?).

Numerical locking problems may also appear for isoparametric finite elements when the two phase material (water plus soil) is quite incompressible, i.e. for very short time steps with respect to the fluid diffusion time scale. Specific elements have to be developed for such problems.

11.3.2.2 Flow of Two Fluids in Rigid Porous Media

The coupled flow of two, different, fluids in a partly saturated rigid media is now considered. Unsaturated soils provide a common example, where the two fluids are water and air. Often, the air phase is considered to be at constant pressure, which is generally a relevant approximation as air pressure doesn't highly affect the water flow. In such an arrangement only one d.o.f. per node is sufficient, and the classical diffusion equations (see Eq. 6.10) are relevant, with parameters depending on the suction or saturation level.

Mixing between different fluids is sometimes possible. Then two or more d.o.f. per node are to be considered. The permeability and storage equations of each phase depend on the suction or saturation level, and so the problem may be highly nonlinear. However, it is not difficult to numerically develop coupling, as the formulation is similar for each phase.

11.3.2.3 Diffusion and Transport Coupling

Heat and single fluid flow in a rigid porous media are considered here. The fluid specific weight and viscosity are dependant on the temperature or salt concentration, and the heat or salt transport by advection–diffusion process is dependant on the fluid flow. Then a diffusion process and an advection–diffusion process have to be solved simultaneously. In this case there are 2 d.o.f. per node: fluid pore pressure and salt concentration or temperature.

11.3.3 Thermo-Hydro-Mechanical Coupling

The phenomena considered in this section are much more complex as they associate multiphase fluid flow and hydro-mechanical coupling (cf. relevant sub-sections of Section 11.3.2) as well as temperature effects. All the features described in the preceding section are to be considered here, associated with some new points.

Heat diffusion has to be modelled. Temperature variation affects fluid flow, by a modification of the fluid specific weight or viscosity. Moreover, if the two fluids concerned are a liquid and a gas (e.g. water and air), then equilibrium between the phases has to be modelled: dry air – vapour equilibrium.

Heat transfer is governed not only by conduction but also by advection by the liquid and gas movements. Similarly, transfers of vapour and dry air in the gas phase are not only governed by diffusion and gradient of species density, but also by advection of the global gas movements.

In such an analysis, the total number of d.o.f. per node is 5 for a 2D problem: 2 displacements, 2 fluid pore pressures and the temperature.

11.3.4 Finite Element Modelling: Staggered Approach

The monolithic approach of coupled phenomena implies identical space and time meshes for each phenomenon. This is not always possible, for various reasons. The coupled problems may have different numerical convergence properties, generally associated with different physical scales or non-linearities. For example, a coupled hydro-mechanical problem may need large time steps for the fluid diffusion problem, in order to allow, in each step, fluid diffusion over a long distance (of the order of magnitude of the finite elements). At the same time, strong non-linearities may occur in solid mechanics behaviour (strong elasto-plasticity changes, interface behaviour, strain localisation...) and then the numerical convergence will need short time-loading steps, which should be adapted automatically to the rate of convergence. Then, it is quite impossible to obtain numerical convergence for identical time and space meshes.

Research teams of different physical and numerical culture have progressively developed different modelling problems. As an example, fluid flow has been largely developed using the finite difference method for hydrogeology problems including pollutant transport, and for oil reservoir engineering (see Section 11.2.3) taking multiphase fluid flow (oil, gas, condensate, water,...) into account. Coupling such fluid flow with geomechanics in a monolithical approach would imply implementation of all the physical features already developed respectively in finite elements and finite differences codes. The global human effort would be very large!

Coupled problems generally present a higher non-linearity level then uncoupled ones. Thus, inaccuracy in parameters or in the problem idealisation may cause degradations of the convergence performance. How can we solve such problems and obtain a convincing solution? First of all, a good strategy would be to start with the uncoupled modelling of the leading process, and to try to obtain a reasonable first approximation. Then, one can add a first level of coupling and complexity, followed by a second one ... until the full solution is obtained.

However such a "trick" is not always sufficient. Staggered approaches may then give an interesting solution. In a staggered scheme, the different problems to be coupled are solved separately, with (depending on the cases) different space or time mesh, or different numerical codes. However, the coupling is ensured thanks to transfer of information between the separated models at regular meeting points. This concept is summarised in Fig. 11.3. It allows, theoretically, coupling of any models.

When using different spatial meshes, or when coupling finite elements and finite differences codes, the transfer of information often needs an interpolation procedure, as the information to be exchanged is not defined at the same points in the different meshes.

The accuracy of the coupling scheme will mainly depend on the information exchange frequency (which is limited by the lowest time step that can be used) and by the type of information exchanged. The stability and accuracy of the process has been checked by different authors (Turska & Schrefler, 1993; Zienkiewicz et al., 1988). It has been shown that a good choice of the information exchange may highly improve the procedure efficiency.

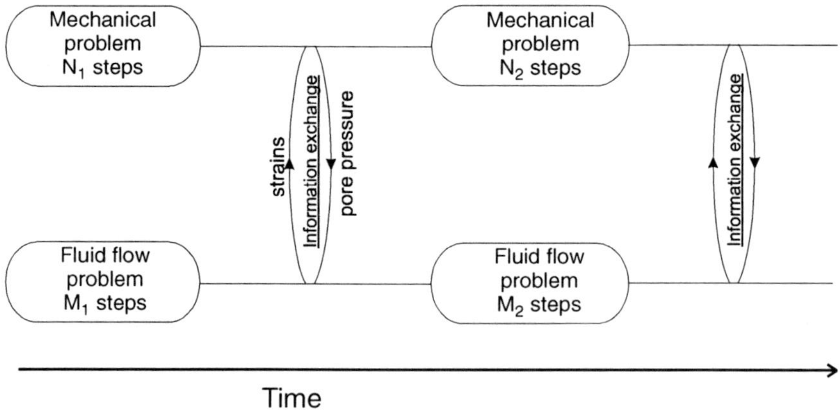

Fig. 11.3 Scheme of a staggered coupling

11.4 Examples

11.4.1 Modelling of Moisture Movements

Alonso (1998) presents, on the basis of in-situ measurements, relevant aspects of the water content development of pavement layers and its effect on the mechanical characteristics of granular bases and subgrades. A general model for the coupled analysis of transfer processes (water, heat) and stress-strain behaviour of unsaturated compacted soils is then presented. A review of some representative properties of compacted soils has been carried out from the perspective of modern concepts of unsaturated soil mechanics. As an application of the methods described a full simulation of a modern pavement structure under the effects of a Mediterranean climate has been developed. The last section of Alonso's paper is devoted to collapse and swelling phenomena of subgrades. Collapse is found in some natural lightly cemented soils but it is more common in embankments compacted on the dry side of the optimum water content. A real case involving severe collapse deformations and the role of described models to analyse the problem is presented. Finally, the available techniques to design and analyse the behaviour of highways on expansive soils are presented.

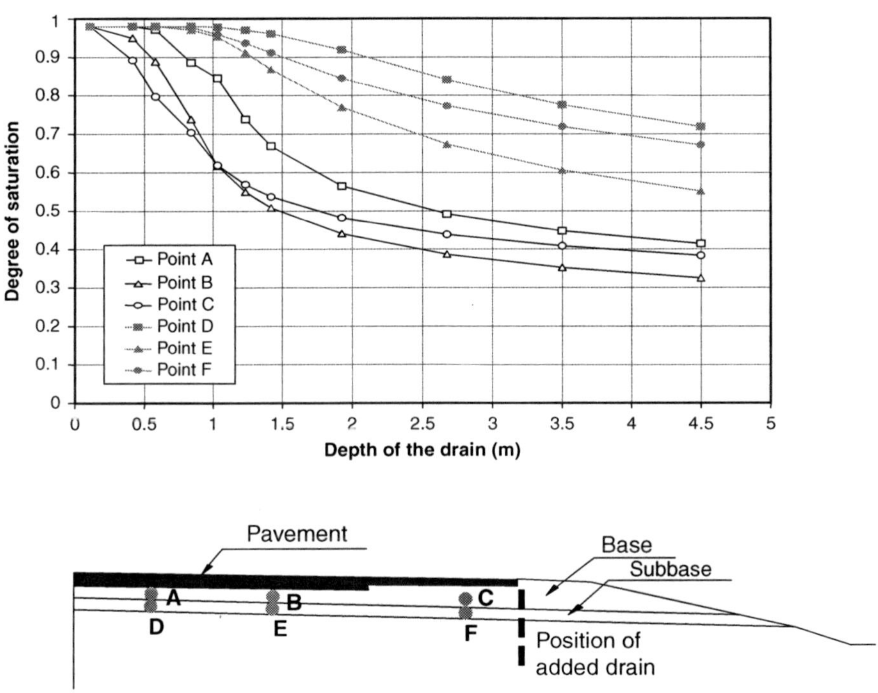

Fig. 11.4 Degree of saturation at equilibrium when a drain is installed at the pavement – shoulder contact, adapted from Alonso et al. (2002)

Alonso et al. (2002) also presented an analysis of the optimum position and depth of longitudinal drains in pavements. An analysis of different climates on the overall pavement behaviour is then given. Three climates have been defined: Tropical, Mediterranean and Sub-alpine (see Chapter 1, Section 1.11), which were defined on the basis of actual data involving rainfall, temperature and relative humidity records. Five years of climate were simulated and the reaction of the selected pavement structure and drainage position were computed and discussed. Figure 11.4 shows some of the computed results. It can be seen that the addition of longitudinal drains has a profound effect on the granular base and sub-base saturation over time, whereas their effect on the subgrade, Fig. 11.5, was found to be more limited. Longitudinal

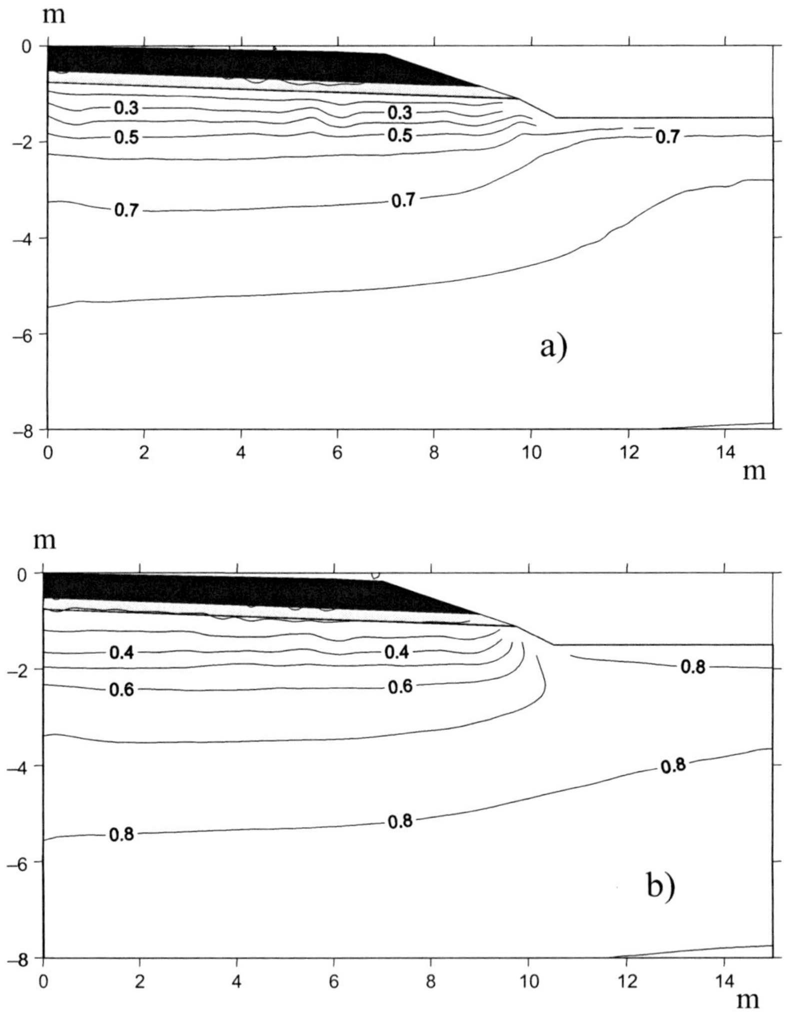

Fig. 11.5 Distribution of degree of saturation for a pavement in a Mediterranean climate. Evaporation through pavement allowed. (**a**) 1st July, (**b**) 1st December, adapted from Alonso et al. (2002)

drains are capable of maintaining a fairly well drained subgrade platform under a Mediterranean climate (as illustrated) whereas their effect was found to be more limited under sub-alpine or tropical climates.

11.4.2 Simulating the Infiltration and Percolation in a Road After Rainfall

Hansson et al. (2005) made an attempt to illustrate the effect of a rain shower and fracture zone permeability on the subsurface flow pattern using a two-dimensional computer model; thus making the simulation domain more like reality (Fig. 11.6). The properties of the materials used in the various layers of the model road fulfil the requirements of the Swedish road design guide. In addition, it was assumed that the asphalt layer of the road had plenty of fractures over a relatively short distance, "a fracture zone", which thus enabled the use of an equivalent homogeneous porous media model. More details about material properties, driving data etc. can be found in Hansson (2005).

Notice (Fig. 11.6) that the fracture zone captures most, if not all, of the upstream surface runoff for the light rainfall event. The heavier rainfall causes a significantly larger infiltration in the road shoulder since the infiltration capacity of the fracture zone, or the granular base layer beneath it, was exceeded. As a consequence, a larger fraction of the total surface runoff reached the road shoulder, and the region of the roadside where both rainfall and surface runoff infiltrated was considerably expanded laterally. This result is qualitatively supported by the findings of Flyhammar and Bendz (2003) who measured concentrations of various solutes in the shoulder and beneath the asphalt cover in a Swedish road partly built with alternative materials, generated from waste and residuals. These materials contained plenty of solutes, and the leaching pattern is similar to the simulated water flow pattern, although the leaching patterns exhibited a large variability between solutes.

11.4.3 Freezing Induced Water Flow

The significance of the coupling between heat and water transport will be illustrated using a freezing experiment performed by Mizoguchi (1990). He packed four identical cylinders with Kanagawa sandy loam. Each cylinder was 20 cm long and had an internal diameter of 8 cm. The samples were prepared for the freezing test by bringing them to the same initial state involving a uniform temperature of 6.7 °C and a close to uniform volumetric water content of 0.33 throughout the cylinders. Water and soil in each cylinder was subjected to freezing from the top down, since their top covers were exposed to a circulating fluid with a temperature of −6 °C. One cylinder was used to obtain initial values, and the other three were removed from freezing after 12, 24 and 50 hours respectively. The cylinders were then cut into 1 cm thick slices for which the total water content (ice + liquid water) was

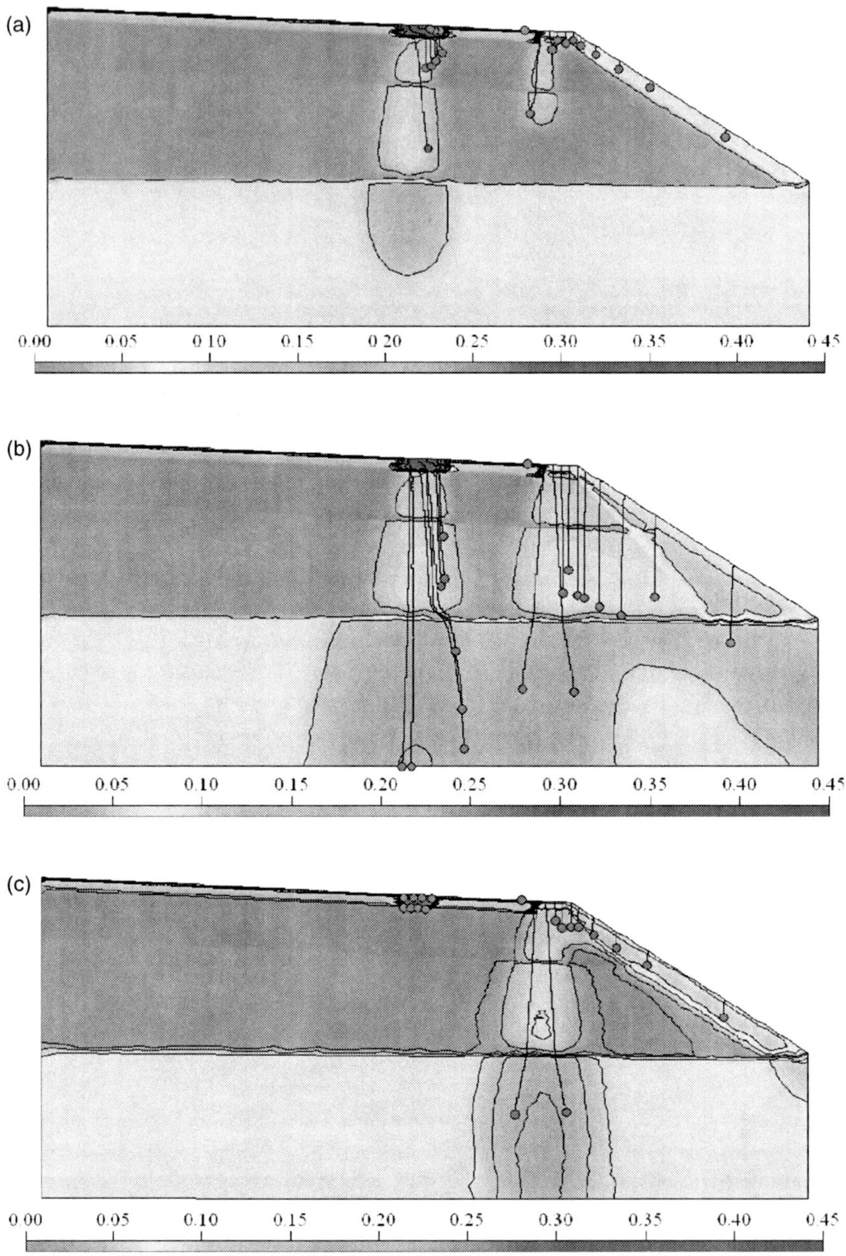

Fig. 11.6 The simulated volumetric water content, θ, in a model road (**a**) after a light rainfall event (*top*), (**b**) after a heavy rainfall event (*middle*), (**c**) after a moderate rainfall event using small fracture-zone permeability (*bottom*) The particles illustrate the flow paths of the infiltrated rainwater. The vertical scale is exaggerated for clarity

determined. The experimental procedure thus described was then reproduced in a computer model in order to test the model.

As described in Chapter 4, Section 4.6, water flows towards freezing fronts where it changes phase from liquid to solid. This process is clearly evident in Fig. 11.7 where the total water content in the upper half of the cylinder increases as the column freezes (Hansson et al., 2004). Since freezing is a relatively quick process, extremely high hydraulic gradients emerge and can lead to sometimes very rapid upward flow of water. The freezing front is clearly visible in Fig. 11.7 as the depth interval where the total water content decreases rapidly. The calculated results are in fair agreement with the measured values. Specifically, the rapid decrease in the total water content at, or immediately below, the freezing front and the gradual recovery deeper in the columns is well predicted.

It is the dramatic redistribution of water caused by the freezing that causes frost heaving, which may damage roads even though the largest problems occur in connection with thaw weakening. It should, however, be pointed out that the computer

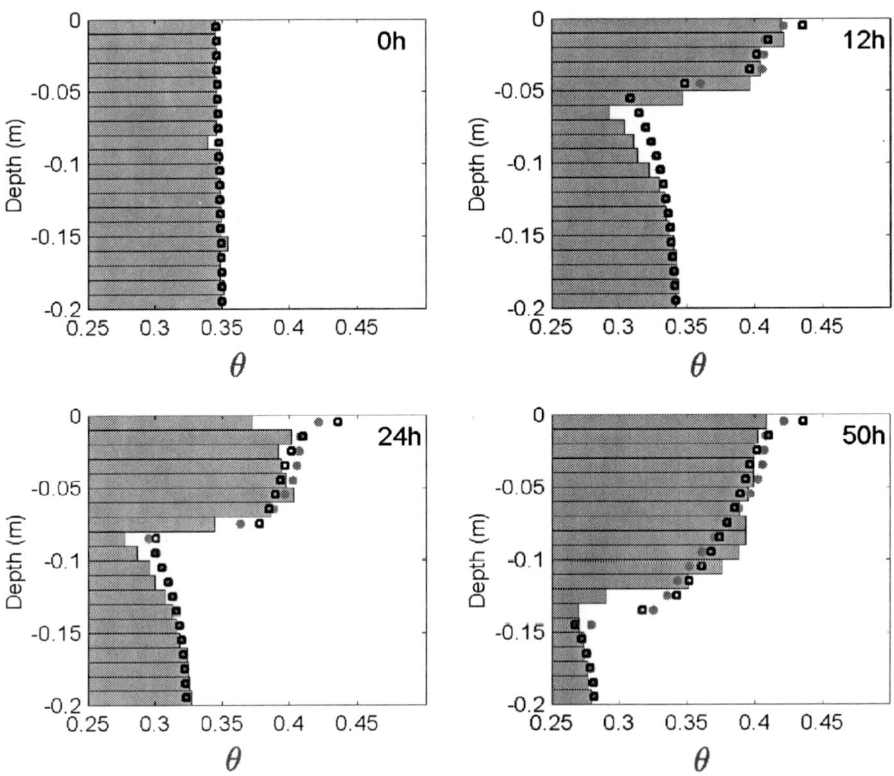

Fig. 11.7 Simulated (*symbols*) and measured values (*horizontal bars*) of the total volumetric water content 0, 12, 24 and 50 h after freezing started. A variable convective heat transfer coefficient, h_c, was used for the first simulation (*solid circles*) and a heat leakage bottom boundary for the second (*open circles*). Simulated values were averaged over 1-cm intervals

model used here neglects effects of frost heave. If the conditions for frost heave had been met during the simulation, the result had been different since the liquid pressure head would have changed as an effect of a relative ice pressure not equal to zero.

11.4.4 Numerical Simulation of Pavements Behaviour from Accelerated Tests

Erlingsson (2007) describes two thin pavement structures that were tested in accelerated testing by using a Heavy Vehicle Simulator. Both were surface dressed structures, one with 20 cm thick unbound base course layer and the other with the base course divided into a 10 cm bitumen stabilized base over 10 cm unbound base. Both structures were instrumented to estimate deflections, strains and stresses in various locations inside the structure. A numerical analysis was also carried out to simulate the response behaviour of the structure that could be compared with the actual measurements. The simulation was performed using different techniques: 3D and 2D axi-symmetric analyses, finite element and multi layer elastic theory, linear elastic and non-linear elastic base behaviour. The results were further used to model the permanent deformation development in each layer. A cross section of the two structures is shown in Fig. 11.8.

Figure 11.9 shows the induced vertical stress under the centre of a single tyre load for both structures where the axle load is 120 kN, or close to one conventional axle load (11.5 ton in the EU), and the tyre pressures is 800 kPa.

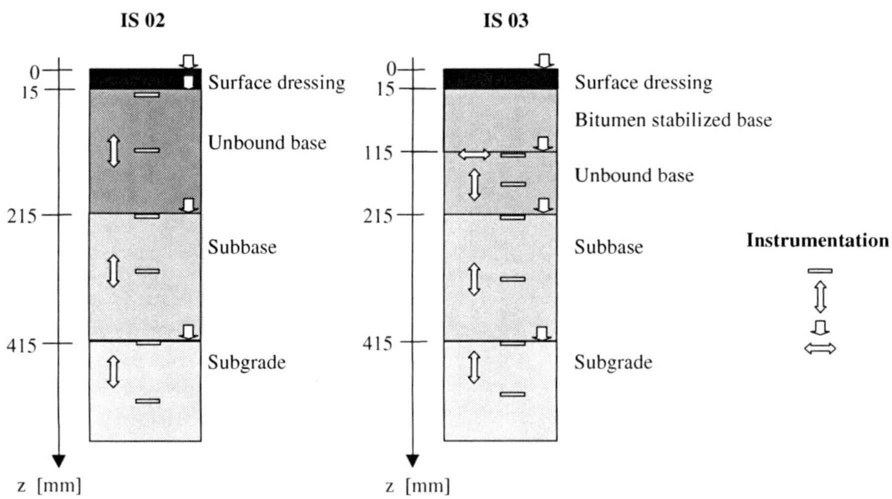

Fig. 11.8 Two test pavement structures: (**a**) IS02 is an unbound structure and (**b**) IS03 is a bitumen stabilized structure
Note: The instrumentation used for the response measurements is shown as well.

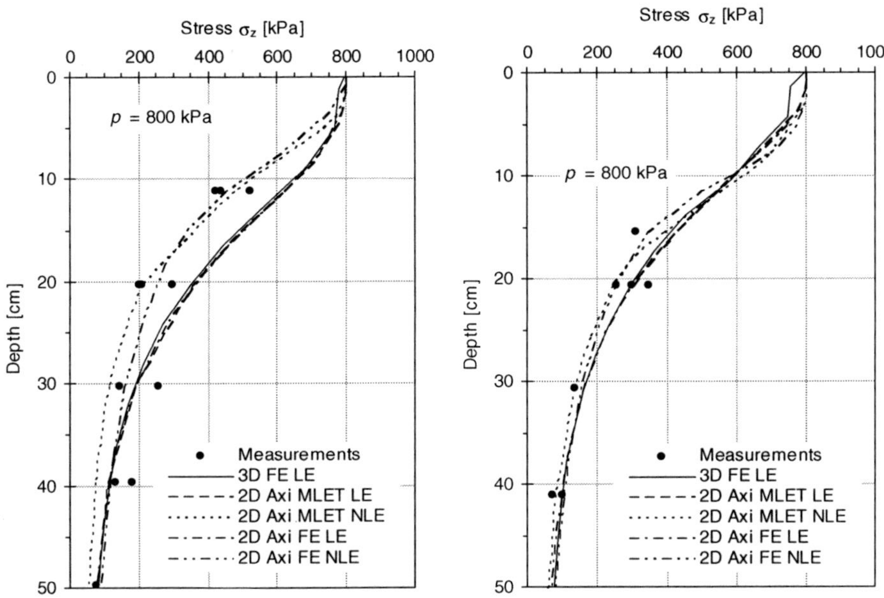

Fig. 11.9 Comparison of measured and calculated vertical induced stresses under the centre of a single tyre as a function of depth for both pavement structures IS02 and IS03
Note: (a) structure IS02 is the unbound structure; (b) structure IS03 is a bitumen stabilized structure. The numerical simulation is carried out using different techniques where 3D = three dimensional analysis, 2D Axi = two dimensional axi-symmetric analysis, FE = finite element, MLET = multi layer elastic theory, LE = elastic behaviour and NLE = non-linear elastic base behaviour.

One can see in Fig. 11.9a the importance of taking into account the non-linear base behaviour for the unbound structure IS02. The linear analyses overestimate the stresses in the upper part of the structure, compared with the two non-linear analyses. In the structure with the bitumen-stabilized base, Fig. 11.9b, this is not as prevalent and both the linear as well as the non-liner analyses capture the overall response of the structure quite reasonably.

Finally Fig. 11.10 shows the results of the predicted as well as the measured accumulated permanent deformation for the base, sub-base and the subgrade layer as a function of load repetition for both pavement structures. The response over the first 300 000 load repetitions are shown.

A simple power law assumption was used in calculating the permanent deformation. This seems to give a satisfactory agreement between the numerical simulations and the measurements for both structures. The largest deviation took place during the early part of the test but, thereafter, the rate of increased permanent deformation was quite similar between the analyses and the actual measurements.

It is also interesting to compare the measured and calculated permanent deformation of the two structures. Adding the three curves of Fig. 11.10 together gives the total permanent deformation, i.e. rutting, in the unbound part of the structure.

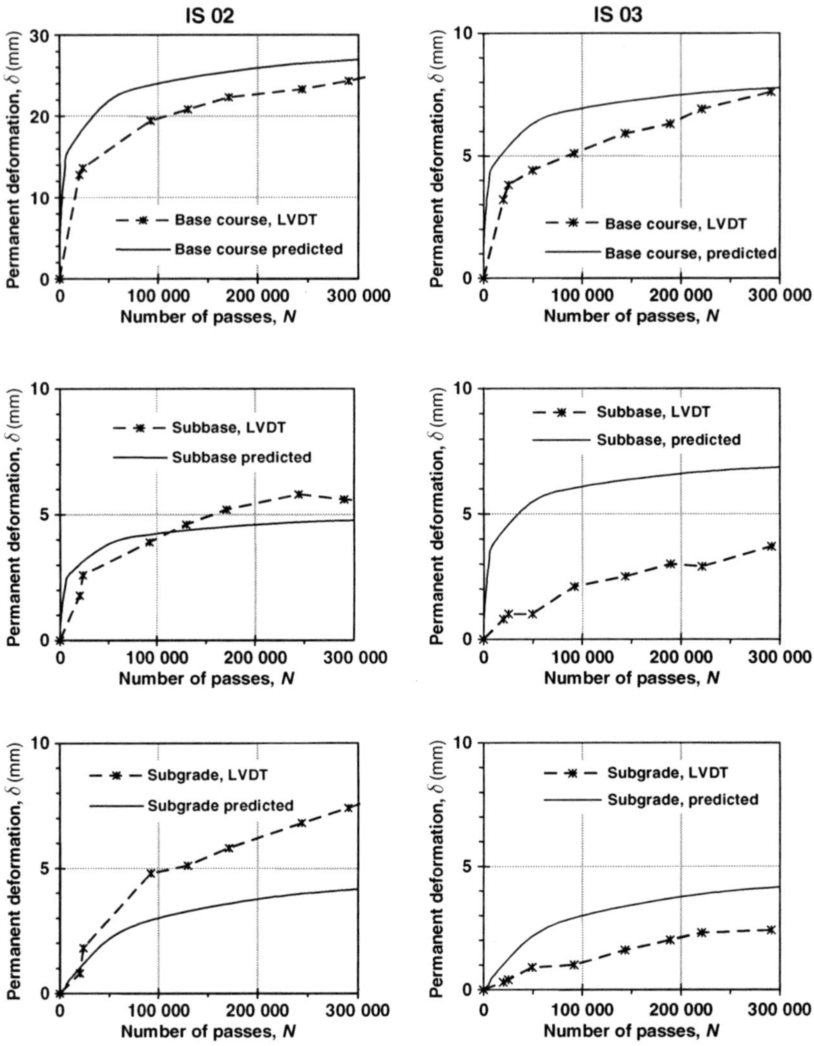

Fig. 11.10 Prediction versus measurements of permanent deformation development for the three unbound layers as a function of load repetition for both pavement structures IS02 and IS03

The one with the upper part of the base stabilized with bitumen shows a total of about 14 mm of deformation after ca. 300 000 passes, but the other with unbound base shows almost a 40 mm deformation after the same number of passes. This indicates quite a different "lifetime" of the two structures. This difference in "lifetime" does not prevail in the measurement and calculation of vertical stresses, where stresses at the top of the subgrade are almost the same for both structures (see Fig. 11.9).

11.4.5 Example of Modelling of the Resilient Behaviour of Pavements

To model the resilient behaviour of pavements, the French pavement laboratory, LCPC, has developed a finite element program called CVCR, which is a part of the finite element code CESAR-LCPC (Heck et al., 1998; Heck, 2001a,b). This program allows the modelling of the response of pavements in 3D, under moving wheel loads, and incorporates the following material models:

- Linear elasticity
- The Huet-Sayegh visco-elastic model for bituminous materials.
- Two non-linear elastic models for unbound granular materials: the Boyce model, modified to take into account anisotropy (Hornych et al., 1998) and the well known k-θ model (Hicks & Monismith, 1971). These models have been described in Chapter 9, Section 9.4.1.

The example below (Hornych et al., 2002) presents an application of CVCR to the modelling of a low traffic pavement with a granular base, tested on the LCPC pavement test track. In this study, the objective was, in particular, to evaluate the ability of the model to simulate experimental pavement response for different load levels and different water contents of the unbound granular material.

11.4.5.1 Experimental Pavement

The experimental pavement structure is presented in Fig. 11.11. It had a length of 28 m, a width of 6 m and consisted of:

- a bituminous concrete wearing course, with an average thickness of 85 mm;
- a granular base (0/20 mm crushed gneiss) with an average thickness of 430 mm;
- a subgrade soil consisting of 2.5 m of mica-schist with a low modulus (around 30 MPa).

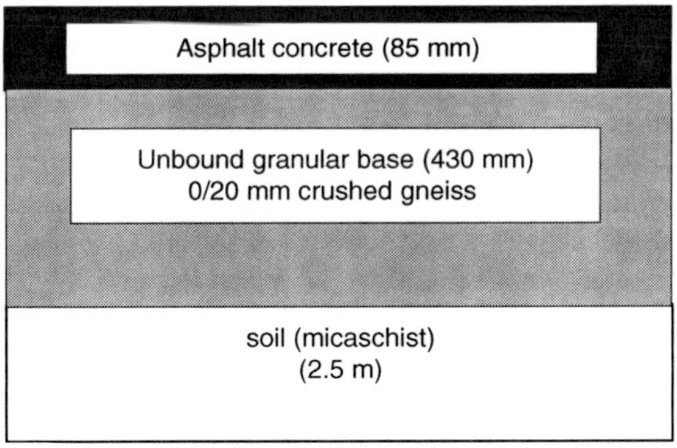

Fig. 11.11 Structure of the LCPC experimental pavement

The instrumentation installed in this experimental structure included:

- strain gauges to measure longitudinal and transversal strains at the bottom of the asphalt layer;
- displacement transducers to measure vertical strains in the top 100 mm of the granular layer and of the subgrade;
- vertical pressure transducers at the top of the subgrade;
- thermocouples in the asphalt layer; and
- tensiometers, to measure suction in the granular base and in the subgrade.

The pavement was subjected to dual wheel loads. Different load levels (45, 65 and 75 kN), and different loading speeds (3.4–68 km/h) were applied during the experiment.

11.4.5.2 Modelling Hypotheses

The pavement was modelled in 3D, considering visco-elastic behaviour for the bituminous material, and the non-linear elastic Boyce model (Eq. 9.10) for the unbound granular material and the soil. The material parameters for the bituminous layer were determined from complex modulus tests and the *in-situ* temperature of the bituminous layer was taken into account in the modelling. The parameters for the unbound granular material (UGM) and for the subgrade were determined from repeated load triaxial tests. For the UGM, tests were performed at 3 different water contents: 2.3%, 3.8% and 4.8%, corresponding to the water content variations observed on the site.

11.4.5.3 Modelling of the Pavement Response for Different Water Contents

A series of calculations was performed for the 3 load levels and the 3 moisture contents of the UGM at a constant loading speed of 68 km/h.

Figures 11.12 and 11.13 show comparisons between measured and calculated maximum longitudinal strains at the bottom of the bituminous layer ($\varepsilon_{xx\ \mathrm{BB}}$), and maximum vertical strains at the top of the UGM layer ($\varepsilon_{zz\ \mathrm{GNT}}$), for the 3 load levels. The results show that:

The model predicts relatively well the strains in the granular layer ($\varepsilon_{zz\ \mathrm{GNT}}$), and their non-linear increase with load level. The strains in the bituminous layer ($\varepsilon_{xx\ \mathrm{BB}}$) are slightly over-predicted.

The water content of the granular layer has a strong influence on the vertical strains in the UGM layer. Increasing w from 2.3 to 4.8% increases the strains by about 60%. The calculations with $w = 3.8\%$ lead to the best predictions, close to the mean of the experimental measurements.

The calculations with the 3 water contents led to a range of variation of the vertical strains in the UGM layer similar to the scatter of the experimental measurements.

Figures 11.14, 11.15 and 11.16 present additional examples of prediction of signals of longitudinal and transversal strains at the bottom of the bituminous layer

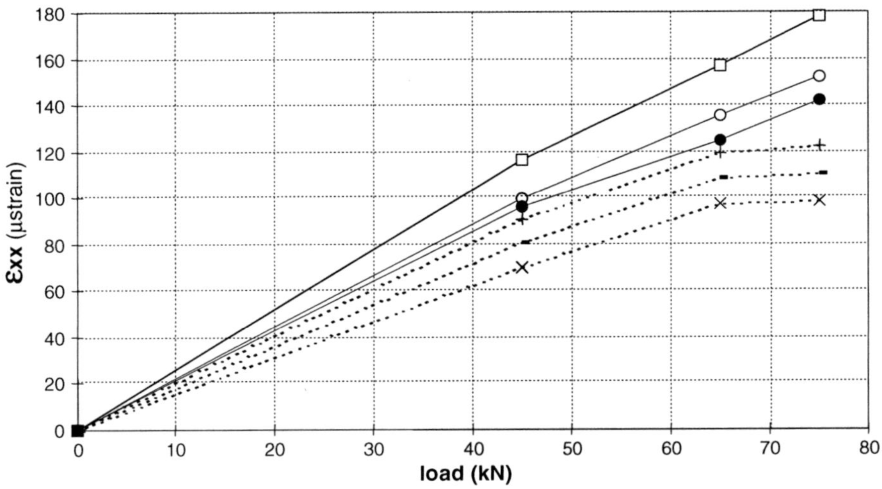

Fig. 11.12 Comparison of experimental and predicted maximum longitudinal strains at the bottom of the bituminous layer, for 3 load levels

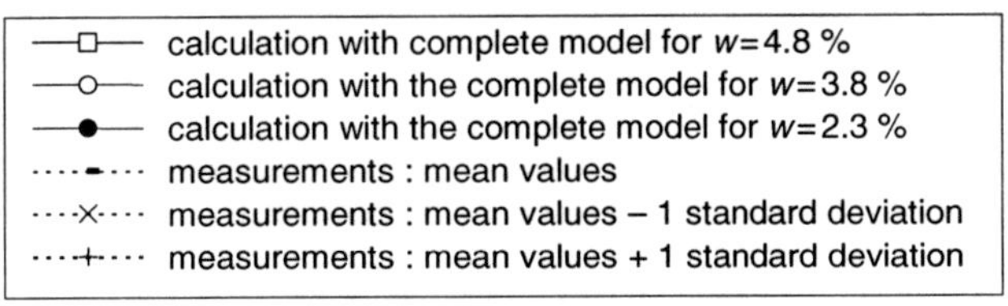

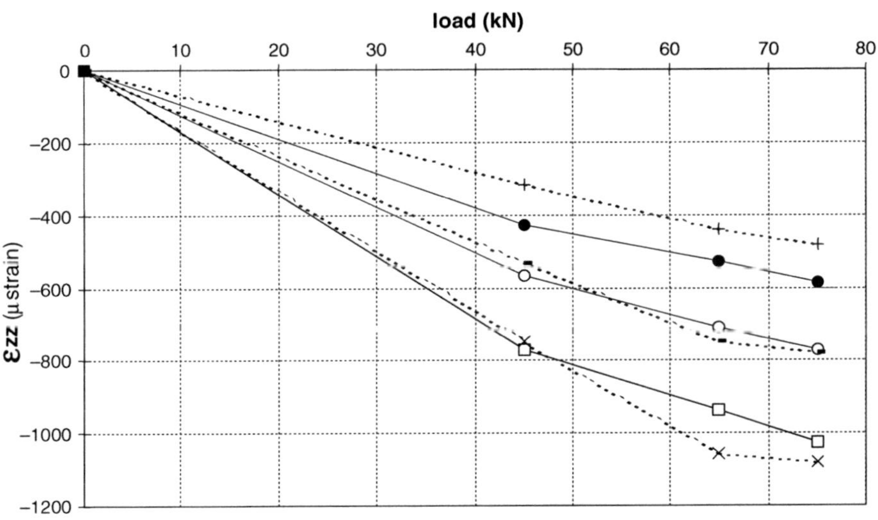

Fig. 11.13 Comparison of experimental and predicted maximum vertical strains at the top of the granular layer, for 3 load levels

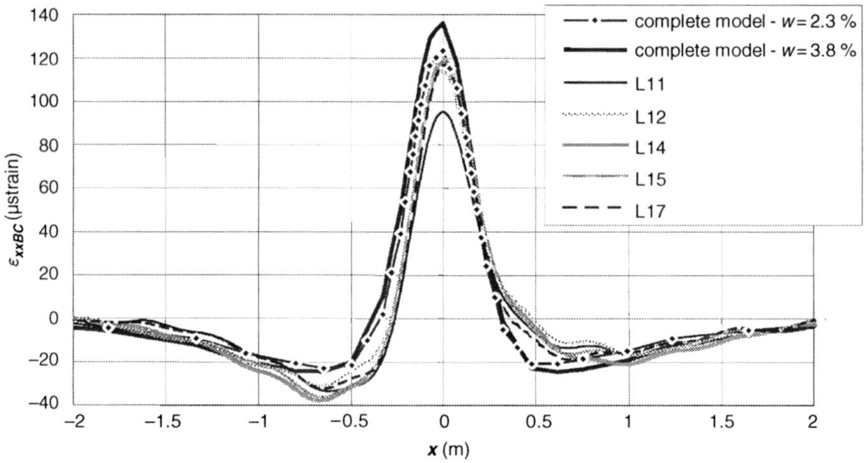

Fig. 11.14 Comparison of experimental and predicted longitudinal strain ε_{xx} at the bottom of the bituminous layer – load 65 kN

(ε_{xx} and ε_{yy}), and vertical strains at the top of the UGM layer (ε_{zz}), for the load of 65 kN. The results show that CVCR (with $w = 3.8\%$ for the UGM) predicts well the strain signals (strain variations when the load moves in the x direction). The experimental curves of ε_{xx} and ε_{yy} at the bottom of the bituminous concrete are non-symmetrical, due to the viscosity of the material, and this is well predicted by the visco-elastic model.

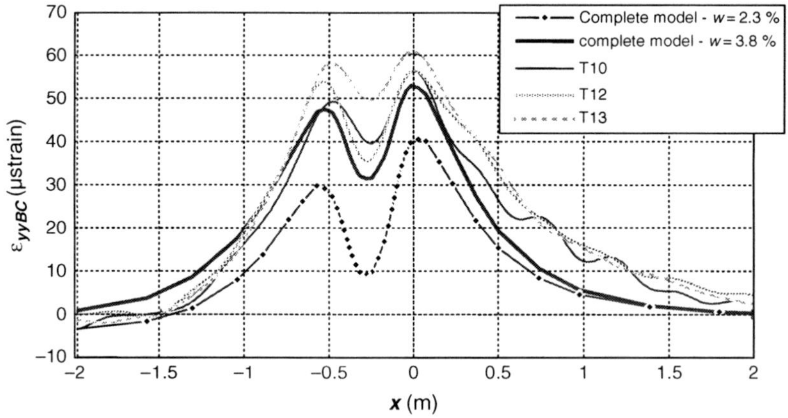

Fig. 11.15 Comparison of experimental and predicted transversal strain ε_{yy} at the bottom of the bituminous layer – load 65 kN

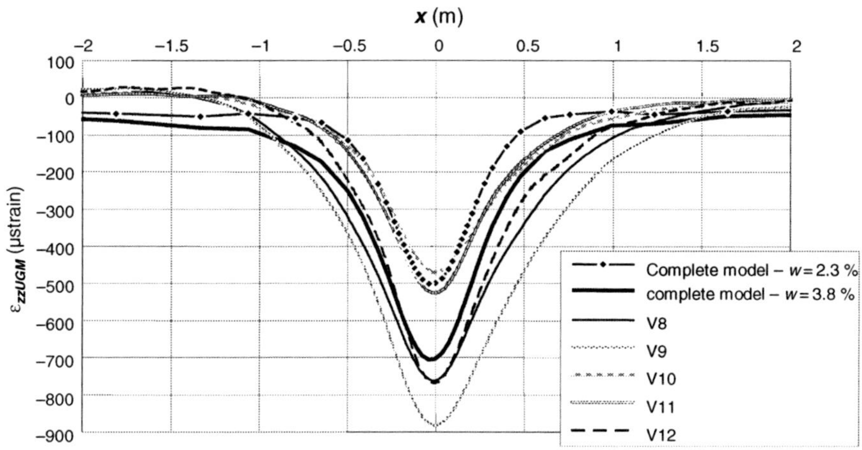

Fig. 11.16 Comparison of experimental and predicted vertical strain ε_{zz} at the top of the granular layer – load 65 kN

11.4.6 Example of Modelling of Permanent Deformations

The modelling of permanent deformations of pavements is more complex than the modelling of the resilient behaviour because it is necessary to simulate the response of the pavement to large numbers of load cycles (typically 10^5–10^6 cycles), with variable loading and environmental conditions.

A programme for the prediction of rutting of low traffic pavements, called ORNI, is also implemented in the finite element code CESAR-LCPC (El Abd et al., 2005; Hornych & El Abd, 2006). To determine the permanent deformations due to large numbers of load cycles, this programme proposes a simplified approach, based on a separate calculation of the elastic response and of the plastic strains. It comprises 3 steps:

i) The first step consists in calculating the resilient response of the pavement, for the different loading conditions considered (different types of loads, different temperature, etc. . .). The resilient response is calculated in 3D, using the programme CVCR.
ii) Then, the resilient stress fields are used to calculate the plastic strains produced by the successive application of the different loads. The permanent strains are calculated locally, at different points in the pavement structure, in 2D (in the plane (0,y,z) perpendicular to the direction of displacement of the load).
iii) Finally, the third step consists in calculating the displacements in the pavement structure. The plastic strains being calculated locally, at different points, do not derive from a displacement field. It is thus necessary to determine the total strains, ensuring their continuity and integrability, and the corresponding displacements.

Two permanent deformation models are implemented in ORNI: the empirical model of Gidel et al. (2001), and the elasto-plastic model of Chazallon (Chazallon et al., 2006). These models have been described in Chapter 9.

11.4.6.1 Experimental Pavement and Modelling Hypotheses

In the European project SAMARIS, the predictions obtained with ORNI have been compared with the response of a low traffic pavement tested on the LCPC pavement test track. The pavement consisted of:

- a bituminous concrete wearing course, with an average thickness of 66 mm;
- A 50 cm thick granular base and sub-base (crushed gneiss);
- A clayey sand subgrade (thickness 2.20 m), resting on a rigid concrete slab.

As in the previous example (Section 11.4.5) the pavement was instrumented to measure strains, temperatures and water contents in the various layers. The loading consisted in applying 1.5 million heavy vehicle loads (dual wheels, with a load of 65 kN).

In the modelling of the resilient behaviour (with CVCR), the bituminous concrete and the soil were assumed linear elastic, and the UGM was described using the anisotropic Boyce model (Eq. 9.13). The bituminous concrete moduli (function of temperature and loading frequency) were determined from complex modulus tests, the UGM and soil parameters from repeated load triaxial tests.

In the modelling with ORNI, it was assumed that no permanent deformations occur in the thin bituminous layer. The permanent deformations of the UGM and of the soil were described using the empirical model of Gidel et al. (2001), with model parameters determined from repeated load triaxial tests, at two water contents for the UGM ($w = 4\%$ and 5%), and one water content for the soil ($w = 11\%$).

Figure 11.17 shows the finite element meshes used to determine the resilient behaviour with CVCR (in 3D), and then the permanent deformations (in 2D).

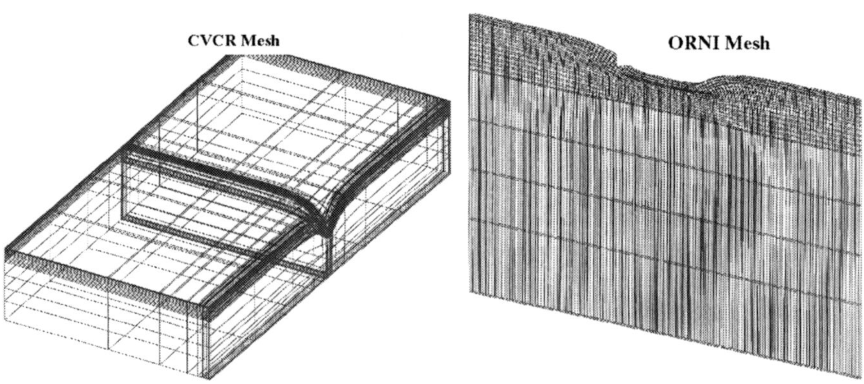

Fig. 11.17 Finite element meshes used for the modelling of the experimental pavement

11.4.6.2 Rut Depth Predictions

A first series of ORNI calculations was performed considering only the rutting of the UGM layer, and assuming different temperatures in the bituminous wearing course (between 15° and 35°, corresponding approximately to the range of temperatures measured in-situ). The results are presented on Fig. 11.18. It can be seen that the temperature in the bituminous wearing course has a large influence on the permanent deformations of the UGM (the temperature affects the modulus of the bituminous material and, therefore, the stresses transmitted to the granular base).

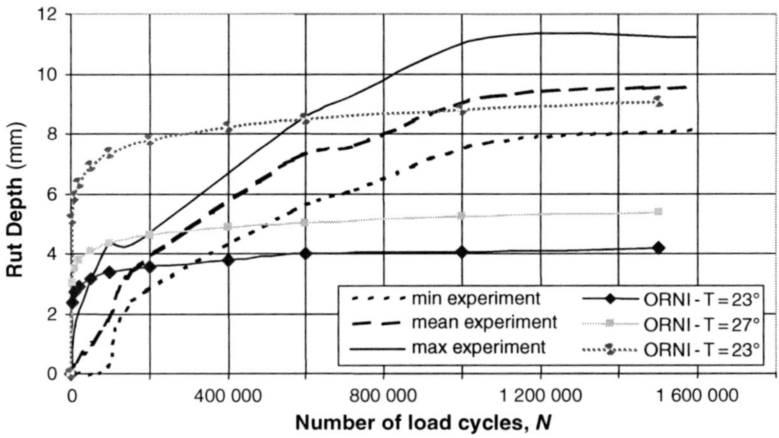

Fig. 11.18 Comparison of maximum rut depths measured on the experimental pavement and predictions with ORNI (rutting of UGM only, different temperatures)

Figure 11.19 presents the results of a second series of calculations where the rutting of the subgrade soil was also taken into account. The contribution of the subgrade to the total rutting is important, representing about 40% of the total rut depth. The final rut depths obtained after 1.5 million loads, with the contribution of the subgrade, are close to the experimental measurements, especially for the temperatures of 23° and 27°, which are close to the average in-situ temperatures. However, the model predicts a too rapid stabilisation of the permanent strains in comparison with the experimental measurements.

11.4.7 Example of the Pollutant Transport Modelling in the Pavement and Embankment

The transport model of pollutant leaching from the secondary road construction material was developed by the Environmental Research Group from the University of New Hampshire, USA (Apul et al., 2003). Water flow in a Minnesota highway embankment was modelled in one dimension for several rain events and calibrated to the field condition (Fig. 11.20). The test facility consists of 40 and 152 m-long

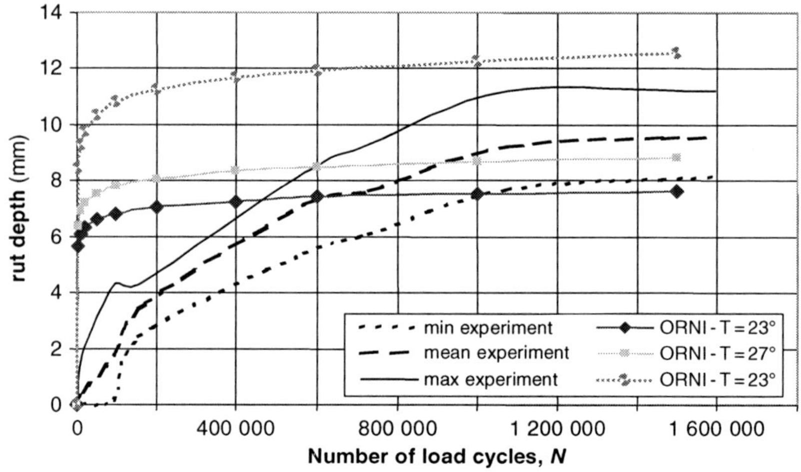

Fig. 11.19 Comparison of maximum rut depths measured on the experimental pavement and predictions with ORNI (rutting of UGM and subgrade, different temperatures)

hot mix asphalt and Portland cement concrete test sections with varying structural designs. Each test section is instrumented to monitor strength and hydraulic properties. The hydraulic properties of the embankment were predicted from water content measurements made in the embankment, a Portland cement concrete pavement with an asphalt shoulder. The hypothetical leaching of Cadmium from coal fly ash was probabilistically simulated in a scenario where the top 0.50 m of the embankment was replaced by coal fly ash. The groundwater table was set at 1.9 m below ground level (b.g.l.), which is within the range (1.3–4.6 m b.g.l.) observed at test site. An entire year's precipitation data repeated 10 times was input as the variable flux boundary condition. The molecular diffusion coefficient of Cadmium in free water

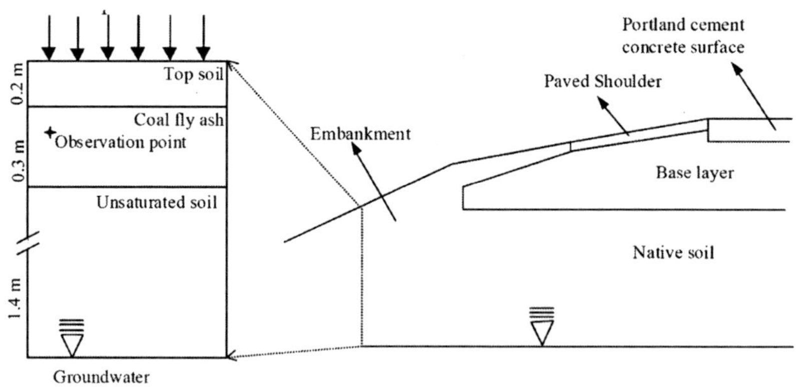

Fig. 11.20 (**a**) Cross section of MnROAD test Section 12; (**b**) conceptual model of the MnROAD embankment (Apul et al., 2003). Reproduced by permission of ISCOWA

was input in the model as a constant (6.2×10^{-5} m^2/day) and tortuosity factor was calculated within the finite element code, HYDRUS2D, as a function of the water content.

The probability distributions of unsaturated hydraulic properties of the embankment were determined from parameter posterior probabilities obtained from embankment infiltration simulations. The probability distributions were used to fit to the four parameters of the van Genuchten SWCC model (see Chapter 2, Section 2.7.3). Weighted moment equations were applied to calculate the means and standard deviations for the normal distributions. Saturated permeability and saturated water content were assigned joint log-normal distributions. To account for the variability of partition coefficient, k_d, uniform distributions were assigned. In the study k_d was considered as lumped parameter. The temporal and spatial variability of k_d that would be expected in the field was incorporated in the modelling approach by probabilistically varying k_d values of the subgrade and the coal fly ash for each simulation.

The average percentage of initial available mass leached after 10 years, as observed 0.01 m below ash, is presented, idealised, in Fig. 11.21, for a point a short way into the ash layer (point marked on Fig. 11.20). No significant Cadmium fluxes were observed 0.25 m below the coal fly ash or at the groundwater table depth. After 10 years, the fraction of initial available mass leached was 5×10^{-6} percent at 0.25 m below the coal fly ash, and 0% at the groundwater table depth (at the 90th percentile of uncertainty). The cumulative release at the 90th percentile of uncertainty and the appropriate probability distributions, was 2.65×10^{-3} mg Cd/kg ash after 10 years. The mean of the release estimate was 1.15×10^{-3} mg Cd/kg ash. Further details

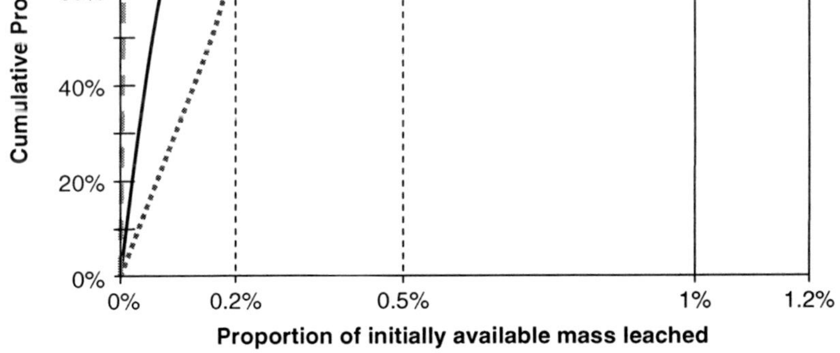

Fig. 11.21 Cumulative probabilities of percentages of initial available mass leached (as observed 0.01m below coal fly ash) after 1, 5 and 10 years. (Apul et al., 2005). Courtesy of D. Apul

are available from Apul et al. (2005) but it is clear that infinitesimal leaching occurs in real pavement/earthworks sections arranged in a similar way to the construction studied in Minnesota.

11.5 Conclusions

The partial differential equations that govern solid mechanics, water transfers, heat transfers and pollutant transfers have been restated. The specificities of the finite element method when dedicated to such non-linear phenomena and their coupling have been summarised. Then a number of numerical simulations have been presented. They cover moisture transfers, freezing, mechanical strains and permanent deformations. It appears that, for the most part, realistic numerical modelling is today available, at least for advanced research teams. But progress is still needed, for example to couple changes of moisture level and changes of the mechanical behaviour.

References

Alonso, E., 1998, "Suction and moisture regimes in roadway bases and subgrades", *Proc. International Symposium on Subdrainage in Roadway Pavements and Subgrades*, Granada, Spain, pp. 3–56 & 57–104.

Alonso, E.E., Cañete, A. & Olivella, S., 2002, "Moisture transfer and deformation behaviour of pavements: effect of climate, materials and drainage", *Proc. 3rd International Conference on Unsaturated Soils*, Recife, Brazil, **2**, pp. 671–678.

Apul, D., Gardner, K., Eighmy T., Linder, E., Frizzell T. & Roberson, R., 2005, "Probabilistic modeling of one-dimensional water movement and leaching from highway embankments containing secondary materials", *Environmental Engineering Science* **22**, pp. 156–169.

Apul, D., Gardner, K., Eighmy T., Frizzell T., Linder, E. & Roberson, R., 2003, "Use of Bayesian methods to estimate long term contaminant leaching in roadways", in *Proc. WASCON 2003*, ed. G. Ortiz de Urbina & H. Goumons, ISCOWA/INASMET, San Sebastien, Spain, pp. 253–262.

Charlier, R. & Radu, J.P., 2001, "Rétention et transfert des polluants chimiques solubles: mécanismes fondamentaux et modélisation numérique", Traite de Mécanique et Ingénierie des Matériaux – MIM, in *Géomécanique environnementale: risques naturels et patrimonie*, ed. Schl1ler, B. & Delage, P., Editions Hermes.

Charlier, R. & Habraken, A-M, 1990, "Numerical modellisation of contact with friction phenomena by the finite element method", *J. Computers & Geotechnics*, **9**(1&2), pp. 59–72.

Chazallon, C., Hornych, P., Mouhoubi, S., 2006, "An elastoplastic model for long term behaviour modelling of unbound granular materials for flexible pavements", *Int'l J. Geomechanics*, ASCE, July/August, **6**(4), pp. 279–289.

Coussy, O. & Ulm, F.J., 2001, "Basic concepts of durability mechanics of concrete structures", *Revue Française de Génie Civil*, **5**(6), pp. 897–919.

Detournay E. & Cheng A.H.D., 1991, "Fundamental of poroelasticity", in *Comprehensive Rock Engineering, Practice and Projects*, **2**, J.A. Hudson ed., Pergamon Press, 1991.

El Abd,. A., Hornych, P., Breysse, & D., Denis, A., 2005, "Prediction of permanent deformations of unbound pavement layers", 7th *Int. Conf. Bearing Capacity of Roads, Railways & Airfields*, Trondheim, Norway.

Erlingsson, S., 2007, "Numerical modelling of thin pavements response and distress development in a accelerated HVS test," *Int'l. J. Road Materials & Pavement Design*, **8**(4), pp. 719–744.

Gens, A., 2001, "*Fundamentals of THM phenomena in saturated and unsaturated materials. General formulation. Thermal and hydraulic constitutive laws*", Paper given to ALERT Geomaterials Graduate School, ENSHMG, Grenoble, France.

Flyhammar, P. & Bendz, D., 2003, "*Vatten- och masstransporter i kantzonen av en vägkropp. Fält- och laboratoriestudier*", (Delrapport I. In Swedish). ISRN LUTVDG/TVTG-7026-SE, avdelningen för Teknisk Geologi, LTH.

Gidel, G., Hornych, P., Chauvin, J.J., Breysse, D. & Denis, A., 2001, "Nouvelle approche pour l'étude des déformations permanentes des graves non traitées à l'appareil triaxial à chargements répétés", *Bulletin des LPC*, N°233, pp. 5–21.

Hansson, K., Lundin, L-C & Šiminek, J., 2005, "Modelling water flow patterns in flexible pavements", *Transportation Research Record* 1936, TRB, National Research Council, Washington, D.C., pp 133–141.

Hansson, K., 2005, "*Water and Heat Transport in Road Structures*", Ph. D. Thesis, Uppsala Universitet.

Heck, J.V., Piau, J.M., Gramsammer, J.C., Kerzreho, J.P. & Odéon, H., 1998, "Thermo-viscoelastic modelling of pavements behaviour and comparison with experimental data from the LCPC test track", *Proc. 5th Conference on Bearing Capacity of Roads and Airfields*, **II**, Trondheim, Norway, July, pp. 763–772.

Heck, J.V., 2001a, "*Module CVCR version expert*", Documentation du code de calcul César-LCPC

Heck, J.V., 2001b, "Modélisation des déformations réversibles et permanentes des enrobés bitumineux – Application à l'orniérage des chaussées", PhD Thesis, University of Nantes, France.

Hicks, R.G., & Monismith, C.L., 1971, "Factors Influencing the Resilient Response of Granular Materials", *Highway Research Record 345*, TRB, National Research Council, Washington, D.C., pp. 15–31.

Hornych, P. & El Abd A., 2006, "*Development and validation of a method of prediction of structural rutting of unbound pavement layers*", Report N° SAM 05-DE27, European Project SAMARIS, March, 94pp.

Hornych, P., Kazai, A. & Piau, J.M., 1998, "Study of the resilient behaviour of unbound granular materials", *Proc. 5th Conf. Bearing Capacity of Roads & Airfields*, **III**, Trondheim, Norway, July, pp. 1277–1287.

Hornych, P., Kerzrého, J.P. & Salasca, S., 2002, "Prediction of the behaviour of a flexible pavement using finite element analysis with non-linear elastic and viscoelastic models", *9th Int'l Conf. Asphalt Pavements*, **I**, Copenhagen, August, p. I-84.

Laloui, L., 2001, "Thermo-mechanical behaviour of soils", *Revue Française de Génie Civil*, **5**(6), pp. 809–843.

Li, X., Radu, J.P. & Charlier, R. 1997, "Numerical Modeling of Miscible Pollutant Transport by Groundwater in Unsaturated Zones", *Computer Methods & Advances in Geomechanics*, Yuan (ed.), pp. 1255–1260.

Mizoguchi, M. 1990, "*Water, heat and salt transport in freezing soil*", Ph.D. thesis. (in Japanese), University of Tokyo.

Pruess, K., Oldenburg, C. & Moridis, G., 1999, "*TOUGH2 users's guide*", version 2.0, Laurence Berkeley National Laboratory, Univ. California, Earth Sciences Div'n.

Schlumberger, 2000, "ECLIPSE, *Technical Description*", Schlumberger.

Thimus, J.F., Abousleiman, Y., Cheng, A.H.-D., Coussy, O. & Detournay, E., (eds.), 1998, "*Poromechanics. A tribute to Maurice Biot*", A.A. Balkema, Rotterdam, 666pp.

Turska, E. & Schrefler, B.A., 1993, "On convergence conditions of partitioned solution procedures for consolidation problems", *Comp. Meth. in Appl. Mech. & Eng.*, **106**, pp. 51–63.

Zienkiewicz, O.C. & Taylor, R.L., 1989, "*The Finite Element Method*", MacGraw-Hill Book Company, **2**, ch. 12, 4th ed'n.

Zienkiewicz, O.C., Paul, D.K. & Chan, A.H.C., 1988. "Unconditionally stable staggered solution procedure for soil-pore fluid interaction problems", *Int. J. for Num. Meth. In Eng'g.*, **26**, 1039–1055.

Chapter 12
Pollution Mitigation

Mihael Brenčič*, **Andrew Dawson, Lennart Folkeson, Denis François and Teresa Leitão**

Abstract There is often a risk of pollution entering or moving in the road environment. This may give rise to problems of various severities dependant on the local environment around and under the pavement. Therefore the risks have, first, to be assessed and then appropriate action taken to minimise the movements and/or the impacts. This chapter describes the criteria to be applied when considering pollution mitigation schemes and the constraints that must be taken into account. Both traffic considerations (which often form the driver for pollution supply) and economic considerations are included in the coverage of the chapter together with some comments on site sensitivity. In particular, the chapter provides a framework for considering alternative mitigation strategies against a background of the benefits and limitations of each. Pollution mitigation measures are only mentioned where they are identifiably different from conventional drainage measures which are covered more fully in Chapter 13.

Keywords Pollution control · impact mitigation · flow disruption · site sensitivity

12.1 Introduction

Roads and road traffic can act as serious sources of various types of pollution. Pollutants spread to the environment through different pathways, with different transport agents and mechanisms. Once pollutants are transported away from the road and traffic sources they can reach various environmental compartments where they can have detrimental effects. Pollution from roads and traffic must be managed and its harmful affects prevented at all stages, especially in environmentally sensitive areas.

The objective of this chapter is to describe general principles of prevention and mitigation of pollution that originate from road and traffic operation and that can influence the water environment. Consideration was taken mainly of mitigation of

* *Co-ordinating Author:*
✉ M. Brenčič
Geološki zavod Slovenije/Geological Survey of Slovenia
e-mail: mihael.brencic@geo-zs.si

A.R. Dawson (ed.), *Water in Road Structures*, DOI 10.1007/978-1-4020-8562-8_12,
© Springer Science+Business Media B.V. 2009

deleterious effects caused by seepage water. Pollution prevention and mitigation is associated with several constraints that can be classified in five major groups; site sensitivity and vulnerability, risk and hazard to pollution, traffic characteristics, economic and legislation constraints.

In the second part of the chapter general principles of mitigation methods are described. A new classification of mitigation approaches based on the pollutant fate model that consist of the chain sources – pathways – targets is described. Classification is described based on the ex-situ and in-situ mitigation methods and descriptions of intervention and non-intervention mitigation measures are also introduced.

The content of this chapter is very much connected with the next chapter "Recommendation for the control of pavement water" where design and technical measures for pollution prevention are described.

12.2 Mitigating Pollution from Roads

The level of pollution originating form roads depends on several factors that can also have an influence on pollution mitigation and prevention. These factors can be divided into two general groups:

- natural factors that depend on the environmental characteristics of the road's surroundings; and
- technical factors that are connected with road design, construction and traffic characteristics.

Together with air, water is the main transport media for pollution dispersal from roads to the environment. Water is a very efficient solvent and, during its path through the road construction and surrounding environment, it dissolves and transports many pollutants – some of them in large quantities and over long distances. Pollutants are transported in water by the mechanisms described in Chapter 6 with solution and suspension of solids being important. Preventing pollution impacts through technical measures represents a considerable challenge to planners, designers and operators of roads.

Road construction is an activity that very directly affects the hydrologic and geo-environment. Due to their linear nature, roads characteristically divide the hydrologic and geo-environment into two or more parts. The degree to which these parts are separated from one another depends on

- the road category;
- the road's topography related to the surrounding area;
- the kind and density of traffic; and
- the existence of connecting structures between the two sides of the road (hydraulic or for biota).

The influences of road operation on water bodies may be classified as direct and indirect. Direct influences include water pollution that can be treated by various treatment methods. Also, the road construction may disrupt sub-surface and/or surface water flow paths. Most of these direct influences are measurable if one is prepared to monitor the impact of the road. Indirect influences are not usually detectable at first sight being mainly connected with activities that are induced by the road operation – for example, new industrial zones development or new residential areas. Ultimately, these consequences can be more severe than direct pollution of water body from road run-off. Nevertheless, we are addressing only the former, the hydrological impact of new developments being far beyond the scope of this book.

During planning, construction, operation and maintenance of roads, protection and mitigation measures for the water environment play an important role. These activities have to be established on the basis of conceptual models that enable the proper implementation of the appropriate technical measures. Natural conditions such as sensitivity and vulnerability of water bodies are among the most important influences to be considered by these conceptual models.

Various constraints can influence the selection of procedures to mitigate the threat of pollution from roads and traffic to many of the environmental compartments. For example, water bodies are very important as they represent an important habitat, may provide a water resource for public water supply and last, but not least, water is one of the most important transport agents for spreading pollutants. These constraints are the consequence of natural site characteristics, usually defined as the sensitivity and vulnerability of water bodies, they depend on geotechnical and hydro-technical conditions in the environment and on road traffic characteristics, especially the share of heavy vehicles in the total daily traffic flow. Legislation is another important constraint to pollution mitigation. To protect the environment as much as possible, there is an increasing trend for laws and other legal instruments to impose upon planners, designers, operators and users of roads many obligations that should be carefully studied and implemented. The knowledge of all constraints can help establish proper criteria for preventing and mitigating harm from road pollution.

In the past extensive monitoring of road infrastructure and its influence on various environmental compartments has been performed. Among other results of monitoring that have been extensively reported are those affecting the hydro- and hydro-geological environment (e.g. Hamilton & Harrison, 1991; Bruen et al., 2006). The most profound and known effects of roads on the water environment, especially in northern European countries, are the influences of winter maintenance practice where salts and some other agents are used as de-icing agents (Amrhein et al., 1992; Oberg et al., 1991). Also reported are the influences of heavy metal contamination on soil micro-structure and, consequently, on the physical and chemical conditions of soils and water percolated through the unsaturated zone in the groundwater (Folkeson 2000). Literature reports on oil spills on Slovene karstic areas reveal that spilt oil sinks very fast into the karstified underground and very soon appears in springs used for water supply (Kogovšek, 1995 & 2007).

Some case studies recording the impact of alternative road construction materials on groundwater are reported in the literature. As an example of this, a road was constructed on a new alignment in Cumbria, Lake District, UK in 1975 over a somewhat acidic bog. Soft marshy material was removed and the area was back-filled with approximately 300,000 tones of Blast Furnace Slag from iron production. The slag was also used below the water table. The slag material contained large quantities of calcium sulphide. Soon after construction, sulphur related pollution began to be observed in surrounding water courses, apparently as a consequence of groundwater seepages from the construction. Sulphide has a directly toxic effect on many life forms but its oxidation to sulphate can be a more serious problem as this can cause severe oxygen depletion, effectively asphyxiating many life-forms within the affected water bodies. In winter the problem was not so severe since there was plenty of water in the local receiving streams to enable dilution to occur. However, as groundwater continued to flow into receiving streams whose flow rates had decreased during the summer, and as water temperature increased, excessive algal blooms occurred. After construction was finished, the records indicate that samples of water discharging from drains from the roadway were found to be contaminated with sulphides and hydrogen sulphide. Biological and fisheries investigations showed that there was a marked impact on the invertebrate fauna and fish populations in the two streams receiving the drainage from the new road (Taylor, 2004).

Many alternative materials are often mentioned by reference to their leaching potential as a potential risk regarding the pollution of surface and groundwater. However, this risk must also be considered regarding some natural materials. Acid drainage is one of the phenomena that can appear in some natural materials. As an example we can refer to the construction of a capping layer consist of hornfels metamorphic rock aggregates that resulted in percolating waters with a pH between 3 and 4 (Odie, 2003). From the construction phase fish-kill was observed in the neighbourhood of the road. Such an acidification was induced by the oxidation of solid and dissolved pyrite (FeS_2) originally present in the aggregate, and resulted in the production of H_2SO_4 and $Fe(OH)_3$. The presence of the latter at the outlet of the drain was testified by a dense orange colour.

12.3 Criteria and Constraints for Pollution Mitigation

12.3.1 Consideration of Site Sensitivity and Vulnerability

Analysis of the natural conditions of a road course is usually the second stage in the planning of the road construction where the first stage is defined by legislation and socio-economic factors. The protection level of the water and water environment from road influences depends on factors that are connected with:

i) road and traffic characteristics;
ii) natural sensitivity and vulnerability of the existing environment; and
iii) presence of the areas with special public interest (e.g. public water supply resources, special areas for vineyards, locations of rare flora or fauna, Natura 2000, etc.).

For example, the protection of the water environment on a low permeable clay stratum will need to be totally different from the protection needed on a high yield intergranular aquifer that represents an important source of drinking water. At the same time the protection required against a highly trafficked road will be totally different from the protection needed against a low traffic road with only a few vehicles per day.

To establish sound and cost efficient measures for water protection from negative road influences, the natural conditions must be correctly characterised and well understood to enable the proper planning and design of protection measures. The natural conditions of the road corridor are analysed in the field using classical hydro- and hydrogeo-logical methods (O'Flaherty, 2001; DoE, 2005; Brassington, 1998). Results of those investigations are used to determine the level of water protection.

When designing technical measures for groundwater protection, the goal of proper protection of hydro-environments from a road's influence will, usually, only be achieved by developing, defining and adopting a classification scheme. Such a scheme must be based on a conceptual model of the road's course above water bodies and a conceptual model of the hydro-logical system's source-pathway-target arrangements. These conceptual models can be divided into two main groups; those for surface water bodies and those for groundwater bodies.

The degree of protection required for a surface water body from negative road influences depends on its ecological, qualitative (chemical) and quantitative status. The most important relation in designing protection measures for surface water bodies is the ratio between run-off water discharged from the paved surface of road and the discharge of flowing water in the receptor river/stream bed. In general, the principle is that the quality of the receiving water should not be deteriorated by the arriving water. This is minimised by ensuring that the road-water: receiving-water ratio should be strongly in favour of the receiving surface water body.

Special care is needed when water from a road is discharged into a standing water body. In this case the mass of contaminant entering the static water body becomes more important rather than concentration (which is the most important in stream and river situations). A mass-balance calculation will typically be required to determine the amount of contaminant that can be handled by the static water body. This requirement is illustrated in Fig. 12.1 for a man-made wetland environment comprising an, essentially, static water body with small and intermittent surface water inflows and outflows. Special care is also needed when dispersal of road run-off is planned. Dispersal is possible only in the case when average daily traffic is relatively low. A less effective, but more controllable solution is provided by "underground" wetlands (Fig. 12.2).

When a road crosses a water catchment area (especially small ones) road constructors are often tempted to concentrate the small upstream watercourses in a single water structure (e.g. a pipe) in order to cross the road. Such a concentration of upstream catchment area run-off modifies the normal flow-rate of the downstream part of the watercourse, leading to an increased erosion over a variable distance unless the stream bed is adapted to the new hydraulic conditions (SETRA, 1993). Upstream, in case of runoff exceeding the discharge capacity of the pipe (e.g. after a

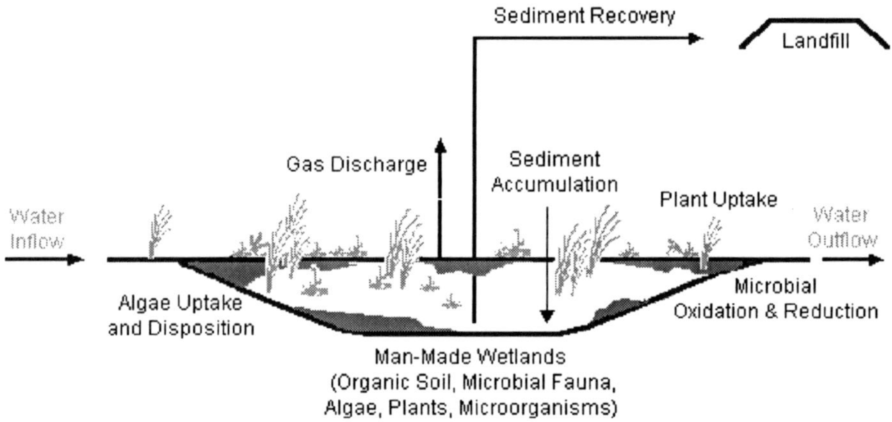

Fig. 12.1 Contaminant mass transfer considerations required for a man-made wetland (Van Deuren et al., 2002). Reproduced by permission of U.S. Army Environmental Command

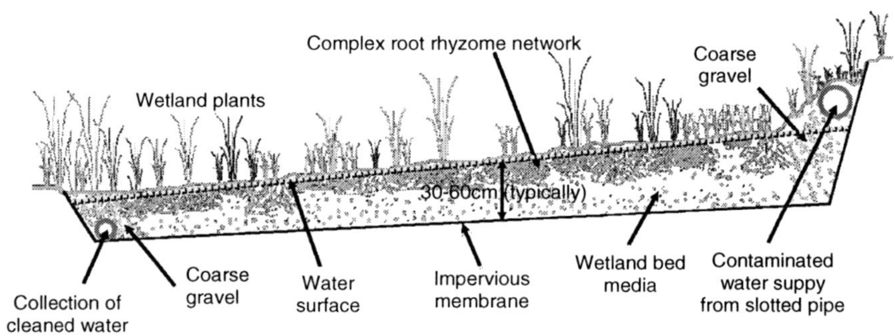

Fig. 12.2 A horizontal subsurface-flow constructed wetland (NAVFAC, 1998)

storm), the road body may temporary act as a dam, inducing flood and stagnant waters. Lastly, all the downstream parts of watercourses that were diverted upstream, toward the main course, are no longer fed by the catchment area. All these hydraulic perturbations induce impacts on aquatic habitats and organisms. They can be avoided thanks to an improvement of the road structure's hydraulic transparency, by means of sufficient upstream-downstream connections across the road structure (Fig. 12.3).

To define the influence of road pollution on groundwater bodies, the groundwater vulnerability concept can be applied. The most useful definition of vulnerability is one that refers to the intrinsic characteristics of the aquifer, which are relatively static and mostly beyond human control. It is proposed, therefore, that the groundwater vulnerability to pollution is defined as the sensitivity of groundwater quality to an imposed contaminant load, which is determined by the intrinsic characteristics of the aquifer. Thus defined, vulnerability is distinct from pollution risk. It is important to recognise that the vulnerability of an aquifer will be different for

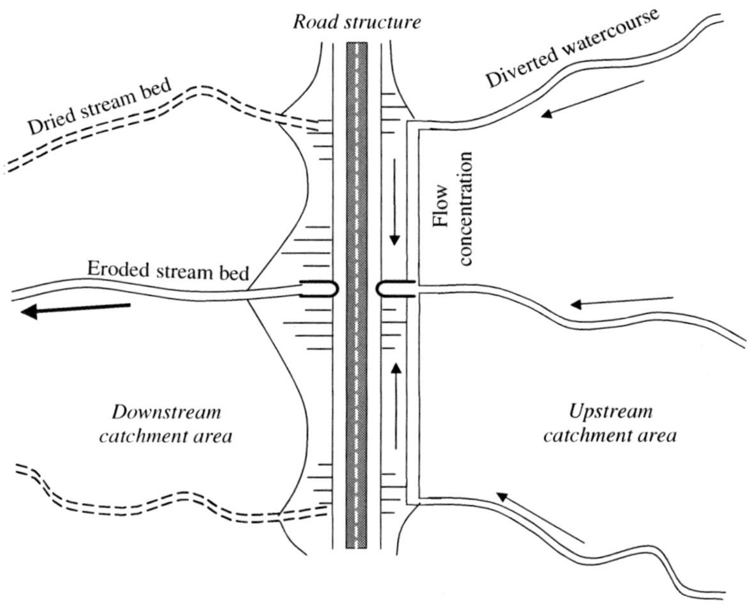

Fig. 12.3 Interrelation between road and streams

different pollutants. For example, groundwater quality may be highly vulnerable to the loading of heavy metals from road runoff and, and yet be less vulnerable to the loading of pathogens originating from the same source. In view of this reality, it is scientifically most sound to evaluate vulnerability to pollution in relation to a particular class of pollutant. This point of view can be expressed as specific vulnerability (COST 620, 1998).

An example to illustrate this approach is a conceptual model adopted in road construction in Slovenia (Brenčič, 2006). For groundwater bodies, protection is defined by a simple conceptual model of an aquifer that is divided into unsaturated and saturated parts. Protection requirements follow from:

- the estimation of transit time of the probable pollutant as it takes its route from the source on the road through the unsaturated zone; and
- on the interaction of the probable pollutant with the soil and water through which it passes *en-route* to the groundwater.

The greater the spreading of pollutant in the vertical direction, the higher is the vulnerability of the water resource. On the basis of the estimation of the pollutant spread and progression in the direction of the water resource e.g. borehole, spring, etc., the classification of arrival times from the spill location, or permanent pollution point due to the road operation, to the water resource is defined. Based on the cross-classification between vertical and horizontal spreading velocity of the pollutant, mitigation guidelines for the selection of suitable technical protection measures of the groundwater from negative highway influences were defined (Brenčič, 2006).

12.3.2 Risk and Hazard for Pollution

Pollutant emissions from roads and traffic present risks and hazards to water bodies where roads are in their recharge area or when they are in direct contact with road environment. Risk is defined as the probability that a particular adverse event will occur during a stated period of time, or it results from a particular challenge (Adams, 1995). Similarly 'hazard' is defined as the attribute that is the consequence of the probability of an adverse event and the degree of harm that can happen if this event occurs. A high hazard is present where the potential consequences to water bodies are significant.

Pollution risk depends not on vulnerability but on the existence of pollutant loading entering the subsurface environment. It is possible to have high aquifer vulnerability but no risk of pollution, if there is no pollutant loading; and to have high pollution risk in spite of low vulnerability, if the pollutant loading is exceptional. It is important to make clear the distinction between vulnerability and risk. This is because risk of pollution is determined not only by the intrinsic characteristics of the aquifer, which are relatively static and hardly changeable, but also on the existence of potentially polluting activities, which are dynamic factors which can in principle be changed and controlled.

The hazard of polluting the proximal road environment is the consequence of three types of emissions that are very much related to overall road and traffic characteristics:

- permanent emissions;
- incident emissions; and
- seasonal emissions.

Permanent emissions are mainly the consequence of vehicle operation on the road and their interaction with the pavement. This type of emission can also be the consequence of the interaction between materials used for road construction, maintenance and their surrounding environment.

Seasonal emissions are the result of the climatic seasonality, which influences circumstances that exist at a particular time on pavement and in the embankment. Typical seasonal emissions are connected with road salting. In northern European countries and Alpine countries during the thawing of snow and ice, a significant amount of chloride is emitted from the roads and their near surroundings. Similar seasonality is connected with higher summer temperatures when pollutants are more bound to asphalt surfaces then during the colder periods of the year.

Accidental spillages of liquids and gases hazardous to water bodies are typical incident emissions on roads. Roads, where the potential hazard of spillage of environmentally dangerous goods is high, should be treated more rigorously than roads where such potential is small.

Protecting the water environment from different types of emissions requires different mitigation measures. Therefore, it is necessary that during the planning and design of roads, a basic knowledge about potential risk and hazards must be established.

12.3.3 Traffic Considerations

The structure of the traffic greatly influences the road run-off pollution load. As a general rule, pollution on low-density roads is smaller than on high-density roads. However, the relation between pollution and traffic density is not linear and it is very difficult to predict the run-off pollution from road traffic characteristics, although these characteristics have a large role in controlling contaminant fluxes. The pollution of road run-off is also highly dependent on the climatic regime.

Traffic characteristics on roads are defined according to the several criteria. The most general parameter is called annual average daily traffic – AADT – however this parameter doesn't define the structure of the traffic. Usually, this parameter is further defined through the passenger car equivalent – PCE – or by the proportion of the AADT that is heavy commercial vehicles (HCV). These approaches allow consideration of the number of different types of vehicles.

In low volume roads, run-off treatment procedures usually differ from those used with high volume roads. For mechanical reasons, the traffic volume and road design criteria are, of course, connected. Consequently, the pollution potential of all roads is linked not only to the traffic levels, but also to the design criteria and approach adopted to match that traffic level. If the traffic volumes are low, the road's design is likely to be thin and the attention to detail to handle runoff and seepage waters is likely to be brief or even absent altogether. Thus, while the risks and volumes of contaminants arising may be less than on a heavily trafficked road, there may also be greater opportunity for these pollutants to enter the local water and ground environments.

If there is no exact legislative demand for run-off and seepage water treatment, the designer must define the treatment procedure according to the traffic density forecast and to the structure. Some national legislations (e.g. in Central Europe) attempt to define technical protection measures based on traffic characteristics (e.g. Brenčič, 2001). Criteria are then related to AADT and PCE. However, these criteria are not very satisfactory, over-simplifying the situation, and should be reconsidered with new parameters based on potential pollutant equivalents of different types of vehicles being applied instead. In Slovenia, the technical legislation dealing with road run-off pollution was implemented defining classes according to PCE. The legislation defines classes of PCE according to the aquifer types and surface water bodies that are crossed by roads. If the prescribed limit of PCE is exceeded then road run-off treatment should be performed. On the highly vulnerable karstic aquifers the limit is set to 6,000 PCE, on intergranular aquifers the limit is set to 12,000 PCE and on rocks and sediments with low permeability the limit is set to 40,000 PCE.

12.3.4 Economic Considerations

During the planning, construction, operation and maintenance of roads, economics plays an important role with pollution mitigation measures providing constraints that have a significant influence on the final cost of the road and its operation. Water protection measurements can represent an important proportion of the total road cost. For example, in Slovenia (a country in which groundwater is a very valuable resource) it was estimated that, over the groundwater sensitive areas, the protection costs represent between 10 and 50% of the total road construction costs.

Roads are constructed due to socio-economic demands and local communication needs. They are among the most important infrastructure objects provided by society's development. Therefore, it often happens that environmental criteria for their construction and operation take second place to the construction criteria, especially in transition economies. During planning and construction, costs for environmental protection measures are very often treated as direct expenses that cause an unjustified rise in the price of the road. Consequently, it can happen that removing these measures is seen to be a source of savings in the project. Damage to the water environment caused by road construction and operation can be very difficult to evaluate in terms of cost and revenues. However, experience shows that indirect costs caused by incorrect (or omitted) protection measures, although difficult to measure, are very high with long term consequences that can be very difficult to remediate. Roads across drinking water 'safe-guard' zones are a particular example illustrating the high costs or high impact that may occur.

A very important economic dimension of protection measures is their operational cost. These costs can represent a large proportion of the total cost of the ongoing road maintenance. Protection measures have to be properly maintained – especially active ones where the run-off is treated before being released to the wider environment. The high operational cost of some run-off treatment systems can lead to incorrect or incomplete maintenance and, consequently, in the generation of a new pollution point at the treatment outlet. Therefore, run-off treatment systems should be carefully designed and costed for all the potential problems that could occur during the maintenance processes.

12.4 Mitigation Methods

The implementation of mitigation and prevention measures from roads and road traffic should follow the pollutant fate in the environment. Before planning and designing of protection against pollutants from road and road traffic, a conceptual model of the pollutant fate in the particular environment should be established. This should help to estimate potential risks and hazards to water bodies' pollution. The model usually consist of three main parts that are represented by definition of pollutant sources, pathways of pollutants through the environment and targets that receive pollution from the sources in road environment. The concept of pollutant fate in the environment is described in greater detail in Chapter 6.

As in every environmentally-driven decision, care must be taken that the benefit to the water body is not offset by an equal or worse disbenefit to another environmental compartment. For example, the higher fuel consumption of cars using a longer road could be evaluated in terms of non-renewable energy impact and

Table 12.1 Classification of pollution mitigation approaches and methods

Mitigation approach	Mitigation method	
	Ex – situ	In – situ
Mitigation at Source	Prevention	Prevention
	Avoidance	Reduction
Mitigation along Pathway	Reorientation	Interception
Reduction at Target	Compensation	Remediation

greenhouse gas emissions and these compared with the lower risk achieved to the groundwater. How to make such a comparison is beyond the scope of this book, but interested readers are referred to (Falcocchio, 2004).

It will never be possible to prevent all deleterious impacts of the road on the hydro-environment. However, there are many actions that can be taken to significantly reduce impacts. In general, the pollution management policy is that protection of the environment should be preformed in a way that the source concentrations of contaminants are reduced as much as possible and to limit, or prevent completely, the appearance of contaminants in the targets. To reach this goal several mitigation approaches and mitigation methods can be adopted (Table 12.1).

Mitigation approaches are divided according to the pollutant fate model: source – pathway – target.

Mitigation at source can be performed with:

- prevention methods;
- avoidance methods; and
- reduction methods.

Prevention methods are in general applied to stop emissions of pollutants in the environment or at particular environmental sensitive areas (e.g. on Natura 2000 areas of sensitive water habitat). A typical general prevention approach is banning of leaded fuel or banning the use of road de-icing agents on the environmental sensitive areas.

Avoidance methods, in general, can be defined as special design procedures, mainly connected with road alignment, that avoid crossing environmentally sensitive areas. They seek to prevent a problem from arising in the first place (or minimize the problem). Often these design options are very costly and they very often interfere with the goals of the road. For example, a road may be longer in order to avoid a particularly sensitive groundwater body leading to greater construction costs and ongoing fuel consumption costs.

Reduction methods are those that are implemented when emissions from roads and the road environment cannot be stopped. They can be implemented by various traffic restrictions such as travel velocity reductions (e.g. on groundwater safe-guard zones) or reduction of traffic flow (e.g. embargo of dangerous goods transport over environmentally sensitive areas). Also, among reduction methods, the proper selection of construction materials can be included (e.g. alternative material use for sub-grade that do not interact with the soil environment).

Mitigation along the pathways can be achieved with

- interception methods; and
- reorientation methods.

Reorientation methods divert water that was polluted at the road surface, or inside the pavement, out of the area sensitive to water pollution perhaps to runoff treatment facilities where water is intercepted and treated. A watertight drainage system that diverts runoff water is a typical example of this.

Interception methods are technical measures that enable interception of pollutant flux. These interceptions can be defined as run-off treatment facilities (e.g. detention ponds) or absorption barriers (e.g. reactive barriers).

Mitigation at the target is achieved when pollutant reaches the target and its deleterious impact is reduced by

- remediation methods; and
- compensation methods.

Remediation methods are only feasible when some deleterious and adverse effects appear at an environmental target (e.g. damage to local fish habitat as the consequence of leakage from alternative material built into a sub-base). A typical remediation method in the pavement and embankment domain could be the replacement of contaminated granular base and sub-base materials by earth works. These methods would be extreme and only used in situations when previous mitigation measures were not successful. The use of these methods should not be implemented as an integral method for permanent pollution protection. However, in environmentally sensitive areas they could be planned as a part of the intervention measures.

Compensation methods are economic measures or replacement measures. The latter are applied in the case that road construction, and all the consequences of it, damage a particular habitat or water body. In this case a new, substitute, habitat or water body is included as part of the construction cost in the area where previously the zone was of lower ecological value. As an economic measure, compensation methods are applied as indemnity to the owners of the land crossed by the road or who are influenced by it. From the environmental point of view, compensation methods for pollution mitigation should be avoided. This approach implements the principle that loss of environmental values can be compensated by economic measures. Remediation and compensation mitigation methods are usually applied outside of the embankment zone, so they are not covered in full detail here.

Mitigation methods can be further divided into:

- **ex-situ methods**. Ex-situ methods are implemented externally as non – technical measures or as technical measures performed in places that are not part of the near-road environment.
- **in-situ methods**. In-situ methods can be defined as mitigation methods implemented on the road or in the near vicinity. These methods are further divided into:
  - ○ intervention measures,
  - ○ non-intervention measures.

    Intervention measures are those that involve intervention by human action, either when a problem is detected or on a regular basis (e.g. to maintain a pumping system). Active approaches are the least desirable for a number of reasons:

    - Their success depends on continued human attention... which is often difficult to guarantee;
    - They continue to require funding after construction, both in terms of payment to the personnel involved and, in many cases, in terms of the running costs of electrical or other energy consuming equipment. In the future there may be pressure on funding and a lack of appreciation of a problem that is in focus at the present time. This can lead to less attention at some future date than is necessary; and
    - Detection of a problem is necessary in many cases for the active approach to be implemented. Some problems will be readily detected – e.g. those resulting from a spillage during a traffic accident – but many will not be easily detectable in which case it is difficult to incentivize the search for a problem which could conceivably (but probably doesn't) exist.

  Non-intervention measures rely on the installation of some constructed element that continues to function over a large part, or all, of the life of the project in which it is installed. They are often more costly than active ones if expenditure is only considered over a year or two. However, in the long term the ongoing costs of providing active control will usually make passive approaches seem more economic.

The constructed element is designed to achieve one or more of the following:

- that any actual or potential contamination pathway is blocked;
- that there is a purpose-installed receptor for the any potential or actual contaminant that will prevent the contaminant from reaching a natural receptor to which it would present a hazard; and
- that the water regime adjacent to the road is maintained in an acceptable manner.

It is always best to attempt to avoid pollution problems rather than to intervene after the event. Data published by the UK's Highways Agency (2006) for 5 British roads reveals the high variability of success of different techniques. Sometimes 99%

reduction in contaminant concentration was achieved, sometimes there was even an increase in concentration after use of a "clean-up" technique due, presumably, to remobilization of previously arrested contaminant.

12.5 Conclusions

No road construction can ever have a zero impact on the environment in which it is placed. The materials of which it is constructed will yield a different response to the hydrological situation than did the soils that they have replaced. The construction interrupts the preceding natural flow regime (Fig. 12.3). The traffic on the road generates various pollutants that fall on the road (Chapter 6, Section 6.2). For these reasons the road designer must assess the potential impact of each aspect, compute the risk of unacceptable pollution and put in place mitigation measures that will address each unacceptable risk in a technically and economically satisfactory manner – this is a major challenge, especially as regulatory regimes become more and more demanding.

References

Adams, J., 1995, "*Risk*", Routledge, 228pp.

Amrhein, C., Strong, J.E. & Mosher, P.A., 1992, "Effect of deicing salts on metal and organic matter mobilization in roadside soils". *Environmental Science & Technology*, **26**(4), pp. 703–709.

Brassington, R., 1998, "*Field Hydrogeology*", John Wiley & Sons, Chichester, UK, 248pp.

Brenčič, M., 2001, "An analysis of standardized measures for the protection of drinking water resources from the negative impacts of roads", *Acta Hydrotechnica*, **19**(31), pp. 137–153.

Brenčič, M., 2006, "Groundwater and highways interaction: past and present experiences of highway construction in Slovenia", *Environmental Geology*, **49**(6) pp. 804–813.

Bruen, M., Johnston, P., Quinn, M.K., Desta, M., Higgins, N., Bradley, C. & Burns, S., 2006, "*Impact Assessment of Highway Drainage on Surface Water Quality*", Environmental Protection Agency, Ireland, Synthesis Report, ISBN: 1-84095-198-2, available on-line from http://www.epa.ie, 39pp.

COST 620, 1998, "*Definition of Terms 'Risk', 'Intrinsic Vulnerability' and 'Specific Vulnerability'*", Report of Working Group & Decisions of 3rd Management C'tt'ee Meeting, Liege, available on-line at http://www.hggeomac.ulg.ac.be/recherches/cost620_website/wg0.htm.

DoE, 2005, "*Ground Water Protection Programs Implementation Guide*", U.S. Department of Energy, Document N°: DOE G 450.1-9, 36pp.

Falcocchio, J.C., 2004, "Performance Measures for Evaluating Transportation Systems: Stakeholder Perspective", *Transportation Research Record*, 1895, Transportation Research Board, pp. 220–227.

Folkeson, L., 2000, "*Pollution of groundwater and soil by road and traffic sources. Dispersal mechanisms, pathways and mitigation measures*", CD Rom from Work Package 1, POLMIT Project, Swedish Road & Transport Research Institute (VTI), Linköping.

Hamilton, R.S. & Harrison, R.M. (eds.), 1991, "*Highway Pollution*". Studies in Environmental Science **44**, Elsevier, Amsterdam 522pp.

Highways Agency, 2006, "*Surface and Sub-Surface Drainage Systems for Highways*", HD 33/06, **4**(3) of Design Manual for Roads and Bridges, 33pp.

Kogovšek, J., 1995, "Izlitja nevarnih snovi ogrožajo kraško vodo: onesnaženje Rižane oktobra 1994 zaradi izlitja plinskega olja ob prometni nesreči pri Obrovu", (in Slovene with English summary, "Threats of dangerous liquids spills on karstic water: traffic accident pollution with gasoline fuel of Rižana spring on October 1994 near Obrov") *Annales, Series Historia Naturalis*, **5**, pp. 141–148.

Kogovšek, J., 2007: "Vpliv prometa na kraške vode", (in Slovene with English summary, "Influences of traffic on karstic waters"). In: Knez, M. & Slabe, T. (eds.) "*Kraški pojavi, razkriti med gradnjo slovenskih avtocest*" (in Slovene, "Karstic phenomena unroofed during Slovene highway construction"), ZRC SAZU (Scientific Research Centre of the Slovenian Academy of Sciences & Arts), pp. 225–238.

NAVFAC, 1998, "*Constructed Treatment Wetlands, Naval Facilities Engineering Service Center, Port Hueneme, California*", presentation given as part of the Remediation Innovative Technology Seminar series available on-line at https://portal.navfac.navy.mil/pls/portal/url/ITEM/200508AC22E54DE6E0440003BA8FC471, part of the US Naval Facilities Engineering Command.

Oberg, G, Gustafson, K. & Axelson, L., 1991, "*Effektivare halkbekampning med mindre salt*", MINSALT-projektets huvudrapport, Rapport 369, Swedish Road & Transport Research Institute (VTI), Linköping.

Odie, L., 2003, "*Rennes' ring road: impact of acid drainage from embankment materials*", (in French "Rocade de Rennes: Impact des eaux acides issues des matériaux de remblai"), CETE Ouest, Saint-Brieuc, France, 10pp.

O'Flaherty, C.A., 2001, "*Highways*", 4th Ed'n., Butterworth Heinemann, 576pp.

SETRA, 1993, "*Water and road: Vol. 4, Damages to aquatic environments*", (in French "L'eau et la route Vol. 4: Les atteintes aux milieux aquatiques"), Service d'Etudes Techniques des Routes et Autoroutes, Bagneux, France, 37pp.

Taylor, J.V., 2004, "*Migration of Contaminants Associated with Pavement Construction*", PhD thesis, School of Civil Engineering, University of Nottingham.

Van Deuren, J., Lloyd, T., Chetry, S., Liou, R., & Peck, J., 2002, "*Remediation technologies screening matrix and reference guide*", 4th ed'n. Federal Remediation Technologies Roundtable, available on-line at http://www.frtr.gov/.

Chapter 13
Control of Pavement Water and Pollution Prevention

José Santinho Faísca*, José Baena, Susanne Baltzer, Beata Gajewska, Antero Nousiainen, Åke Hermansson, Sigurður Erlingsson, Mihael Brenčič and Andrew Dawson

Abstract This chapter sets out the requirements, possible problems concerning surface and subsurface water flow for pavements and offers some technical solutions to control these waters. It presents the general principles for the design and choice of a drainage system, the measures to adopt during construction and maintenance phases and considers the control of surface and subsurface water contamination, in order to minimize the possible detrimental effect to existing aquifers and habitats. This is achieved by a thorough review of available drainage measures, including many illustrations.

Keywords Road drainage · sub-soil drainage · filter criteria · drainage layers · trenches

13.1 Introduction

The objective of this chapter is to set out the requirements, possible problems and to give some technical solutions to the control of surface and subsurface water flow for pavements.

The proper management of a pavement is needed to maintain the strength of the road structure, to provide long service life, safe traffic conditions and the environmentally acceptable treatment of pavement water. The increase of moisture in the pavement and in the subsoil or in the pavement foundation can decrease the bearing capacity (as discussed in Chapter 8) and stability, and contributes to physical and chemical phenomena, which modify the pavement's structure and further may increase erosion, expansion, dilution, cracking, risk of collapse and frost damage (Hall & Correa, 2003).

This chapter presents the general principles for the design and choice of a drainage system, the measures to adopt during construction and maintenance phases, and considers the control of surface and subsurface water contamination, in order to

* *Co-ordinating Author:*
✉ J.S. Faísca
Estradas de Portugal/Roads of Portugal, Portugal
e-mail: jose.faisca@estradasdeportugal.pt

A.R. Dawson (ed.), *Water in Road Structures*, DOI 10.1007/978-1-4020-8562-8_13,
© Springer Science+Business Media B.V. 2009

minimise the possible detrimental effect to existing aquifers and habitats. In particular, the aim should be to guide excess water out of the higher construction layers in such a way that the water beneath the pavement doesn't weaken the pavement structure or allow subterranean water to enter the structure.

13.2 Objectives

There are two aspects, which must be addressed in answering the question: "Why is road drainage so important?". They are:

- A road's infrastructure is an engineering work, aimed at the establishment of a platform on which vehicle circulation is possible under safe conditions, with proper traffic flow, utility, and economy, independent of the region's climate conditions; and
- Water, along with heavy traffic, is one of the greatest causes of road deterioration. As previous chapters have shown, even relatively small increases in water content can often result in significant reductions in the mechanical properties of the aggregate and soil layers in and under the road, thereby speeding up pavement failure.

The overall objective is, therefore, to keep the pavement and the subsoil dry enough to avoid any potentially harmful effects of the water and to control the environmental effects of that drainage water. This is done by decreasing the infiltration through the pavement platform's surface by adopting integrated solutions. These should simultaneously

- permit the re-establishment of natural groundwater patterns,
- avoid the access of runoff water from nearby land areas to the road platform,
- reduce the risk of erosion due to surface and subsurface flow on/in the nearby slopes, and
- provide preventive measures against soil and/or the aquifer's contamination either as a consequence of an accident or due to regular traffic and the construction.

13.3 Conception and Drainage Criteria

13.3.1 Road Alignment and Routing

For new roads, before drainage can be considered, the routing of the road must be fixed. Road routing is a complex procedure that involves very many factors including social, economic, engineering and environmental criteria as well as public acceptance. However, proper selection of the road corridor is one of the most important factors in the protection of water bodies. Technical measures aimed at mitigating a problem provide a poor substitute for efficient routing and alignment which could have avoided the occurrence of the problem altogether. Nevertheless, in those cases where conflicts between roads and water bodies cannot be avoided,

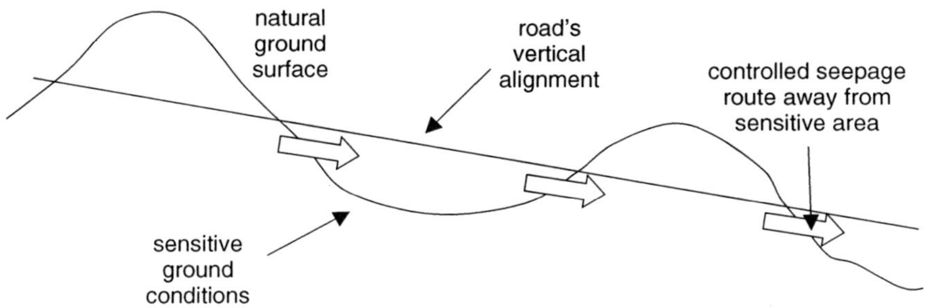

Fig. 13.1 A vertical alignment used to lead water away from a sensitive area

design of the road's elements, together with technical measures, plays an important role. The design of the road's alignment has a major influence on the inclination of the road surface. Although the main criterion for road surface dewatering is usually traffic safety (e.g. to prevent aquaplaning) water protection criteria must also be included into design practice. The selection of the longitudinal and transversal cross fall of the road can

- Direct the water from different parts of the highway surface and sub-surface to different drainage outlets, thus allowing "isolation" of spillage-affected run-off to localised collection points rather than allowing the contaminant to affect an entire drainage system,
- Lead water to less sensitive ground- or surface-water zones where it is more easily handled,
- Be an influence on the recharge area and, at the same time, on the total pollutant loads flowing from the particular section of the road.

Horizontal alignments can be selected to take a road away from a place where it may be expected to have an undesirable impact. Vertical alignments can permit the drainage of water collected from above the pavement's formation level (e.g. water deriving from cutting slopes and from pavement runoff) to be led out of a local catchment where it might be undesirable, to another location where it can be handled with less risk to the environment (Fig. 13.1). Less obviously, but more practical in many situations, horizontal curves may be introduced into a road's alignment in order that the pavement has to built with some super-elevation. This will then provide a cross-fall across the pavement to the margin on the inside of the bend (Fig. 13.2). By carefully choosing the road's horizontal and vertical alignments, both surface and sub-surface waters can be moved to a desired outlet point where the impacts will be minimised.

13.3.2 Fundamental Drainage Considerations

The drainage system employed in a road construction will depend on factors such as:

- The importance of the road;
- The amount of traffic;

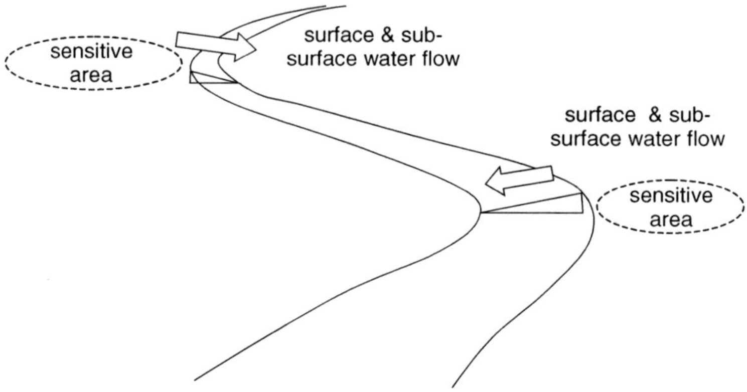

Fig. 13.2 A horizontal alignment designed to lead water away from sensitive areas

- The zone (rural or populated);
- The sensitivity of the groundwater; and
- The sensitivity of streams, rivers and lakes.

Drainage systems can be classified as follows:

- Surface systems – these involve ditches and open channels in the surface of the ground;
- Subsurface systems – these are not directly accessible from the surface. Water is collected from water in the ground's pore space and conveyed in trenches and pipes. Subsurface systems can be divided into shallow (interceptor) drains that collect percolating water above the water table and deep (water table lowering) drains.

In a very permeable soil, separate drainage may not be necessary, only provision of some capacity for immediate runoff and for snow and snowmelt.

Open channel drainage systems are favoured because of their low cost (compared with capacity), easy rehabilitation and maintenance. However, deep open ditches and steep inner road slopes can be dangerous for wandering vehicles, especially on a road with high traffic speeds. Steep slopes may also increase cracking along the centre of narrow roads, and, possibly, increase erosion and the need for channel cleaning.

Subsurface systems have advantages in special circumstances and can be efficient tools in road rehabilitation projects, especially when there is limited space available, limited support of slopes or a need to use deep drainage.

The correct depth for the drainage depends on the pavement thickness, the road layout (cutting or embankment), the type of subsoil and the climatic conditions (intensity of runoff, frost depth and snowmelt conditions). The drainage depth is usually the depth of all the layers of the road structure that contribute to the bearing capacity of the pavement.

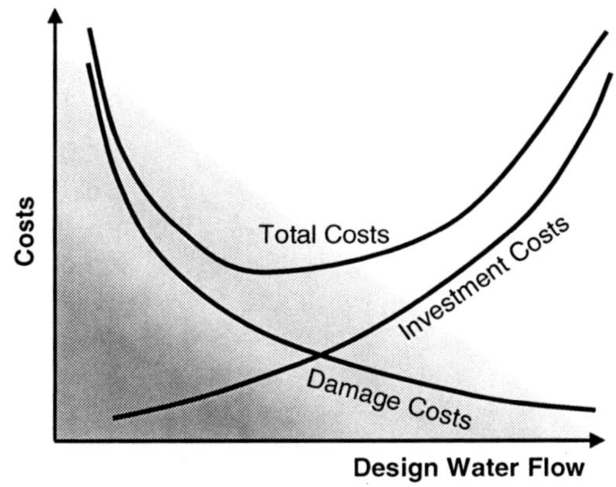

Fig. 13.3 Drainage economics

The design of drainage is a part of the design of the whole road, so all aspects of the road can have an effect on it. Drainage design is, simply described, the action of finding an optimum balance between the total costs (investment and maintenance costs) compared to the possible advantages and disadvantages (pavement life, traffic safety, easy maintenance and wider environmental aspects) – Fig. 13.3.

Considering all these aspects, there are a number of steps which, in general terms can be considered fundamental. They are listed in the Portuguese Road Drainage manual as follows (IEP, 2001):

1st Step – Gather all the relevant information:

- Road importance (traffic flow values, status and whether a "lifeline" road);
- Geometric characteristics of the road (layout and profiles);
- Drainage areas and existing drainage systems;
- Geology;
- Meteorological data (precipitation, temperature, frost, etc.);
- Hydrological and hydrogeological conditions in the area surrounding the road (ground-water conditions); and
- Identification of specific constraints (technical, social, economic or environmental);

2nd Step – Identify critical/sensitive areas:

- Vulnerable areas with particular conditions, geological, environmental or ecological;
- Areas with a high frost formation probability;
- Specific areas of the road, such as high and low level points;
- Extreme gradient and cross fall situations; and
- Cutting/embankment transition areas.

3rd Step – Adopt standard/typical layouts, where possible, for road segments with similar characteristics.

4th Step – Define the basic data for the water flows in each layout, using existing methods, tables and software to perform necessary calculations.
5th Step – Analyze the possible and adoptable solutions, based on standard drawings and typical dimensions used in each region or country.
6th Step – Perform the hydraulic calculations so as to obtain drainage sizes. If the estimated amount of water is small, exact hydraulic calculations may not be needed.
7th Step – Consider the location of the discharge points, as well as the need to design retention and/or treatment basins, which may be associated with individual drainage systems.

Roads are normally constructed with two types of drainage systems, the surface and the subsurface drainage systems, each taking care of their separate sources of water and moisture.

13.3.3 Surface Drainage System

The surface drainage system should remove all flow of rainwater from the road's surface, and from the highway slopes as well as the runoff from adjacent land. Surface drainage systems are also important in the proper management of polluted runoff and in minimizing environmental impacts. The surface drainage can be divided into transverse drainage and longitudinal drainage.

Transverse Drainage – Typically used to allow existing water courses to pass under/over the road which would, otherwise, form a physical barrier. These are normally constructed as aqueducts or culverts (see Chapter 12, Fig. 12.3).

Longitudinal Drainage – The main objective is the fast collection and removal of the rainwater that falls upon the road's immediate surroundings, and of the water from the adjacent areas, edges, excavation slopes and central reserve. This is fundamental for maintaining the safety of traffic by eliminating water films and puddles from the road surface (which can result in aquaplaning) at the same time reducing the possibility of water infiltration into the pavement's layers or foundation, which may reduce its load carrying capability.

Longitudinal surface drainage systems include gutters, channels, ditches, swales, galleries and collectors, complemented by their respective manholes, catchpits and sumps. Surface drainage is not the main topic of this book, so readers should look elsewhere for detailed information on this topic (e.g. Kasibati & Kolkman, 2006).

13.3.4 Subsurface Drainage System

Subsurface drainage is made up of different parts but all are linked directly with the surface drainage system and all are, fundamentally, taking care of groundwater or water that infiltrates through the pavement surface.

13.3.4.1 Drainage Regime

Part of the rainfall-runoff infiltrates into the ground and continues as subsurface flow. Part of this may, in turn, continue as a sub-horizontal subsurface flow, depending on the permeability of different soil layers. This can lead to increases in the moisture level under the pavement, reducing the bearing capacity. To decrease this phenomenon, the scheme design should note that:

- it is a good practice in embankments in the Mediterranean countries either to keep the thickness between the underside of the pavement and the natural soil to at least 1.0 m, or, if necessary, a drainage layer (see below) as well as other measures should be used, depending of subsoil characteristics; and
- in a cutting, the depth of the lateral drains should allow for an adequate drainage depth to the groundwater level, normally more then 1.0 m.

Because of the different types of subgrade and pavement construction, it is important to be able to differentiate between those pavements where water flow will be largely vertical, wetting the subgrade (with the associated loss of subgrade support strength) and those situations where vertical water flow will be arrested by impermeable layers in the sequence, forcing the water sideways and necessitating different drainage measures. Figure 13.4 shows three types of pavements, A B and C, which are now described.

A. Subgrade layer with low permeability – infiltrated water will flow above the subgrade, at the bottom of granular base/sub-base layers, according to the maximum crossfall;
B. Permeable subgrade layer and impermeable subsoil – infiltrated water flows on the layer between the subgrade and impermeable subsoil;
C. Permeable capping layer and subgrade – the water percolates vertically through every layer.

The basis for choosing whether approach Case A, B or C applies is set out in Fig. 13.5.

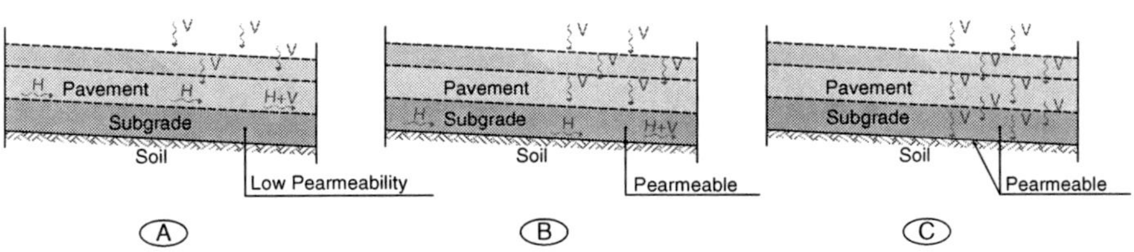

Fig. 13.4 Cases for water infiltration and flow in the pavement sub-surface

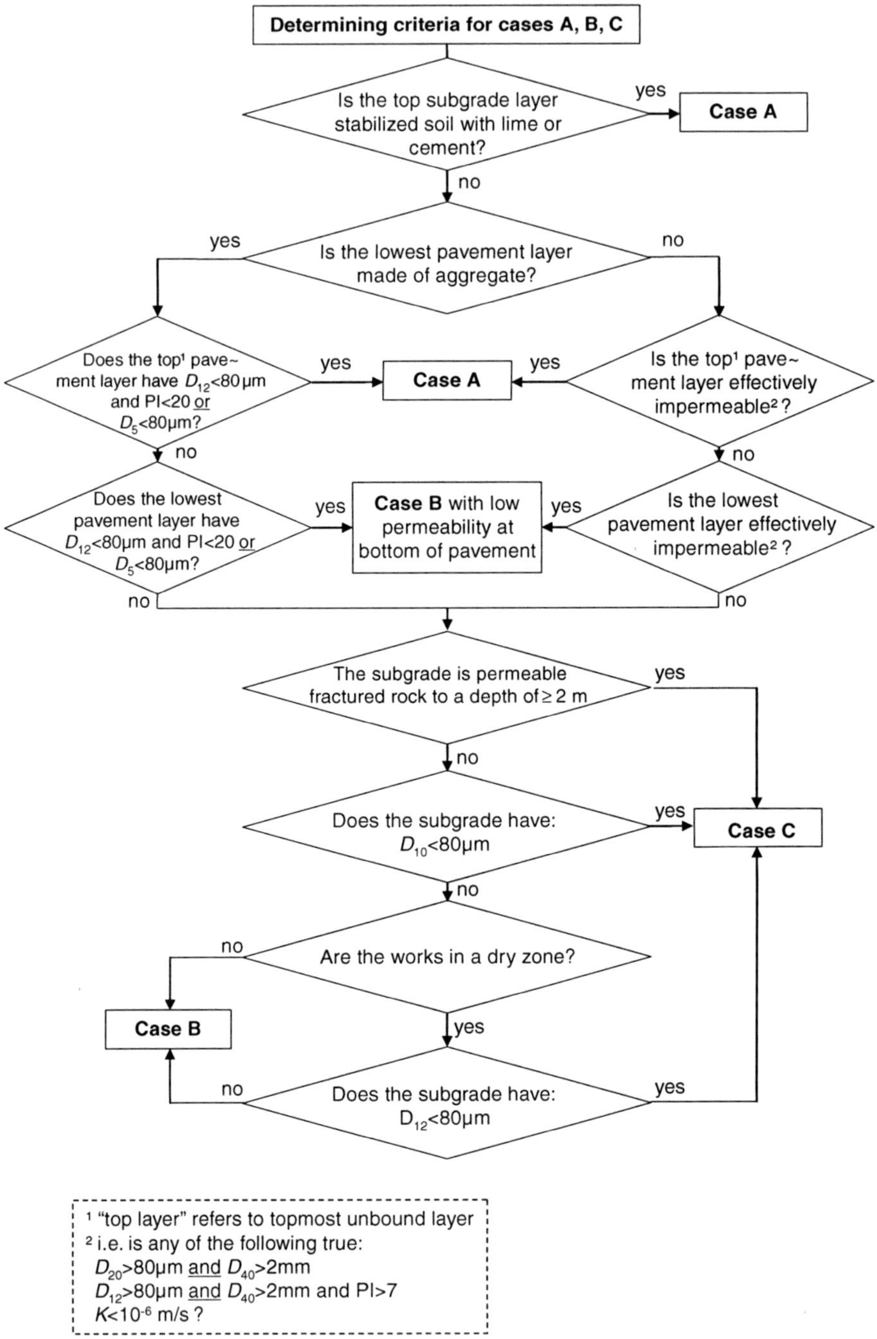

Fig. 13.5 Selection of appropriate drainage approach as per Fig. 13.4. For explanation of the D_n notation, see Chapter 2, Section 2.4.1

13.3.4.2 Longitudinal Drains

In drainage design, an undamaged asphalt surface is considered almost impervious. However, water can infiltrate into the structure through cracks and joints (see Chapter 5, Sections 5.3 and 5.4.2). Also, shoulders and slopes with higher permeability and high water tables can allow significant amounts of water into the structure. Whether the water arrives via cracks in the asphalt and is flowing through the pavement according to regime A or B, is threatening to arrive from the margins, or is simply close to the underside of the pavement due to a high phreatic surface, it is then necessary for longitudinal interceptor drains to be provided or other drains that will lower the water table or keep it in a low position. Many types of longitudinal drain are available as described in Section 13.3.9.

In many countries, especially the Mediterranean ones, where roads have open verges and slopes (especially where there are flat areas and hollows) it is customary to construct the verges/slopes with an impermeable surface cover using soils with a percentage of fines typically more than 25% of its weight ($D_{25} < 80\,\mu m$), with a minimum thickness of 20 cm to limit water seepage into the pavement structure.

13.3.4.3 Drainage Layers

A rather common subsurface drainage system used to remove the infiltrated/seepage water from the pavement structures is by providing a permeable layer. Permeable layers should be at least 10–15 cm thick and extend under the full width of the roadway. They can be used under both concrete (PCC) and asphalt (AC) pavement surfaces. Permeable bases are usually located just above the subgrade and are discussed in more detail in the second part of Section 13.3.6. Permeable unbound granular bases must be separated from high plasticity subgrade soils by mean of geotextiles or impervious materials.

The drainage layer should drain into a longitudinal drainage pipe. In order to encourage the lateral flow of water, a minimum cross-fall should be considered, of, at least, 2%. For curved lengths of road and those with a permeable central reserve, the pavement bed must have a cross-fall of between 2% to 4% inclination, starting 1.0 m away from the paved area (marked with E in Fig. 13.6).

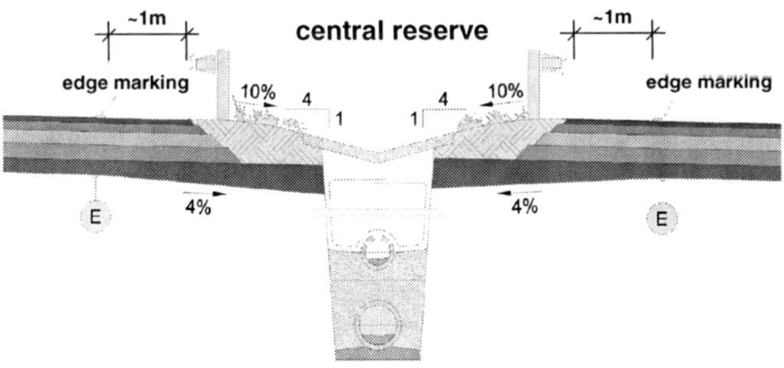

Fig. 13.6 Typical pavement cross-falls. E = position 1 m inwards from the edge of the pavement. Similar cross-falls will exist at road edges

13.3.4.4 Porous Asphalt

Although they may not be considered as part of the drainage system, a pervious type of asphalt treated surface layer, known as porous asphalt, has become common in Europe in recent years. The main advantages attributed to porous asphalt layers are noise reduction, improvement of skid resistance in wet weather, and enhancement of runoff water quality. Asphalt treated drainage layers of this type are discussed further in Chapter 5, Section 5.7.

13.3.4.5 When Drainage is Unnecessary

Some authors suggest that subsurface drainage may not be necessary if:

- annual rainfall is not significant;
- the subgrade has a relatively high permeability value;
- the pavement is structurally adequate without drainage;
- lateral and vertical drainage in the pavement section exceeds infiltration; or
- heavy traffic level is negligible.

For example, Christopher (1998) found that drainage provides no additional benefit if average annual rainfall is less than 400 mm and permeability of the subgrade exceeds 3.5×10^{-5} m/s, however Dempsey (1988) and Forsyth et al. (1987) suggested different values, 3×10^{-6} m/s and 1.7×10^{-4} m/s, for this parameter.

13.3.5 Combined Drains

Sometimes it is desirable to combine surface water runoff collection and sub-surface seepage water into one, combined, drainage system. Typically, such systems are in the form of lateral trench drains (see Section 13.4.1) that are open to the surface. These are the traditional means of draining roads and continue to be used on lower-trafficked roads where run-off contamination is lower and runoff volumes smaller (because the road is narrower).

Their principal drawbacks are:

- Stone scatter – aggregate in the drain can be disturbed by over-running vehicles leading to a safety concern from loose stones being thrown into the air by vehicle tyres;
- Contamination of seepage waters – contaminated runoff is introduced almost immediately into the sub-surface without any mitigation measures thereby spreading the pollution (c.f. Chapter 12, Section 12.4);
- Size – it is necessary to build the drain large enough to handle storm events – a size that would not be necessary for most sub-surface drainage requirements; and
- Backing-up – the possibility of the drain being temporarily filled during a storm event allowing water to seep back into the subsoil and road construction – the very place from which it is supposed to be draining.

The Highways Agency (2006) publishes a table that indicates when it is sensible to consider a combined system and when not to do so.

13.3.6 Drainage Layers

13.3.6.1 Drainage Layers in the Pavement

The sub-surface drainage system often includes a (permeable) drainage layer in the pavement. Its function is to quickly remove water entering the pavement layers, either through infiltration to the groundwater or to a sub-surface drainage system, before any damage to the road can be initiated. It is common practice to include drainage layers where the groundwater level is high compared to the location of the road, where the subgrade soil has low permeability and on high class roads (Summary of replies to the WATMOVE questionnaire, www.watmove.org).

There are situations where a drainage layer is not considered necessary. Responses to the WATMOVE-questionnaire show that a majority of countries assess the necessity of a drainage layer before deciding to include one. If the groundwater level is low compared to the location of the road, the subgrade soil has a high permeability or if it is a low-class road, a drainage layer may not be included.

There are different national traditions for which layer in the pavement works as the drainage layer, when included. Answers to the questionnaire show that some countries only use one layer as a drainage layer, i.e. the sub-base, whereas other countries mention (e.g.) three different layers. Dependent on the available material and the specific construction, any one or more of these is used as a drainage layer. The questionnaire responses of pavement engineers in many European countries are shown in Fig. 13.7.

An advantage of having the drainage layer just above the subgrade is that it can also act as a capillary break which is highly desirable in cold climate areas to prevent frost-generated water movements from the subgrade into the pavement (see Section 13.3.6). If the drainage layer is placed on the top of subgrade, the

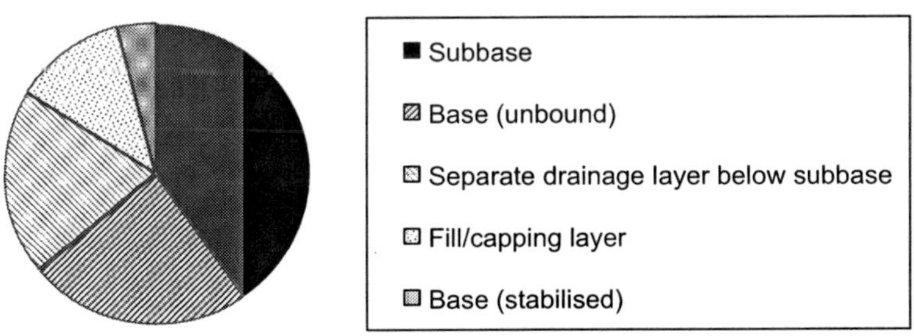

Fig. 13.7 Layer used as primary drainage layer in European pavements. (% of countries indicating that they use this layer. It was possible to indicate more than one layer)

permeability of the granular base and sub-base must be greater than the infiltration rate,[1] so that water can flow freely to the drainage layer.

Dawson (1985) reported that many UK sub-bases (typically containing as much as 10% by mass less than 75 μm) act more like sponges, absorbing water, than as permeable drainage materials. Jones & Jones (1989) report measurements of coefficient of horizontal permeability in the range 1–60 × 10^{-3} m/s for typical aggregate sub-bases and Roy & Sayer (1989) report in-situ measurements (by injection) of 2–110 × 10^{-3} m/s for similar materials. In other granular base course injection tests, Floss & Berner (1989) quote permeability values between 10^{-6} and 10^{-4} m/s for sandy gravel aggregates in-situ. Biczysko (1985) tested broadly graded aggregates in the laboratory for their horizontal permeability, obtaining reliable permeability coefficients in the range 2–50 × 10^{-4} m/s. However, he found that the aggregates with finest gradings (10% of material finer than 75 μm) were difficult or impossible to saturate – yielding an apparent permeability of 3 × 10^{-6} m/s – indicating that the lowest values are likely to over-estimate the in-situ behaviour and that substantial

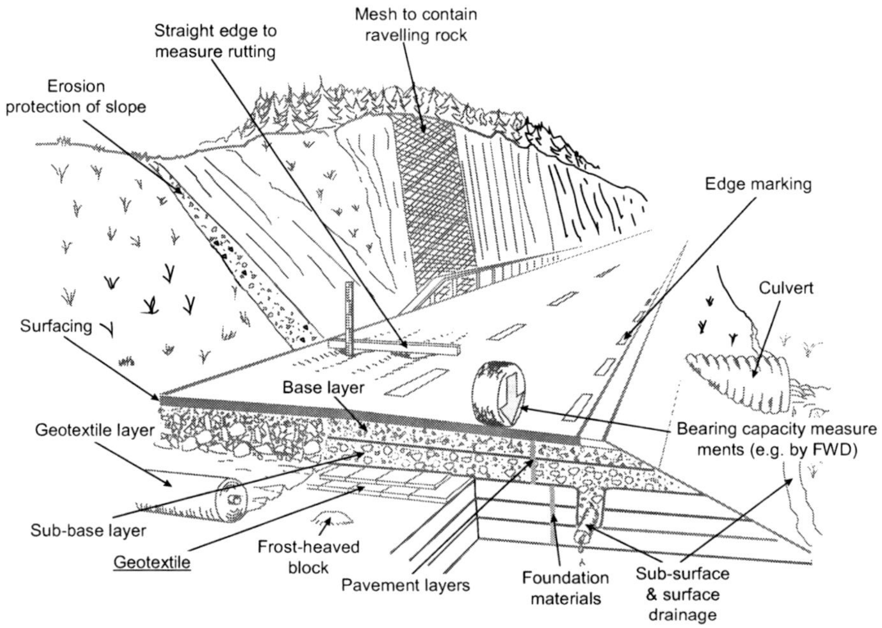

Fig. 13.8 The drainage layer is often directly connected to the longitudinal drains, but can also continue to the open slope side. In this example, surface water runoff also arrives at the drainage layer at the side of the road. Therefore the figure shows a sub-base drainage layer where water can flow either way (from Vägverket, 2004). Courtesy of the Swedish Road Administration

[1] This can be said because, under vertical flow, there is a hydraulic gradient of 1 so the Darcy flow velocity has the same numeric value as the coefficient of permeability.

through flow is unlikely to take place. The laboratory values reported by Jones & Jones (1989) and by Biczysko (1985) were both obtained by a specific permeameter designed for highway aggregate testing (Department of Transport, 1990) as described Chapter 3, Section 3.1 (Fig. 3.6).

Therefore, it is *not* sufficient to have a granular base or sub-base layer and assume it will drain. Instead, when the base or sub-base layer is also to act as a drainage layer as in Fig. 13.8, its material must satisfy both the requirements of strength and the permeability requirement for a drainage layer. When the layer is on top of the subgrade, the material may also have to satisfy requirements to act as a capillary break and filter, so that the fine particles of the subgrade soil do not migrate into the drainage layer (see Section 13.3.9).

An overview, resulting from the questionnaire mentioned above, of the requirements to ensure a successful drainage layer is given in Table 13.1.

Table 13.1 Requirements for drainage layers

Requirement	No. of countries having requirement	Value/type of requirement*
Grading specification	13	Maximum percentage of fines normally limited. In countries with cold climates the percentage is limited to somewhere between 5% to 10% passing 63 μm. In Mediterranean countries the percentage is usually larger as, in these climates, there can be a risk of layers being too dry. The fines ensure some suction. Some countries also set a requirement for the coefficient of uniformity.
Mechanical performance	11	Stiffness (i.e. plate bearing test or CBR) (6) Rate of compaction in-situ (2) Durability (1) Los Angeles value (2) Type of rock (1) Compression strength of rocks (1) Soundness test (1) Atterberg limits (1)
Change of design with increased width of pavement	8	Thickness (4) Permeability and thickness (2) Thickness and crossfall (1) Crossfall (1)
Design permeability (saturated)	4	$K \geq 1 \times 10^{-5}$ m/s (Germany) $K \geq 9.26 \times 10^{-5}$ m/s (Poland) $K \geq 10 \times 10^{-5}$ m/s (Slovenia) $K \geq 10.58 \times 10^{-5}$ m/s for drainage blankets laid beneath or within the pavement structure (Romania)
Design drainage time	4	5 h, defined as time to 15% saturation (Spain) 2 h, −50% saturation (Virginia, USA) 48 h, −40% saturation (Romania)

*Name or number in parenthesis is countries using the requirement. A total of 16 countries took part in the survey.

Care must be taken if a permeability coefficient and a certain grading envelope are both specified. It would not be difficult to specify one and thereby prevent the other from being achievable. While the relationship between grading and permeability can not be precisely defined (see Chapter 2, Section 2.5.1), controlling one will certainly have a large effect on the other.

The performance of the drainage layer does not necessarily stay unchanged with the passage of time. Some countries report that they have noted that the layer might become more or less clogged with time. The fines content might increase caused by degradation of aggregates and/or migration of fines from other layers. This causes decreased permeability and increased frost susceptibility.

13.3.6.2 Open-Graded Drainage Layers (OGDLs)

According to Huang (2003) the placement of a drainage layer directly under the asphalt or concrete pavement surface is preferable, because the water in the pavement, either percolating through cracks or entering from the sides, is quickly allowed to move to a lower level from where it can easily be drained. No pore pressures can develop because of the high permeability and rapid dissipation properties of the OGDL eliminating any chance of pumping occurring. Furthermore, it eliminates the final, negative, effects of water or frost. A properly designed and constructed permeable granular base layer may have a similar structural performance as a conventional base. However, OGDLs have a number of disadvantages:

- the deficiency of fines in the drainage layer may cause stability problems. They are difficult to compact into a stable foundation on which higher pavement layers can be constructed. Perhaps even more problematic is the trafficking of the OGDL by the construction plant that will lay the next pavement layer;
- the water in the sub-base cannot drain readily into the drainage layer;
- if the outlet becomes blocked the drainage layer becomes a reservoir for pore water that can develop pressure pulses under passing traffic thereby causing erosion and loss of bearing capacity. This can be a particular problem under jointed pavements where ingress of water along joints may be rather large once joint sealants have failed.

A recent study in Finland showed that more open-graded sub-base is a satisfactory means of reducing the moisture content of a granular base course, but that such a sub-base should not be used beneath an open-graded base course due to stability issues.

Typically, open–graded unbound granular materials are used as permeable layers, however the use of cement or asphalt treated permeable bases can add some extra strength and stability to the drainage layer (and, hence, the pavement) if needed. The resulting material is some kind of buried "no-fines" concrete or porous asphalt. Normal OGDLs are relatively expensive to source, due to the wastage of the fine fraction, and expensive to compact due to the stability issue. Treatment only adds to this.

Table 13.2 Example of gradation of unbound granular permeable bases in Spain

Sieve size (mm)	25	20	8	4	2	1	0.500	0.250	0.063
Passing (% by mass)	100	65–100	30–58	14–37	0–15	0–10	0–6	0–4	0–2

Although authors differ over the precise value, pavement layers can be considered as permeable when their coefficient of permeability exceeds approximately 10^{-5} m/s. Therefore, many of the conventional dense-graded unbound granular layers cannot be considered as permeable. Table 13.2, shows a typical grading of permeable granular layers used in Spain, where material passing the 1 mm size sieve is limited to 10% (by mass). Since sometimes segregation problems have been detected when pavers are used to lay the layer down, a uniformity coefficient, C_u, less than 4 is required. In addition, in order to get enough stability during the construction of the layer above, the permeable base must mostly be made of crushed aggregate.

Illinois experimented with the use of thin treated OGDLs (7.5–15 cm thick) beneath both concrete and asphalt surfaced pavements during the late 1980's and early 1990's (Winkelman, 2004). Four projects were constructed to monitor the effectiveness of the drainage layer and the performance of the pavement. Five additional projects were constructed based on the early performance of the monitored projects. However, continued monitoring of the initial projects, and additional projects, indicated two of the pavements were quickly deteriorating. Surface pavement distress, severe lane to shoulder differential settlement, and high pavement deflections for these two projects indicated a failure of the pavement structure.

The immediate response was to stop using OGDLs, although the non-failing pavements built over OGDLs were not removed. Six of the projects were monitored through from construction until 2003 to ascertain the longer term performance of these pavements and to see whether the OGDL had been beneficial. Some contained cement-treated OGDLs, some asphalt-treated OGDLs.

It was concluded that:

i) The use of an OGDL is more expensive than the use of a standard stabilised base material or lime modified soil;
ii) Some limited benefit due to drainage may be achieved, particularly early during the life of the pavement. However, the longer term performance of the monitored pavements was not much better than that typical for other pavements in the area built without OGDLs;
iii) The intrusion of fines from the subgrade and the aggregate separation layer into the OGDL resulted in settlement, faulting, and eventually premature failure of the pavement, therefore, the use of a geotextile fabric or dense graded aggregate filter under the OGDL to prevent the intrusion of subgrade fines is recommended;
iv) The benefits of using cement-treated OGDLs over asphalt-treated OGDLs, or vice versa, could not clearly be determined;

v) For continuously reinforced concrete pavements, the limited benefits of using an OGDL do not outweigh the increased costs, construction difficulties and maintenance requirements;
vi) Segregation is problematic during construction and careful use of plant is needed to minimize it; and
vii) Careful consideration should be given to subgrade soil analysis, topography and surface drainage, and pavement type. OGDLs are not suitable for all situations.

Most of the pavements were quite heavily trafficked and it may be that different findings would have been obtained on more lightly trafficked pavements.

13.3.7 Cold Climate Effects

It is a design objective that snowmelt and rainfall should have fast access to the side of the road and the drainage system. This avoids surface ice formation and skidding, but also protects the structure so that as little water as possible filters through to the pavement. Edges, kerbs, channels and runoff barriers must be kept clear by routine maintenance. Sufficient snow storage (and snowmelt) capacity on the road side must be included in the design. Dry structures and the road bed should be effected less by frost.

During spring-thaw there is considerable increase in the moisture of many unbound materials. The magnitude of the spring-thaw weakening (on bearing capacity and slope stability) depends, very much, on the functioning of the drainage system.

Due to snow cover on the road side, frost depth is typically deepest in the middle of the road and this will cause uneven frost heave across the section of road. The heave depends on how long the freezing front remains in the frost susceptible soil, the soil characteristics and also how much water there is (see Chapter 4, Section 4.6). If the frost heave differential is big enough, it can cause surface roughness, cracks, breakage of the surfacing and lifting stones (causing first roughness and later on surface cracking).

The permeability of most soils decreases when frozen. In springtime the frost can form an almost impermeable water barrier in the soil below the road since thawing will be fastest from the top and in the middle of the road. In that case the melted water will flow in a longitudinal direction until it can find an exit route, at which point it will flow to the road side or into the subsoil causing localised seepage and/or erosion problems.

Low permeability side-slopes decrease the infiltration of rainfall, but they can also act as flow barriers for the seepage of water out of the pavement structure and thus increase frost damage as well as decrease the strength of the structure. In that case openings with gravel filling at certain intervals and on low-lying locations can help the situation as in Fig. 13.9. Another option is a drain. A drain is the only option if the road structure is deeper than the open ditch at the side of the road.

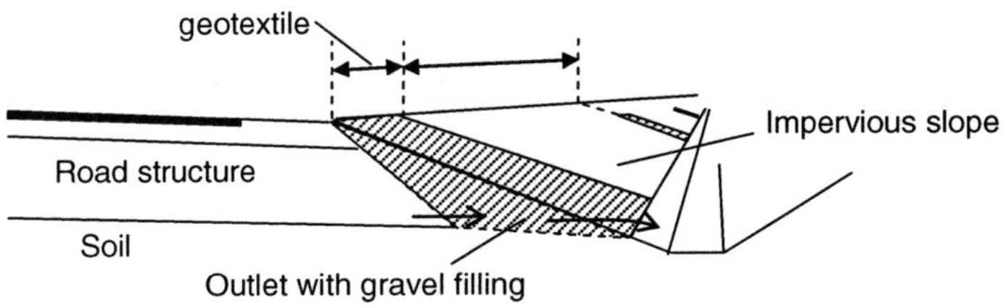

Fig. 13.9 Outlet with gravel filling on a low permeability slope

If there is no outlet ditch or drain available, and the amount of water is small, the water may be soaked into the soil. A transition wedge of coarse material and non water susceptible structure can be used to avoid frost damages at a low lying location where water is periodically accumulated.

In cold climates, where frost heave in winter is a problem, an open-graded layer can be constructed at the bottom of the pavement to prevent water being pulled from the subgrade due to a freezing front in the pavement. This will lower frost heave in the road construction by decreasing the water content compared to pavements without a capillary break. However, frost-susceptible layers will always heave to some degree if there are freezing conditions. For example, Hermansson (2004) reports a field study where an extremely well drained soil heaved 80 mm during a period of two months. The open-graded layer will not affect frost heave in a susceptible subgrade if the freezing front gets that low in winter, but it will significantly reduce the effect on the highway surface by allowing the road to bridge differential subgrade heaves.

More information on drainage of low volume roads in cold climates is given in a report by Berntsen & Saarenketo (2005).

13.3.8 Road Runoff Collection and Treatment

Where there are environmentally sensitive areas or high traffic flows, increasing the risks of accidents and generating contamination from wear, water flowing over the surface of the road and the embankment should be collected before it can soak into the ground in an uncontrolled manner. Water seeping through the earthworks and collected by a drainage layer has to be led, by virtue of a fall in the drainage layer and by shaping it, to collection points. At the collection point the water quality can be monitored and, according to the quality measured, it can be fed to a soakaway (Section 13.4.7) or piped away for treatment.

Water that arrives at an outlet from a drainage system may need treatment to bring the water quality to an acceptable level. In extreme cases, conventional waste-water treatment systems can be installed, but such a level of treatment is seldom required, except where water has passed through some contaminated soil body. If the water could, following a traffic accident or similar, contain spillages of fuel or cargo from traffic, then it will be necessary to install a oil/fuel separator. If these are installed then it will usually be sensible to provide a retention pond on the upstream side of the separator. This lagoon can act as a location to temporarily store the spilled fluid from which it can be extracted and taken away for off-site handling. The aim will be to give the road operator or environmental manager sufficient time to respond to the spill before the fluid is allowed to enter a surface water body or allowed to soak down to the groundwater. Storage lagoons also provide a zone in which solids can sediment from water and provide a means of attenuating the peaks of hydrographs.

Man-made wetlands and reed beds can also be provided as water purification areas (see Chapter 12, Section 12.3.1). They are particularly suited to groundwater outlets as the water arrival will be more consistent, at a low flow volume rate, than for surface water runoff and there will be much less need to provide storage for excess water arriving after a rain storm. Reed growth is much more suited to this consistent supply of water, while the area to be occupied by the water purification planting can remain relatively small. They often have the additional advantage of providing an enhancement to the pre-existing environment. Naturally occurring wetlands are under serious threat in most countries and they should not be used for this purpose.

The WATMOVE questionnaire sent to European road authorities (see www.watmove.org) revealed that settlement of solids and oil separation are the most common treatment methods used in practice, see Fig. 13.10.

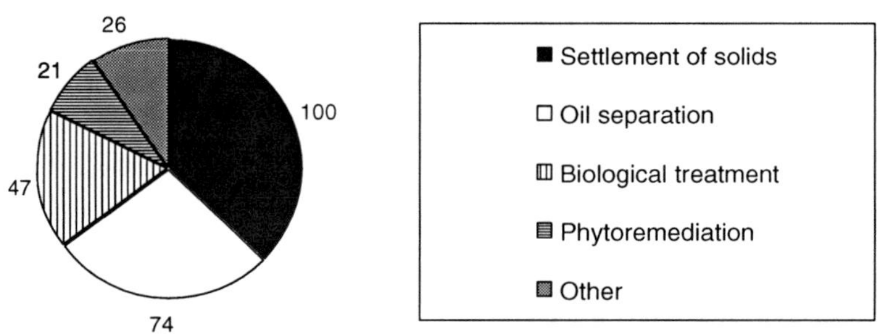

Fig. 13.10 An overview of treatment methods used in Europe. The number is the percentage of countries in the survey using the treatment

13.3.9 Filter Criteria

Aggregates and geotextiles employed in drainage systems have to operate next to soil or aggregate that surrounds them. To achieve good performance, they must remain permeable, retain the surrounding ground or aggregate in place and not clog. These requirements are met by defining specific performance criteria. The first of these is the non-sedimentation criteria, which is usually provided by aggregates with no plasticity and a limited amount of fines (usually no more than 5%).

The following filter criteria should be fulfilled:

- F_{15} (filter material)/D_{85} (soil) < 5
- F_{15} (filter material)/D_{15} (soil) > 5
- F_{50} (filter material)/D_{50} (soil) < 25

where the number, n, following F is the size of the sieve through which n% of a filter soil will pass and the number, m, following D is the size of the sieve through which m% will pass of the soil being drained. This terminology was introduced in Chapter 2, Section 2.4.1.

Typically, literature about geosynthetics also defines the following filter criteria (e.g. Bergado et al. (1996) who give a long list of references on the subject, Christopher & Holtz (1985) or Koerner (2005)). Firstly, to ensure that the filter holds the soil in place and doesn't let it through – the "retention criterion" – the following requirements are, typically, set:

- for fine-grained and erodible soils $O_{90} \leq 10\ D_{50}$
- for fine-grained cohesive soils $O_{90} \leq 0.12\,\text{mm}$
- for problematic (gap-graded) soils $O_{90} \leq D_{90}$
- for coarse-grained and well graded soils $O_{90} \leq 5\ D_{50}\sqrt{C_u}$ or $O_{90} \leq D_{90}$

where O_{90} is the effective opening size – 90% of the openings in the geosynthetic are smaller than this value, and C_u is the Uniformity Coefficient (see Chapter 2, Eq. 2.4).

Next, to ensure that particles of soil don't enter the filter's pores and clog it – the "clogging criterion" – the following requirements are, typically, set:

- $O_{90} \leq 5D_{90}$ although, for cohesive, soils O_{90}/D_{90} may be larger.

There will usually be a "permeability criterion", too, as the geosynthetic should remain at least as permeable as the protected soil throughout its lifetime. For this reason, the permeability of the geosynthetic should be N times greater than that of the soil it is filtering at the outset, where N typically varies between 10 and 100. To achieve this criteria such as the following may be set if permeability values are not obtainable:

- for fine-grained cohesive soils $O_{90} \geq 0.05\,\text{mm}$
- for coarse-grained and well graded soils $O_{90} \geq D_{15}$ and $O_{90} \geq 0.05\,\text{mm}$.

Where a filter is provided between the soil and a grooved, slotted or perforated medium such as a porous plastic pipe, the relationship between the dimension of the filter material and the grooves or the tube perforation size should be as follows:

- $F_{85} > 1.2$ times the grooves' width, or
- $F_{85} > 2$ times the perforations' diameter.

Where it is difficult to know the best design solution, drains can be filled with aggregate wrapped in a geotextile which has a coefficient of permeability 10 times more than the surrounding soil's coefficient of permeability.

The characteristics required of geotextiles and geotextile-related products for use in drainage systems is contained in European Standard EN 13252 (2000). In Europe, all geosynthetics used in drainage systems should be CE marked.

13.3.10 The Pavement as a Water Reservoir

A book on Water in Road Structures would not be complete without a brief description on the use of the pavement as a water store. Pervious pavements (see Chapter 5, Section 5.7) can be taken one step further and not only used to convey water away from the surface, but can also be used to temporarily store the water *in* the pavement. There are two principal reasons for doing this

- hydrograph attenuation and
- water quality improvement.

Here, the use of pervious pavements may only be briefly discussed, but interested readers may find out much more in the book devoted to the topic by Ferguson (2005). Typically they comprise highly permeable surfaces (e.g. of concrete blocks or stone cobbles) with a coarse, open-graded layer underneath that will act as a water storage and transport layer. It may also retain water that will, ultimately, soak away into the subgrade.

13.3.10.1 Hydrograph Attenuation

Normally, rain falling on impermeable surfaces, such as a road surface, quickly enters the drainage system, arriving at the outfall to river or stream very soon after falling from the sky. It is estimated (Interpave, 2005) that in a fully forested, lowland catchment only 5% of rainfall will flow across the ground surface, the remainder will be delayed by the vegetation to such an extent that it will soak into the ground. For agricultural land with less vegetation 30% may flow across the surface. However for an urban environment with piped stormwater drainage systems 95% is carried to the surface water bodies. For this reason, in an urban environment the water arrives much more quickly than if it had taken a natural route (movement over vegetated surfaces and by percolation through the ground). Thus the flow pattern in the river or stream rises and falls much more quickly, and reaches higher maximum flow values and lower minimum flows, than it would in a non-built-up area (see Fig. 13.11, explained in more detail below). This "peaky" flow leads to increased frequency of flooding and to reduced irrigation flows in times of drought as there is less water soaked into the ground to provide dry weather seepage supplies to surface water bodies. For these reasons, if a rapidly filling, but slowly emptying store can be provided in the pavement then this undesirable effect will be reduced.

In the example illustrated in Fig. 13.11, three rainfall events occur within a two-day period. It is assumed that this particular pavement can hold 20 mm of rainfall (i.e. 20 l per m^2). The first storm (0–8 h), which peaks at 5mm/hour, almost causes the pavement to fill with water. The pavement has drained back to half full when

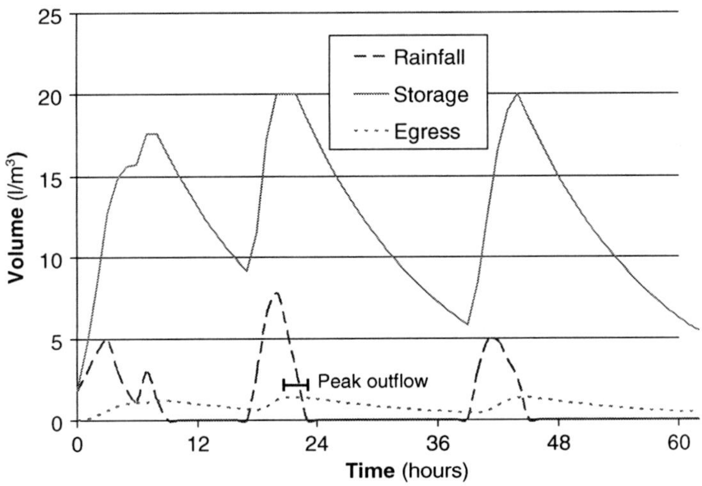

Fig. 13.11 Sample rainfall, storage and outflow hydrograph

the next storm (18–22 h) arrives. This storm, with peak loading of 8mm/hour does cause the pavement to fill for a short time, whereas by the time the third storm arrives (40–44 h) the pavement has drained further and the 5 mm/h peak intensity storm is just handled by the system. Note how the maximum outflow (egress) is about 1.5 l/m^2 when the pavement storage is full (21–23 h) very much less than the peak rainfall of 8 l/m^2 which would otherwise arrive into the drainage system and be fed to a stream or river.

This benefit is well illustrated by the performance of car park drainage pavements six years after initial construction reported by Brattebo & Booth (2002). Two of their pavements had grassed unbound surfaces (sand and gravel) held in a plastic grid arrangement while another two used concrete block surfacing with about 40% and 10% open area. Figure 13.12 shows rainfall and run off for a grass-sand and a reference asphalt pavement. Even for the heavy rainfall event (121 mm in 72 h) only 3% of this ran-off the surface of the permeable pavement whereas run-off from the asphalt pavement closely follows rainfall. The other three permeable pavements gave even less run-off.

13.3.10.2 Water Quality Improvement

The slow percolation of contaminated runoff into the pavement through a porous aggregate layer to an outlet substantially slows water movements, provides the possibility of filtration and allows water to pass by a large surface area of stone. By these means the water drops the suspended, fine, solids that it is carrying into the pore space of the pavement layers. The contaminants tend to adhere to the surfaces of the porous material's particles – particularly to the fine fraction as it provides the largest area of fines. Hydrocarbon contaminants also tend to be sorbed to the solids. For these reasons the water that leaves the pavement is substantially

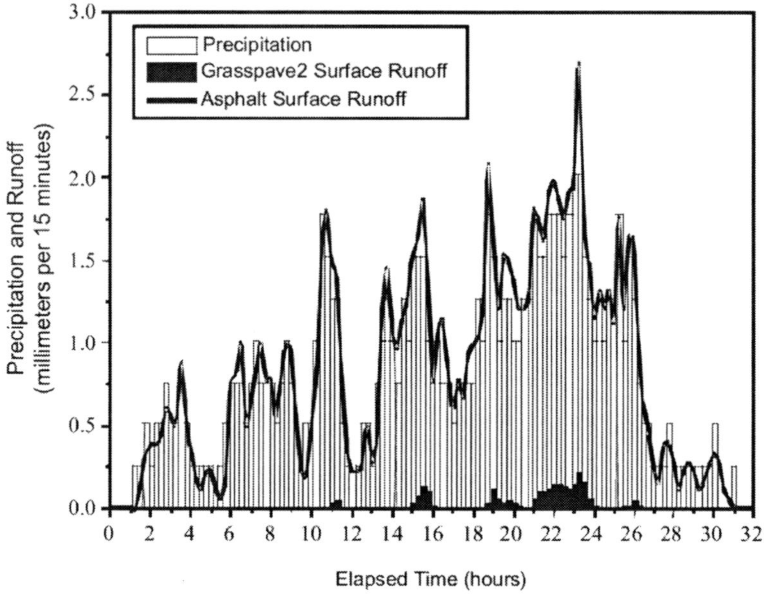

Fig. 13.12 Rainfall and runoff from two car park pavements (after Brattebo & Booth, 2002). Reproduced by permission of Derek Booth

cleaner than when it arrived. The possibility of cleaning runoff flows has been largely observational to date.

In the study performed by Brattebo & Booth (2002) water that percolated into the pavement was caught and its quality measured and compared with that of the asphalt pavement's run-off. Table 13.3 summarises the findings. While hardness (dissolved carbonate) and conductivity (influenced by content of salts) has increased, indicating dissolution of some species from the construction through which it has passed, Cu, Zn and oil concentrations are all reduced. Thus the pavements had some ability to clean the percolating water of chemicals commonly found in pavement run off. Far greater attenuation might be expected in real pervious pavements where the route to outlet (either by soakaway or by overflow pipe) would be far longer and cross more sorptive materials than in the experiments described here.

13.3.10.3 Pavement Constructions

There are several arrangements for pervious pavements. The most common comprise:

- A structural layer of porous asphalt (see Chapter 5, Section 5.7) over an aggregate base or sub-base with high void space,
- Open concrete or polymeric blocks with soil or aggregate infill over a structural and porous aggregate base,
- Permeable concrete block pavers laid on a permeable bedding course underlain by pervious aggregate or by "honeycomb" blocks.

Table 13.3 Mean concentrations of detected constituents in water running off or through 5 experimental pavements (from Brattebo & Booth, 2002)

	Hardness (mg CaCO3)	Conductivity (μmhos/cm)	Copper (μg/l)	Zinc (μg/l)	Motor oil (mg/l)
	Infiltration				
Plastic grid filled with sand. Grassed	23.4	48	1.29 (6)	10.8 (2)	<detection
Plastic grid filled with sand	14.7	38	<detection	14.3	<detection
60% open concrete blocks filled with gravel. Grassed	47.6	114	1.88 (4)	12.2 (3)	<detection
90% solid concrete with gravel	49.8	113	1.7 (7)	8.6 (3)	<detection
	Surface run-off				
Asphalt	7.91	14.1	9.07	22.2	0.183 (1)

The water from 9 storms was measured. The number in parentheses indicates the number of samples in which contamination was at less than the detection limit.

13.3.10.4 Other Considerations

From a pavement point of view, it is desirable to keep the aggregates in the pavement as dry as reasonably possible – so as to promote strength, stiffness and resistance to deterioration (see Chapters 8–10) – but the pervious pavement concept is directly in opposition to the general principle. Therefore other strategies have to be taken to make sure that the pavement provides sufficient strength, stiffness and durability for traffic loading. Normally this requires particular attention to aggregate quality. While, typically, aggregates to a 4–40 mm grading range, or similar, are required to ensure a suitable pore space for water storage, their requirement for durability may be higher than conventionally. There are greater stresses on particles consequent on them having an open structure with fewer contact points (see Chapter 9, Section 9.2). Also, more load spreading should be achieved, if possible, in the overlying pavements. Furthermore the design must allow for softening of the subgrade caused by the water stored in the pavement. Some design information is given in Woods Ballard & Kellagher, (2007) with more information given in Pratt et al (2002) and, for block pavements, Interpave (2005).

13.4 Current Techniques

13.4.1 Lateral Drains

In order to stop water getting into a pavement foundation and to avoid the consequent reduction of its support capability, the construction of ditches or trenches is a common procedure. They are usually filled with highly permeable material, wrapped in geotextile and with a perforated tube or porous material near the bottom. Alternatively a geo-composite based drainage system, known as a "fin" drain, may be used. Fin drains are, typically, only a few centimetres thick. Trenches or ditches are, usually, excavated by digging plant while narrow trenches for geo-synthetic fin drains may be dug or may be saw-cut.

These types of systems, named longitudinal drains, are placed in parallel with the road's centre line, usually at the edge of the pavement structure, and will lie under the surface water channels or gutters whenever these are permeable.

Lateral drains, according to their function, can be divided in two main groups, interceptor drains and water table lowering drains.

Water table lowering drains allow the lowering of the water level in the pavement structure platform, when close to the pavement's under-side. They are normally placed at depths varying from 1.2 to 2 m below the pavement surface so as to reach the water table (or to reach the height to which the water table may sometimes rise). Thus they keep the water table low and help to prevent the water from being pulled up to the pavement layers by capillarity.

Interceptor drains are drains of the same type as the previous ones, although they usually go down less far into the ground. They aim to ensure the internal drainage of the pavement and to intercept percolating water.

13.4.1.1 Trench Drains ("French" drains)

A trench drain consists of drain wrapped in geotextile, see Figs. 13.13–13.15. The drain is made of a mineral material such as a rounded or crushed aggregate. Originally either no carrier pipes or un-jointed pottery pipes were employed at the bottom of such drains. Nowadays, several materials are used for this type of pipe, from perforated or porous concrete, to PVC and fibreglass, the last ones with grooves or perforations. The pores, joints, perforations or grooves are designed to allow water collection. Typical diameters vary from 150 to 200 mm, with a longitudinal gradient that satisfies the self-cleaning condition ($\geq$ 0.25%). Whenever these drains reach

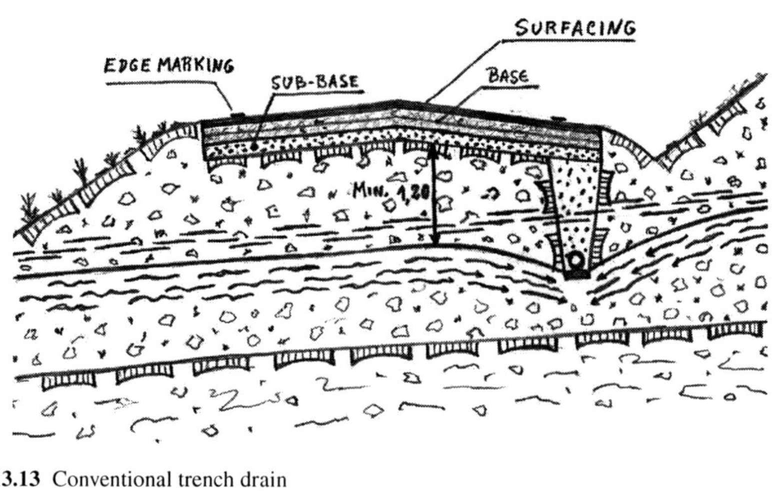

Fig. 13.13 Conventional trench drain

Fig. 13.14 Constructing of a trench (French) drain

their maximum capacity, a lateral pipe with the adequate discharge capability should be placed underneath the drains to take away water from the trench. The geotextile is employed as a filter which prevents migration of fine soil particles into the drain and its silting-up. The water permeability of the geotextile should allow water to flow freely from the surrounding ground into the drain. The characteristic pore opening size, O_{90}, of the geotextile used for the trench drain is selected to prevent mixing of soil with aggregate (see Section 13.3.9). Procedures for selection are provided by (e.g.) Christopher & Holtz (1985), Joint Departments (1995) or Koerner (2005).

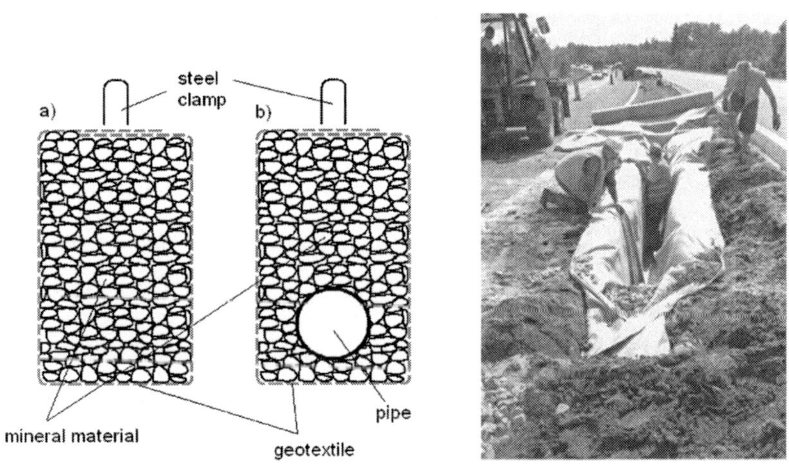

Fig. 13.15 Trench drain (**a**) without pipe, (**b**) with pipe, (**c**) in central reserve[2]. (Credit: B. Gajewska)

[2] Readers may like to identify violations of safe working practice which can be seen in this picture. The inclusion of this photograph does not mean that the authors condone such practice!

The cross-sectional area of drain is determined depending on the amount of water which ought to be carried away and on the grain-size distribution of the mineral material in the drain. Determining the dimensions is usually performed according to empirical procedures as the materials, climate, groundwater conditions and materials all have a pronounced influence upon the water that is to be conveyed and that can be carried by the drain.

The pavement layers must be shaped so as to ensure that water in them moves towards the drain and the base of the trench must be substantially lower than the layer to be drained:

- to ensure a suitable hydraulic gradient towards the drain which will drive drainage action;
- to aid entry across the geosynthetic liner which may require a small head difference before wetting is achieved and water passage possible; and
- to ensure that here is adequate capacity within the drain to hold exceptional water flow events without a risk of water flowing from the drain into the layer.

Thus a distance from the base of carrier pipe to the bottom of the layer to be drained of about 0.5 m is typical. Successfully designed and properly made, trench drains have a lot of advantages. Among other things there are (Użdalewicz, 2001):

- a lengthy life in which it works effectively;
- low construction and operating costs;
- the possibility of managing the "area above the drain", for example as a footway.

The following are the stages of construction of a trench drain:

i) digging of a narrow trench excavation;
ii) cleaning out the narrow trench;
iii) lining of excavation surfaces with geotextile (along the excavation or cut set across excavation) (e.g. Fig. 13.14);
iv) placing a covering of aggregate at the bottom;
v) installing the drain pipe, if needed;
vi) filling the drain volume with aggregate;
vii) closing the drain and jointing of geotextile edges (e.g. with "U" shaped clamps) (e.g. Fig. 13.15a & b); and
viii) covering the closed drain surface with 3–5 cm (or more) of top soil or other low permeability cover (except where surface runoff is also to be collected – see Section 13.3.5).

The minimal length of overlap for geotextiles should be at least 30 cm (Edel, 2002). The longitudinal direction of overlaps should be consistent with the direction of water flow through the drain. The aggregate filling the trench drain should be compacted (in layers).

13.4.1.2 Fin Drains

Fin drains or drainage screens are also longitudinal drains, manufactured from composite materials. Their essential make-up is of two geotextile faces that provide a filter function between the surrounding ground and a rigid plastic core that is sandwiched between the geotextile faces – see Fig. 13.16c. The so-called "drainage core" is, typically, formed of a high-density polyethylene, HDPE, structure. Often this feeds into an integral collector at the bottom. The core permits the water to flow in the plane of the geocomposite (compared with most simple geosynthetics

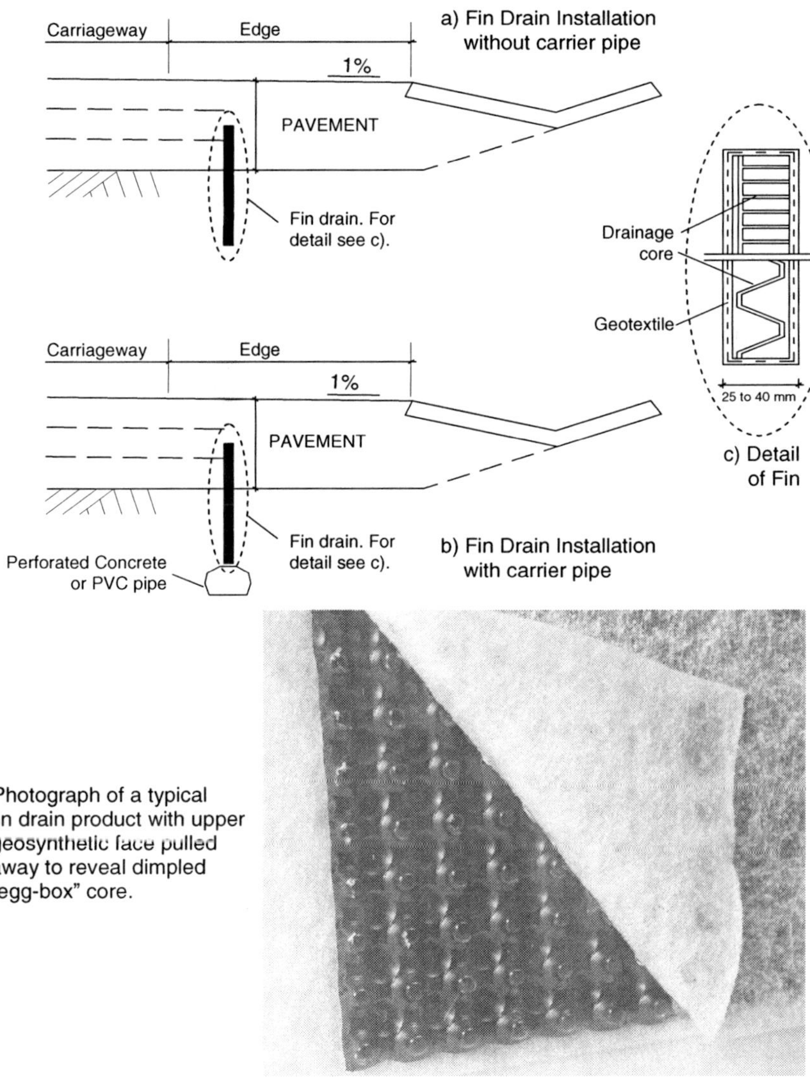

Fig. 13.16 Examples of geocomposites used in fin drainage systems

in which only cross-flow can readily take place). These drains are usually placed at the pavement's edge, allowing the collector of percolating water (Fig. 13.16a & b). Their main purpose is not to lower the groundwater level. The advantage of fin drains is their narrow thickness, allowing the construction of narrow trenches. This is particularly advantageous in improvement works. The disadvantage is that they are less able to carry high volumes of water and, thus, are less suitable where groundwater lowering of permeable soils is to be attempted.

A fin drain is normally supplied in the form of a roll of constant width. Thus, its height in the ground is constant, too. Care must, therefore, be taken that the base of the slot in which the drain is to be placed will fall evenly towards an outlet rather than following the vertical profile of the pavement alongside which it lies (Highways Agency, 2006).

13.4.1.3 Californian Drains

So-called "Californian Drains" are sometimes used. These consist of parallel and closely-spaced tubes, arranged vertically or sub-horizontally. The tubes can be perforated or grooved and are installed into natural ground or fill. The main objective of such drains is the reduction of pore water pressure in a certain area, in order to lower the water level or treat a retained water pocket. A typical application, to stabilise a slope with high pore water pressures, is illustrated in Fig. 13.17.

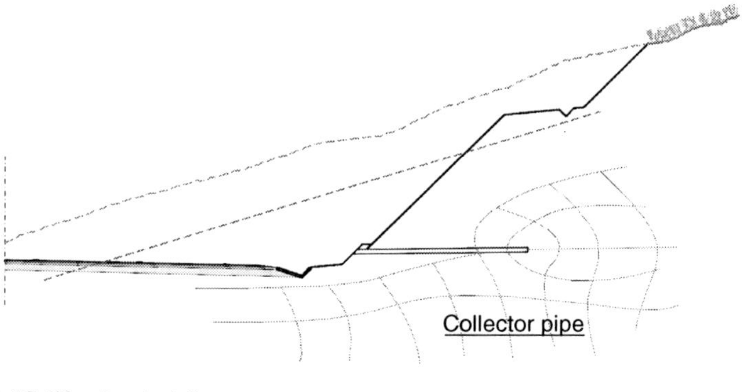

Fig. 13.17 "Californian drain"

13.4.2 Drainage Layers for Rigid Pavements

In a rigid, concrete, pavement structure, water filters mainly through

- open joints between the concrete slabs;
- open pavement-shoulder joints; and
- and the areas between the verges and the pavement.

Unless transverse drains are used (see Section 13.4.5), this filter water is collected by the drainage layers and is directed to longitudinal drainage pipes. The drainage layer should be constructed of a non-sedimentary granular material, placed between the pavement's aggregate base and the subsoil. It is also advisable to place a geotextile to separate the different layers. The minimum thickness of the drainable base should be approximately 0.15–0.20 m.

13.4.3 Transverse Drains in Rigid Pavements

Transverse drains are similar in construction to lateral drains, except that they run perpendicular or slightly skewed to the centreline of the carriageway (Kasibati & Kolkman, 2006). They are mostly used to drain water that may infiltrate into the road bases and sub-bases at joints, see Fig. 13.18 (FHWA, 1980).

It is important to provide drains at joints in rigid pavements as the joints' sealing can deteriorate with time allowing water to flow into the pavement. Perforated pipes are usually used as transverse drains and they may empty directly into the side ditches. Pipes as transverse drains are classified as passive drains, meaning that they do not constrain the water movement but they are placed in the hope that water will find its way into the drain. However, some of the water may move past the drain and cause some problems. Transverse drains should be used with caution in frost heave prone areas as differential frost action may damage the pavement structure.

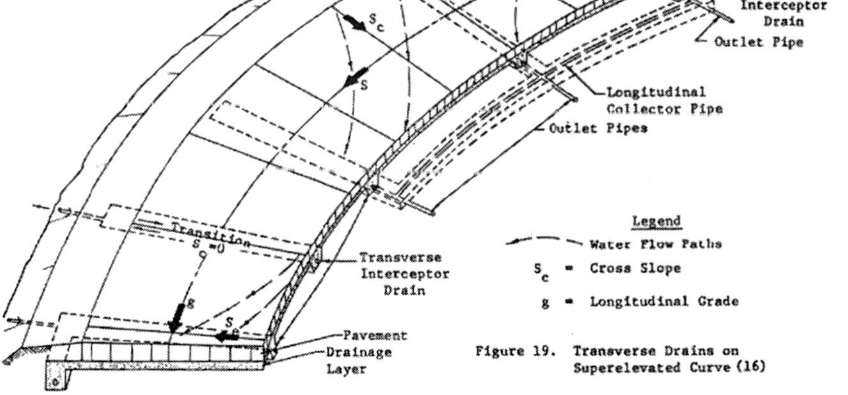

Fig. 13.18 Transverse drains on super-elevated curve. Courtesy of the FHWA (1980)

13.4.4 Earthworks Drains

This kind of drainage is installed with the aim of controlling waters emerging from earthworks, which includes not only the water that appears at the base of the excavations but also the flow coming from excavated slopes. Five types of systems can be used:

13.4.4.1 Drainage Layers

These consist (see Sections 13.3.4 and 13.3.6) of a layer of granular material with constant thickness (normally between 0.40 and 0.60 m) that is spread at the base of the excavation along the formation, or, for an embankment situation, at its foundation (Fig. 13.19). This layer is placed between geotextiles having separation and filter functions.

Fig. 13.19 Drainage layers

At the base of large embankments crossing deep valleys, the main aim may be to keep the water table at its original position in the old valley bottom. For this purpose it may be more desirable to use a series of trenches let into the original valley profile which lead water away before it can rise into the base of the embankment. Due to their appearance in plan these are colloquially known as "Christmas Tree" drainage systems (Fig. 13.20).

13.4.4.2 Drainage Masks

Drainage masks provide control of emerging water (e.g. a spring line) on a slope's face with a geotextile covered by hand-placed rock (Fig. 13.21). The material used for a drainage mask should be angular and should comprise 100–500 mm sized stone. The rock provides improved slope stability both by allowing the reduction of pore water pressures without erosion and by adding mass to the slope.

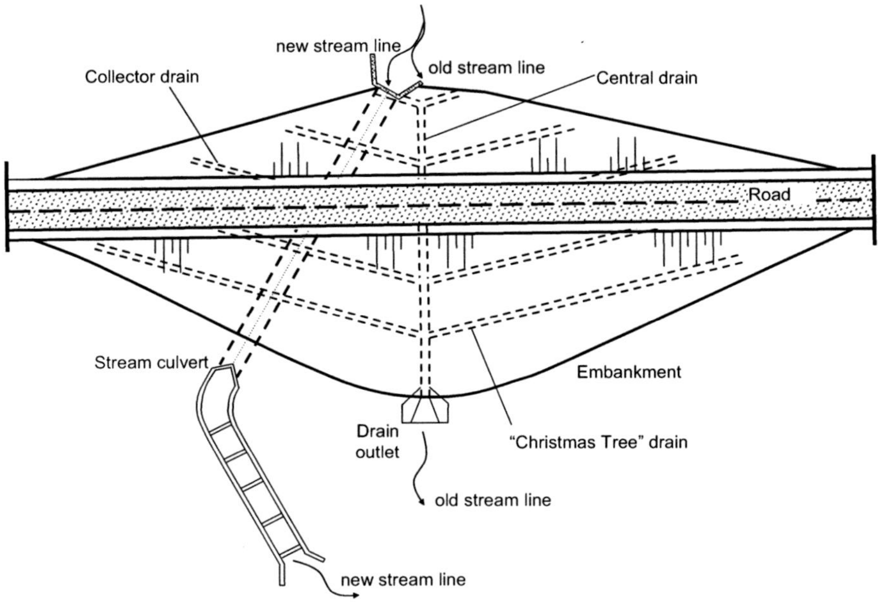

Fig. 13.20 "Christmas tree" drain

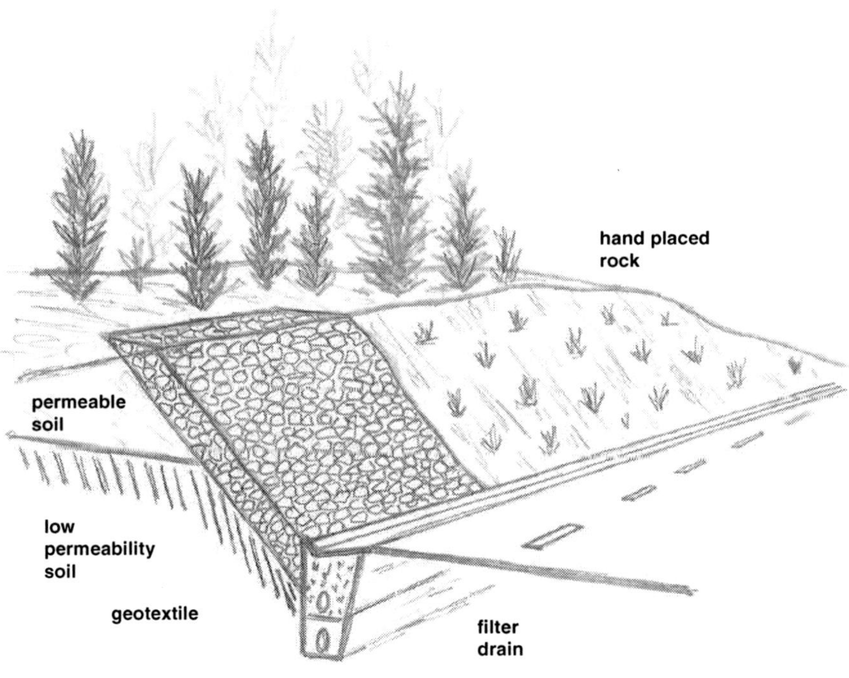

Fig. 13.21 Drainage masks

13.4.4.3 Drainage Spurs

Drainage spurs have the double function of draining and reinforcing the excavated slopes. Spurs are constructed as stone-filled trenches excavated perpendicular to the slope face (Fig. 13.22) to provide both drainage and buttress support. The material to be used in drainage spurs should be coarse and angular (e.g. broken rock in the size range 100–200 mm) and should be contained within some geosynthetic wrap to ensure soil does not migrate into the large void space. A heavy-gauge geosynthetic is indicated given the abrasive nature of the fill.

13.4.4.4 Wells

Wells are vertical shafts of sufficient diameter in order to lower the water level. They are usually associated with a pumping system and, hence, require constant maintenance.

13.4.4.5 Cut-off Drains

Drains can be placed as a trench or fin at the toe of a cutting, between the cutting and the pavement construction (Fig. 13.23). These act to lower the water in the cutting both increasing the stability of the slope and also reducing the water arriving at, and the pore water pressure in, the foundation of the pavement. Typically these cut-off drains also perform the function of lateral pavement drains (see Section 13.4.1). However, they differ from lateral pavement drains because they are designed to collect water from both sides of the trench or fin.

In porous ground where water tables are high, or high rainfall is anticipated, the drain must be designed to carry relatively large, continuous flows unlike normal lateral pavement sub-surface drains. This may have implications for the design dimensions of the carrier pipe at the bottom of the trench, for the regularity of points at which the trench is emptied to a surface water body or other outlet, and for the accessibility for maintenance. The last point deserves emphasis – if a normally operative cut-off drain ceases to function, water pressures will quickly rise in the toe of the cutting slope and in the pavement foundation, leading to rapid pavement deterioration and reduced cutting slope stability. In general, fin drains are less desirable when large flows have to be carried away as their capacity is normally less than that of a comparable trench drain.

Longitudinal cut-off drains may also be installed in sidelong earthworks so as to prevent water from ever arriving at the road's construction. Figures 13.24 and 13.25 give examples.

13.4.5 Pavement Underdrains

In excavated areas and cuttings where the longitudinal slope is more than 3%, a longitudinal water flow may appear fed by water from under the pavement that is

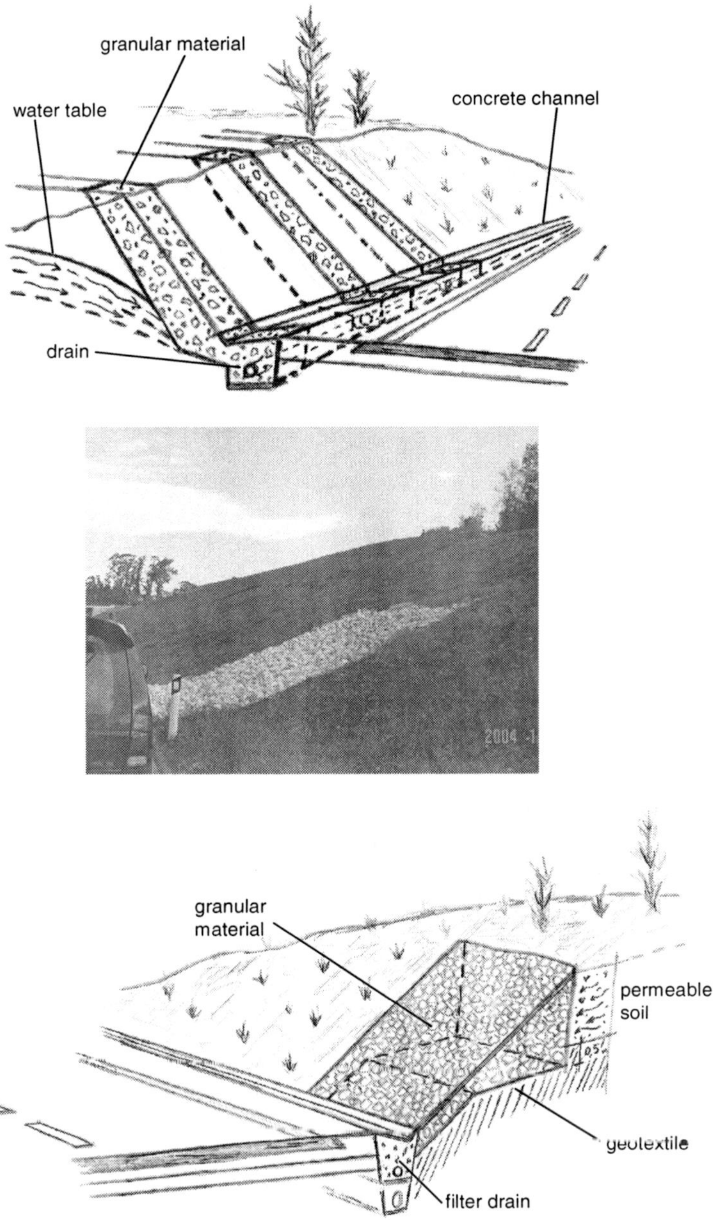

Fig. 13.22 Drainage spurs

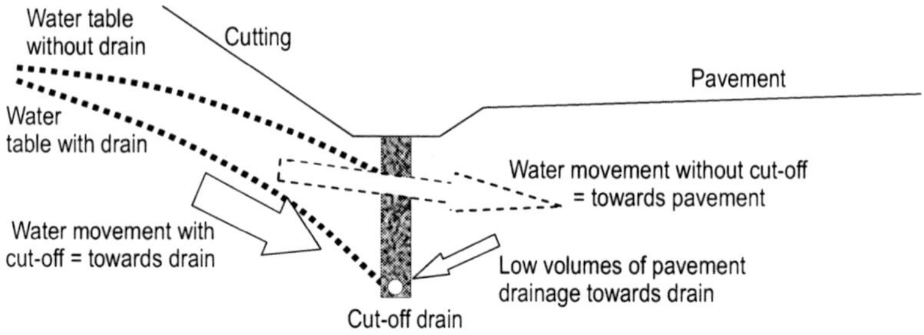

Fig. 13.23 Cut-off drain operation

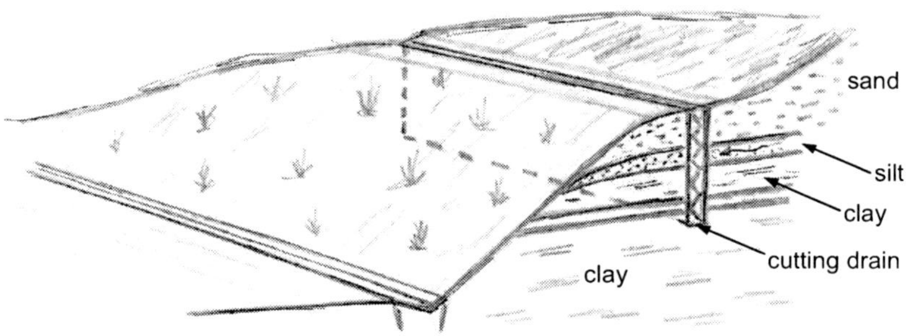

Fig. 13.24 Cutting drain

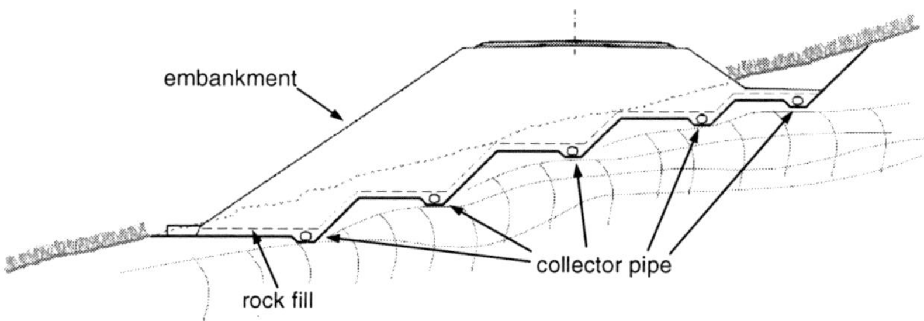

Fig. 13.25 Longitudinal drains in "$^1/_2$ hillside"

separate from the flow in channels, gutters and gully's. In these cases, the inclusion of pavement underdrains (Fig. 13.26), installed transversally under the pavement, can be used, in order to collect all subsurface waters. These kinds of drain are best constructed in transition areas and, in areas of excavation or fill, placed centrally to improve the rapid flow of the infiltrated water. In sandy soils they should be placed with a spacing between 20 and 25 m while, in very clayey soil, these distances should be reduced to about 5 m.

It is advisable that these transverse drains be constructed in a trench, filled with drainage material wrapped in geotextile, or by synthetic filters, constructed right

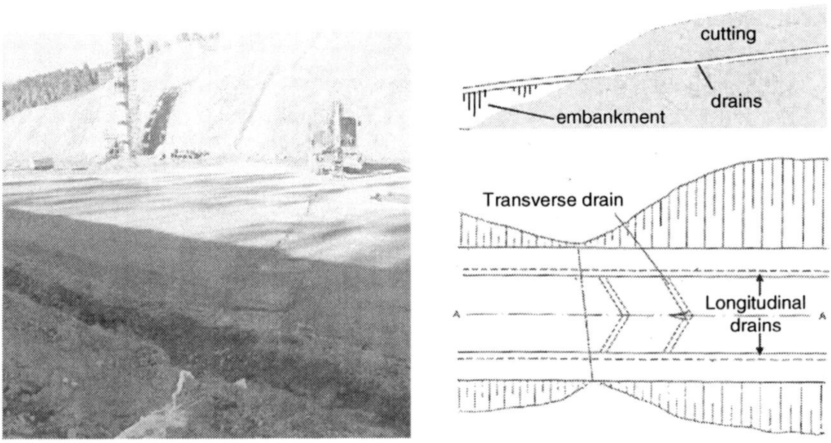

Fig. 13.26 Transverse drains

down below the level to be drained. In cold regions, the dimensions and depth of the network may be greater than in non-frost affected regions.

13.4.6 Deep Drains in Frost-Affected Areas

In colder climates, deep drains ("cut-off" drains) are used to reduce local frost damage by intercepting the flow of groundwater and seepage water under the road structure, usually where there is a crossfall (see Section 13.4.4). The depth is usually at least the design frost depth (e.g. in Finland, this is between 1.5 and 2.2 m). Lesser depths may be used if there is very low permeability soil below. Deep drains can be installed beneath open ditches but then some gravel cuts may be needed to connect the drain and the structure. However, to ensure the fastest drainage of the road structure during the thawing period, the best location is connected to the structure under the inner slope, see Fig. 13.27. A particular benefit over an open trench is that such a drain doesn't get clogged with ice. In most cases the need for a deep drainage or cut-off drain is due to longitudinal variation in the permeability of the subsoil. For example, a rock or belt of clay may change the flow direction to flow under the road bed (see also Section 13.3.7).

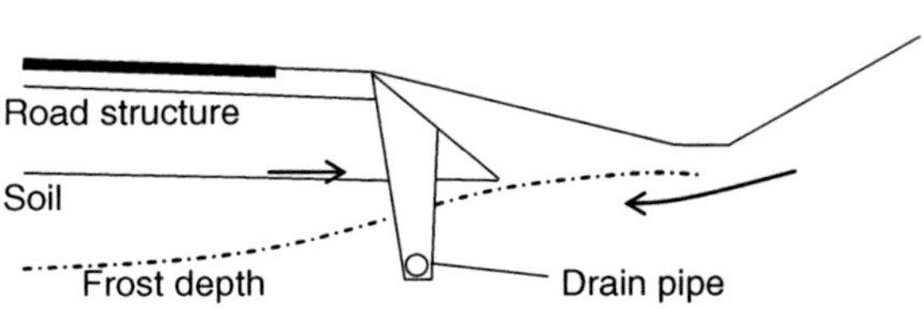

Fig. 13.27 A deep drain example from Finland

In road rehabilitation, the trench for a deep drain is made with a narrow bucket at the toe of the road embankment slope, a drain pipe is laid near the bottom (filter fabric first if necessary), initially filled with drain gravel or crushed stone # 5–10 mm and followed by a top filling with coarse gravel or similar material, or crushed rock. The sides should have a low permeability lining. Narrow channels should be blasted through rock thresholds. Inspection wells can be made, for example, from plastic culvert material at 400 mm diameter with a cover on the top and some silt storage at the bottom, at about 50 m intervals. A cover flap should be used at the drain outlet opening. The water from this outlet is led to a lateral ditch, to a diversion ditch or into a rainwater sewer. A wide gravel outlet can ensure a safe discharge in the case of a local blockage of the outlet especially if the drain opening is located in a low gradient area. A sunny position for the outlet also decreases the risk of freezing. If the collected water has to cross the road, a separate pipe is usually constructed, instead of a culvert, to avoid freezing problems.

Among the possible benefits of deep drainage are:

- decreased growth of vegetation and thus some lower maintenance costs;
- better support of the slopes and pavement edges;
- improved traffic safety since open drains can be shallow and the slope gradients low;
- reduced probability of cracking at pavement joints, especially for narrow roads; and
- sheet ice should be eliminated.

Because deep drainage may also lower the groundwater table in the long term, the risks and disadvantages for the environment have to be evaluated separately.

13.4.7 Dispersal and Soakaways

Water that has been collected from runoff or from sub-surface drainage systems has to be disposed of. The simplest means is to route it to a naturally occurring surface water body (stream or lake). Often a retention pond (see Section 13.4.8.1) is interposed between the water coming out of the highway and the outflow into the surface water body. The retention pond reduces peak flow rates (therefore making the outflow easier for the environment to receive without flooding) and may have environmental benefits, too (see Chapter 12, Section 12.4 and Section 13.3.8 in this chapter). However, such surface water aspects are beyond the scope of this book and interested readers are referred to one of the many texts dealing with surface water drainage. Alternatively, the runoff or seepage flows can be introduced into the ground via a soakaway. This is particularly attractive when there would be little fall between an outlet and the receiving surface water body. Sometimes water can be dispersed into a wetland area. Only purpose-built wetlands (see Chapter 12, Section 12.3.1 and Section 13.3.8 in this chapter) should be used and they should be designed to take this water although, in the past, it has not been unusual for wetland areas to develop around points where sub-surface water seeps to the surface. Areas handling sub-surface seepages should expect low flows over long periods compared with the short, "peaky" hydrographs associated with runoff.

Permitting water to soakaway to the ground is only permissible where regulation, or regulatory authorities, allow. In particular, the use of soakaways in areas where the groundwater is used for drinking water is very restricted or, in some countries, not permitted at all.

Water collected from embankment grips will usually be of acceptable quality for disposing by soakaway as it came from the natural subsurface and is returning to it. The issue is not, therefore, likely to be only one of quality (unless the cutting intercepts an already contaminated groundwater), but maybe more one of volume. Can the disposal soakaway disperse the water supplied to it without surcharging and without causing problems to the receiving groundwater levels? Whether seepage water from the road drainage system can be disposed in the same way is less certain. In many cases the water collected from seepage started as rain that fell onto the road construction, collecting contaminants as it did so. By the time it arrives at the potential disposal route it will have travelled through many pavement layers and a sub-surface drainage system during which sorption and natural attenuation processes may have removed much or all of the contaminant that it once contained (Dawson et al, 2006).

According to answers to the WATMOVE questionnaire (see www.watmove.org), approximately half the countries in Europe that use soakaways have requirements concerning the water that flows from pavements into them. Often the water has to go through some kind of treatment, i.e. as a minimum sedimentation and oil separation, or a minimum vertical infiltration/percolation time must be ensured by the design.

Soakaways need to be positioned to meet several criteria:

- that they encounter a substantial zone of porous ground. For this reason they cannot be used in clay soils. In moderate permeability soils, soakaways may be made more efficient by providing radiating "branches" from a central water store so that slow seepage over a wider area can provide sufficient rate of seepage;
- that they have sufficient capacity to hold most[3] of the water supplied to them through the connected drainage system until it has percolated into the ground. For this reason a soakaway must have void space that is above the maximum groundwater level; and
- that the soakaway space is kept open – either by filling the soakaway hole with coarse aggregate that has a high voids content and lining the sides to prevent wash in of fines, or by supplying a solid wall with openings (see Fig. 13.28).

For a successful design the volume of water to be drained; the return period for this amount of water to come into the soakaway again; and the percolation rate of the soil/sub-soil must all be assessed. Then a water balance calculation can be

[3] Typically soakaways should be able to cope with water flow from a two-year return period 'high' (or other agreed return period).

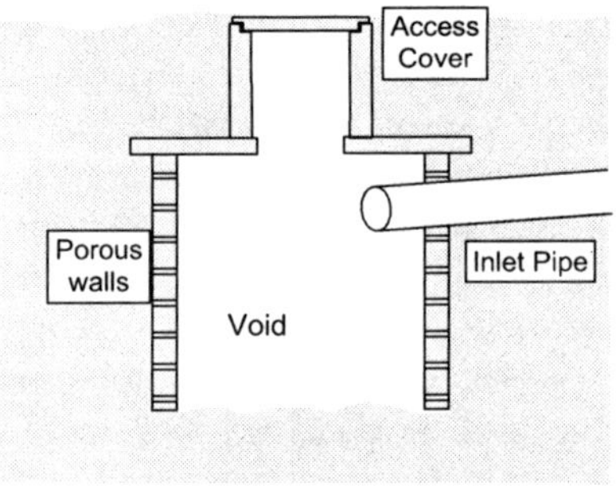

Fig. 13.28 Schematic of a walled soakaway

performed – for example by calculating the volume of water arriving at the soakaway in each hour, m, (from a knowledge of the rainfall intensity, P (mm/h), and the area being drained to the soakaway, A (m^2), and calculating the amount soaking into the ground, IR (m^3/h/m^2) (Eq. 13.1). The excess generated during the storm of length n (h) must never accumulate to a volume larger than the storage capacity, V (litres), of the construction.

$$\sum_{m=1}^{m=n}(P_m - IR_m) \times A < V \tag{13.1}$$

Swales (see Figs. 1.10 and 1.11) are a form of linear soakaway, with water being able to soak into the ground but, hopefully, leaving contaminants behind in the lining and vegetation of the swale. As a system for dealing with surface runoff they are beyond the scope of this book, but the water that they introduce into the ground does need to be considered by the designer of drainage of subterranean waters. They should not be so positioned that they will raise the groundwater levels in areas where this would decrease the stability of the surrounding slopes or where it would feed water into the pavement foundation, thereby reducing the bearing capacity of the pavement. For these reasons, the positioning of a swale for disposal of surface runoff presents the designer with a dilemma – too close to the pavement and it is likely to result in reduced pavement performance, too far away and it will not be easy to make it useful for its primary purpose.

13.4.8 Road Drainage and Treatments Systems

In recent years, it has been demonstrated that diffuse or irregular sources of pollution, like runoff, constitute significant contamination points in the natural drainage system. In most cases one has to take appropriate measures to control this pollution. In general, the environmental aspect of a road project nowadays constitutes an integral and fundamental part of the project, so that there is a strong interdependency between the various aspects considered in design and in the interventions that have to be planned. In particular, it will probably be necessary to perform an integrated assessment of the road project itself, the drainage, the water resources, the quality of the water, the geology and the geotechnical aspects, with special focus on hydrology and landscape studies. Only in this way can one balance need for the road and the impact it will cause and allow the design, dimensioning and implementation of optimal and adequate systems for the conveyance and treatment of water flowing from the road platform.

The choice of treatment system to adopt has to address diverse conditions:

- the type of pollutants to treat (in particular heavy metals);
- the regional climate;
- compatibility with the roads' drainage project;
- the impacts on the landscape; and, also
- the creation of a passive system, without the need of energy which, thereby, has great reliability and low maintenance needs.

Therefore, once the discharge points have been associated with particular, sensibly-sized, areas of the pavement platform, one can compute the discharge and proceed to study which treatment system should be adopted and its interaction with the drainage system.

The implementation of a project of this kind doesn't overcome the need for an effective control of the treatment efficiency and this may be achieved through the application of an adequate monitoring system for the site. This should commence operation immediately after the start of the road use. Data can be obtained by sampling the pollutant charge in the runoff on first use. It will form a baseline against which to check any eventual flaws in functioning of the recovery and treatment system. Monitoring will allow the operator to detect, and thus rectify, any defects in a timely manner. The successful implementation of a sampling program of this nature will be very useful not only for the scheme in view, but especially for new situations, helping to find better project options which can be implemented in subsequent projects.

In parallel one should promote periodic maintenance and conservation action in the drainage network and in the treatment areas. In fact, without adequate maintenance the investment made will be likely to prove fruitless.

13.4.8.1 Road Wastewater Treatment Options

The following figures illustrate some options for treatment systems. Figure 13.29 shows a system where little or no treatment is needed. Figures 13.30 and 13.31 show situations in which progressively more treatment is provided, while Figs. 13.32 and 13.33 show situations in which "hard" treatment solutions with settlement, retention and infiltration tanks are provided in some manner. Sometimes settlement and retention tanks can be entirely fabricated from concrete, on other occasions they can be formed of excavations in soil at a location where settlement of solids onto and in the soil is acceptable as the soil has been carefully selected and prepared to prevent

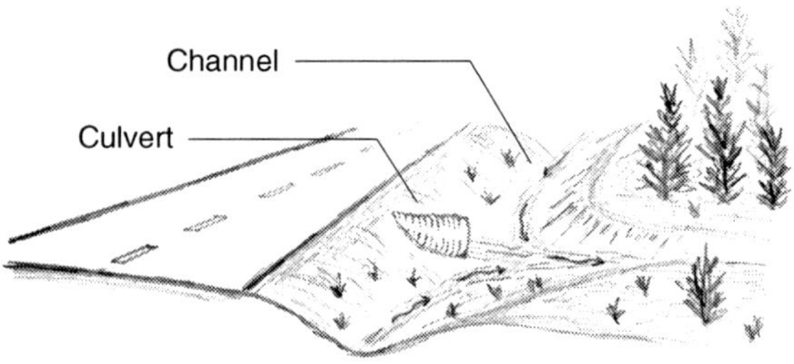

Fig. 13.29 Environment with low sensitivity

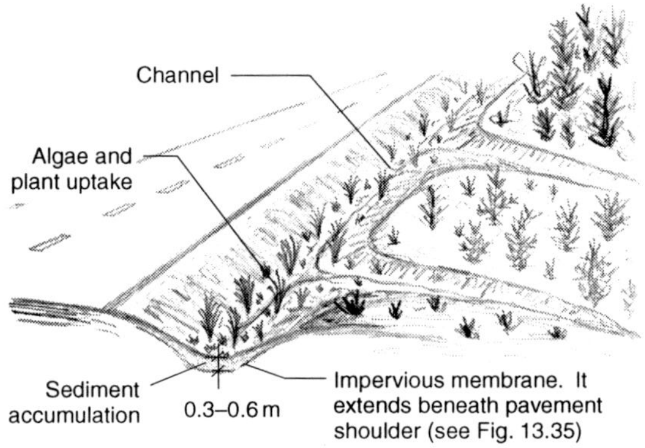

Fig. 13.30 Environment with high sensitivity

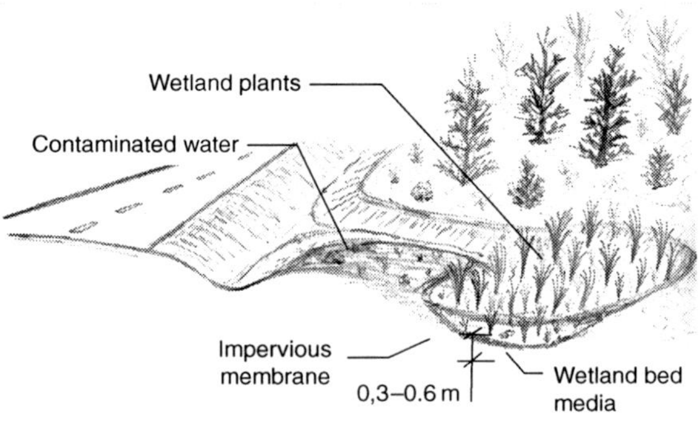

Fig. 13.31 Environment with extremely high sensitivity

long distance movement of the contaminants (e.g. by the use of soils with a high sorptive capacity that have been carefully compacted as a liner).

The potential of wetlands as treatments (Fig. 13.31) was illustrated in Chapter 12 (see Figs. 12.1 and 12.2) and described in Section 13.3.8.

13.5 Sealing Systems for Environmental Protection

13.5.1 Sub-Soil Barriers

Sealing systems can be laid during construction to prevent contaminated water from moving in an undesired direction or to keep natural groundwater separate from contaminated road runoff and road construction seepage waters. In many places in which geomembrane barriers could be placed, spillage of petroleum and diesel from vehicles is a possibility. Sealing systems are used for sealing highways and embankments.

During the design of a sealing system the designer should take into account the sensitivity of the area, crossfall and alignment of the road. When the seal is placed on a slope, a very important part of the design procedure is the analysis of the slope stability as the shear strength between the layers of the sealing system may be much less than found between soil layers, thereby significantly reducing the factor of safety against slippage.

Liners are part of the sealing systems that consist of a base, a sealing layer and a protection layer. The base is that part of the construction on which the sealing layer should be placed and it can consist of natural soil or artificial aggregates placed on the natural soil. Materials selected for the granular base should not consist of sharp or large rock blocks that could damage the sealing layer. The base should be stable and compacted to at least 92% of optimum (Proctor) density. An important aspect is to ensure a planar base.

The sealing layer provides the low permeability of the sealing system. The required thickness depends on the sensitivity of the area that is to be protected and on the quality of the material used to make the sealing layer. The material of the sealing layer also depends on the purpose. Materials for sealing the pavement area will be different from those materials used for sealing the slopes of an embankment.

The protection layer is intended to protect the sealing layer from traffic (e.g. break-through caused by vehicle crashes), damage from the placing of coarse or sharp overlying material and negative climatic influences (e.g. freezing and drying). For this purpose natural materials such as soils, crushed rock and some artificial materials, such as concrete materials, are used. If necessary, the surface of the protection layer should also be designed against erosion due to high water flow velocities above it. Figure 13.34 shows an example where altered land use has increased run-off and flow speed to $> 1\ \mathrm{m/s}$ over a trench so that the existing protection provided by 0/100mm crushed rock is no longer adequate. This could rapidly cut down to an underlying groundwater, damaging its quality. A high performance protection layer would be needed over a sealing layer in such a situation.

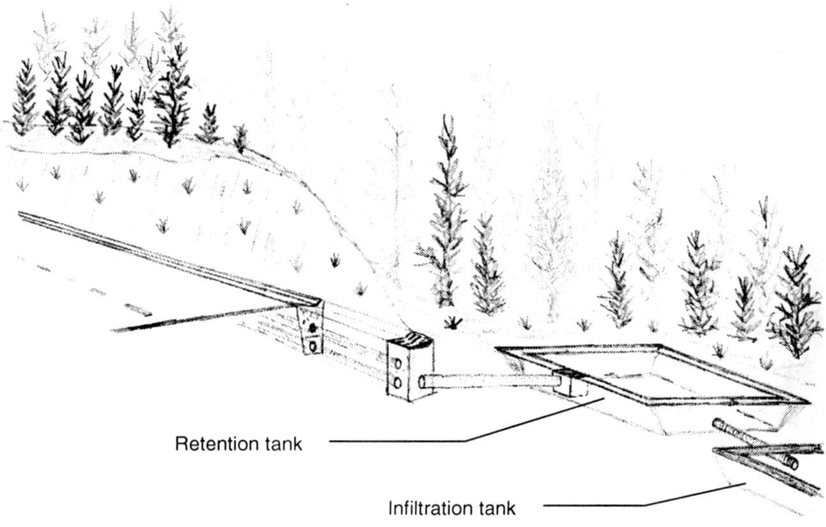

Fig. 13.32 Environment with extremely high sensitivity. A combination of various types of water treatment is included

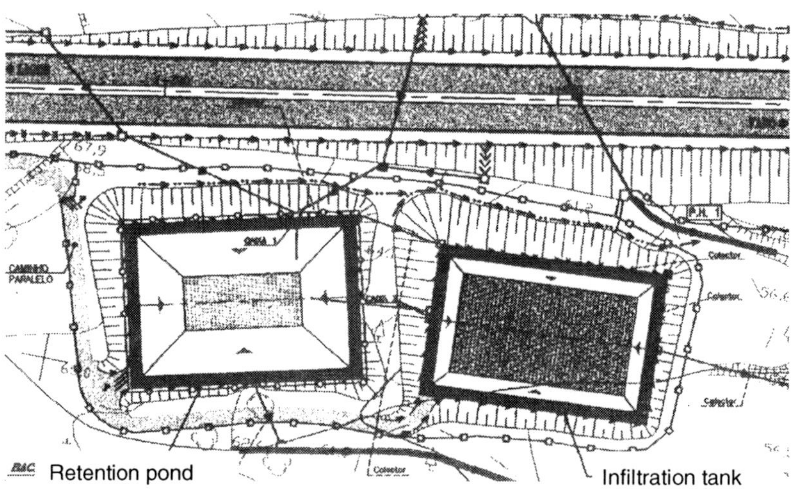

Fig. 13.33 Plan of environment with extremely high sensitivity. A combination of various types of water treatment is included

Fig. 13.34 Excessive run-off/seepage eroding the top of a drain following heavy rainfall

For materials in the sealing layer, natural and geosynthetic barrier materials (GBR) can be used. The most common natural material used for sealing is clay, which is sometimes available on the construction site or in clay pits that are positioned in the vicinity. Materials available at the construction site can be enhanced with the addition of clean bentonite clays.

Geosynthetic barriers (GBR) can also be used. They come in various forms:

- polymeric geosynthetic barrier GBR-P;
- bituminous geosynthetic barrier GBR-B; and
- clay geosynthetic barrier GBR-C.

Four types of geosynthetic barriers application may be distinguished (prEN 15382, 2005):

- deep GBR on side slopes – where the GBR is installed under the drainage collection system and covers the entire slope as well as the ditch area (Fig. 13.35);

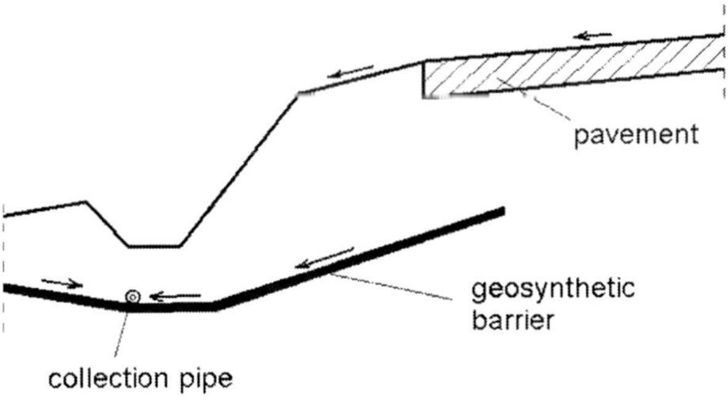

Fig. 13.35 Application of geosynthetic barrier (GBR)

- high GBR on side slopes – where the GBR is installed above the drainage collection system as a high laying sealing system and covers the side slope of the road to prevent an overflow of the road surface runoff;
- deep GBR in central reserve – where the GBR is installed under the drainage collection system and covers the section in the central reserve, where sealing is required; and
- high GBR in central reserve: where the GBR is installed above the drainage collection system as a high level sealing system and covers the section in the central reserve where sealing is required.

Polymeric liners are supplied in rolls and must be joined on-site to form continuous sheets over large areas. This is a specialist task requiring the use of experienced personnel if one desires a reasonable confidence in achieving an effective water barrier.

Geosynthetic barriers are prone to damage by ultra-violet light and by vermin. The first can be overcome by ensuring that the material is covered in soil or other material rapidly after unrolling. Some are more resistant to vermin than others, but it is always sensible to consider ways of preventing damage from animals (perhaps by providing a light steel mesh cover a little above the placed geomembrane as a vermin barrier).

Clay sealing sheets are also available, especially when clay material is not present on the site. Their advantage compared to on-site materials is their precisely defined properties that allow easy design and construction. Typically, these comprise a thin (circa 1 cm) layer of rather dry bentonite formed between two geo-textile sheets. Supplied as a roll, these liners are unrolled on site and overlapped without seaming. Once buried and in contact with water the bentonite sorbs very strongly, causing significant expansion. This expansion develops an effective seal between the liner and the soils around it and between one roll of liner and another. If punctured, the bentonite expansion means that holes self-seal. Bentonite is an excellent sorbent of many species of heavy metals and some organics. Bentonite clay liners should be properly maintained and they should be prevented from drying out. If this happens, cracks up to some centimetres in width can appear and the sheet will no longer act as a the barrier. In that case bentonite layers can be more permeable than a sub-base.

prEN 15382 (2005) does not advise that geosynthetic barriers be connected to drainage systems when embedded in shoulders or slopes. Figure 13.35 shows a typical application of geosynthetic barriers. Details of technical solutions may be found in prEN 15382 and in RiStWag (2002).

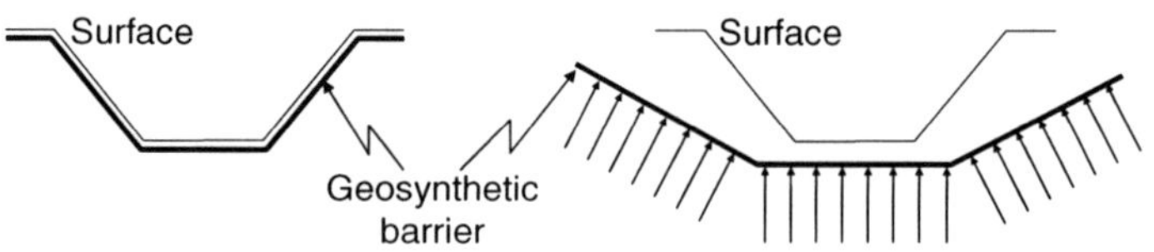

Fig. 13.36 Elementary *misapplications* of geosynthetic barriers

Placing geosynthetic barriers on slopes with a thin cover of soil and lack of sufficient overburden to compensate for uplift pressures are elementary misapplications (Fig. 13.36).

13.5.2 Surface Seals

For pavement seals the following materials are used:

- asphalt layers;
- stress absorption membranes ("SAMIs") (see Fig. 13.37); and
- junction sealing material.

SAMIs act over an old cracked pavement surface, sealing the cracks against water ingress. They often provide a small degree of differential horizontal movement between an old and a new pavement layer so that the relative movements of the parts of the old pavement either side of a crack do not cause a stress concentration in the new, overlying, pavement layer at the same point. Without this ability, a crack in the new layer can quickly form immediately above the old crack – a problem known as "reflection cracking". Water can then, very quickly, re-enter the road and the overlay will not act as an effective seal (e.g. Fig. 13.38).

13.5.3 Examples of Seals

An example of pavement sealing was seen in Slovenia for highways crossing very highly sensitive aquifers. There, the following requirements are used (Ajdič et al. 1999).

Fig. 13.37 A stress-absorbing membrane. In this case a bitumen-rich interlayer is contained between two geosynthetic sheets which have to be sealed to the old pavement and to the new overlay (sample courtesy of MacPave Corporation)

Fig. 13.38 An asphalt overlay over a concrete pavement showing severe reflection cracking. Reproduced by permission of MacPave Corporation

- In the asphalt layer the permeability can be controlled via the air void ratio (see Chapter 5). The wearing course of the asphalt layer should include not more than 5% air voids and the base course not more than 7% air voids.
- A stress absorption membrane should be constructed using a polymer modified bitumen in a layer of 1.5–2.0 kg/m^2 and appropriate fine aggregate. Junction sealing should be performed with bituminous tape.

An example of the use of a lining system beneath an embankment is Highway A-15 at Botlek in part of Europoort, Rotterdam (in the Netherlands). Approximately 400 000 t of (municipal solid waste) incinerator bottom ash was used in an embankment for this major roadway construction. The ash was covered with a compacted sand-bentonite mixture with a minimum thickness of 20 cm to reduce water infiltration. The formation (founding level) of the embankment was shaped so as to bring any water seeping through the ash to a sampling point at which quantity and quality of seeping water could be monitored. The aim of the cover and lining systems was to prevent the contamination of underlying clean groundwater by infiltrating water that would have passed through the ash embankment material, potentially collecting undesirable contaminants on the way. In fact, due to the heavy industrial use of the land in Europoort over many years prior to the embankment's construction, the natural groundwater is degraded at a regional scale, so use of the ash posed few risks of causing unacceptable contamination (Mank et al. 1992).

At another site in the Netherlands, a wind barrier was built at Caland (Stoelhorst, 1991). This project, built in 1985, used more than 650 000 t of bottom ash in an embankment 700 m long and 15 m high. The ashes were covered with a primary cover of 0.5 m of compacted clay with a sand drainage layer (0.5 m thick) and top soil (1 m thick) overlaying the clay layer. The slope of the compacted ash was between 40% and 50%. As at the Botlek site, groundwater quality is monitored, in this case on both sides of the embankment.

13.6 Design of Drainage Systems

The construction of new roads can cause impacts on the water resources of affected regions, causing irreversible effects in some cases.

Surface and subterranean water resources are finite and irreplaceable natural resources for survival, therefore their protection against abnormal flow and against pollution is of great importance, nowadays making their preservation an indispensable part of a sustainable development policy. For this reason it is fundamental that a drainage system be developed that regulates the flow of effluents from the pavement platform, that controls the subterranean drainage and that minimises the hydrological impacts of the road on the environment.

13.6.1 Hydraulic Calculation for Drains (q_L)

In order to estimate the water flow into drainage pipes, one should differentiate between the two distinct situations introduced earlier:

- pipes above the water level (intersection drains); and
- pipes below the water level (groundwater lowering) drains.

When the drainage system is above the water level, the infiltration water from edges, channels and gutters, and from some of the transverse drainage that is covered by permeable surfacing, must also be considered according to the relationship of Eq. 13.2.

$$q_L = R \cdot B \cdot L \tag{13.2}$$

where q_L is the water flow through the pipe (m^3/s), R is the surface runoff water flow ($m^3/(s.m^2)$), L is the section's length (m) (see Eq. 13.3) and B is the width of the section requiring calculation (m) (see Eq. 13.3 as shown in Fig. 13.39).

$$B \cdot L = \sum_{i=1}^{n} b_i \times l_i \tag{13.3}$$

where b and l are individual widths and lengths, respectively (Fig. 13.39). Other non-runoff flows can be added into Eq. 13.2 by simple addition, provided they are expressed in units of m^3/s.

In cases where the drainage system is used not only as an interceptor but also to lower the water level, dimensioning should consider specific calculations for the underground flow into the drain. In this situation the projected flow should be the sum of the aforementioned value and that estimated through the application of Darcy's Law.

Such a flow estimate and the depth of installation for the drain are based on the assumption of specific tests and sophisticated calculations. Nevertheless, in most

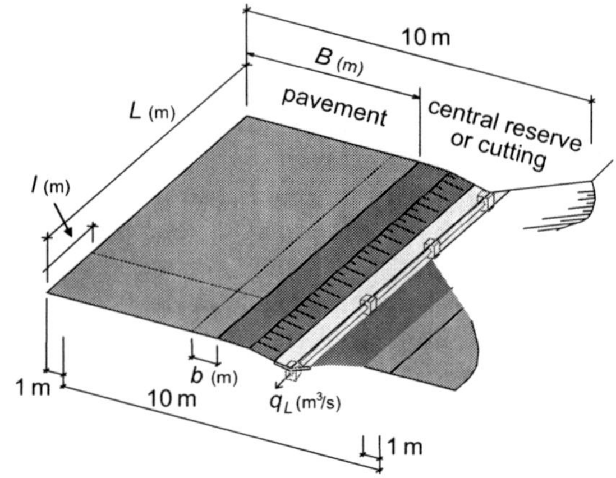

Fig. 13.39 Drainage zones for a section of carriageway and hinterland (adapted from Carreteras (2004))

cases they are revealed to be of limited practical relevance because, in the range of commercial diameters, perforated pipes have a considerably larger capacity for in-flow than is strictly required and the depths at which they are installed usually guarantees the lowering of the water level in the zone between drains.

Having said this, and in order to simplify dimensioning, some authors consider that the in-flow to the drain amounts to approximately 35% of the total flow generated as slope runoff with 20% of the surface runoff from the road pavement being added to cater for flows originating in the road platform, i.e.:

$$q_L = 0.35q_E + 0.20q_P \tag{13.4}$$

where q_L is the water flow to the pipe (m^3/s); q_E is the surface runoff water from slopes (m^3/s) and q_P is the surface runoff water from the platform (m^3/s).

Regarding the depth of installation of the drains, one can make a first estimate using the formula:

$$z = z_w + 0.5 \cdot b \cdot \left(\frac{IR}{K}\right)^{0.5} \tag{13.5}$$

where z is drain depth (m), z_w is the depth at which the groundwater level should stabilize (m), b is the distance between drains (m), IR is the rate of infiltration into soil (m/s) and K is the soil permeability (m/s).

A specific hydro-geological calculation must be done whenever the drainage system aims to lower the water level. When deciding on the transverse profile to use in a new road's project, the details of the subterranean drainage, based on tables and criteria, are very important.

Finally, one should add that in order to satisfy the criteria for self-cleansing and guarantee an adequate geometry, the drains should have a minimum longitudinal inclination of 0.5%, which, in exceptional cases, can be reduced to 0.25%. This inclination should not exceed 20%.

There are various computer software codes on the market that can perform calculations of flow as described, e.g. CANALIS, HYDRA and MOUSE.

13.6.2 Drainage Details

Some typical design details are presented in the following figures. Figure 13.40 shows the typical details for a drainage channel to be installed in the verge between the pavement and a cutting slope, Fig. 13.41 shows the likely details to be employed

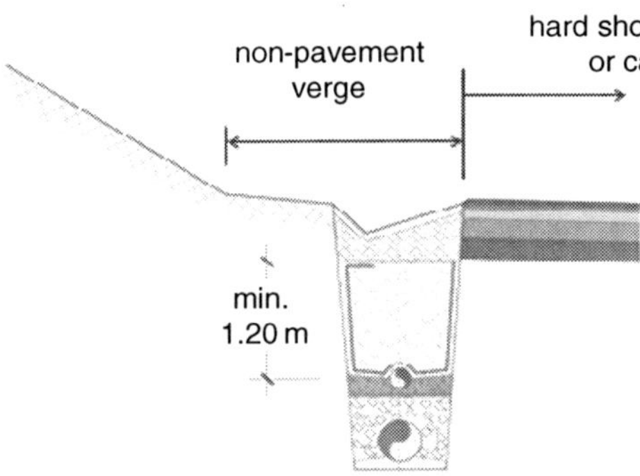

Fig. 13.40 Cuttings – Standard concrete channel in verge with drain and pipe

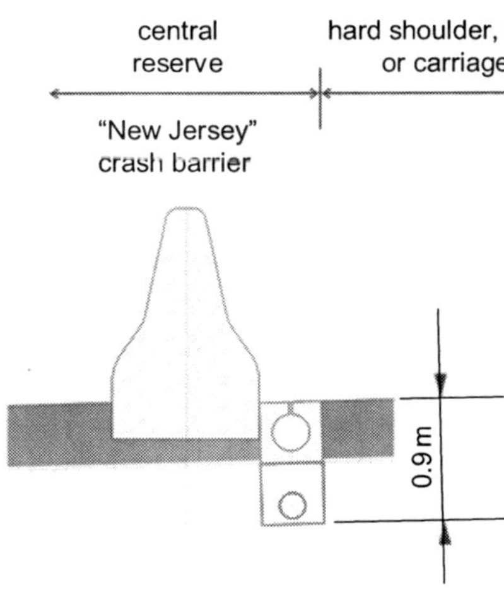

Fig. 13.41 Drain for use in conjunction with concrete barrier and linear slot drainage channel

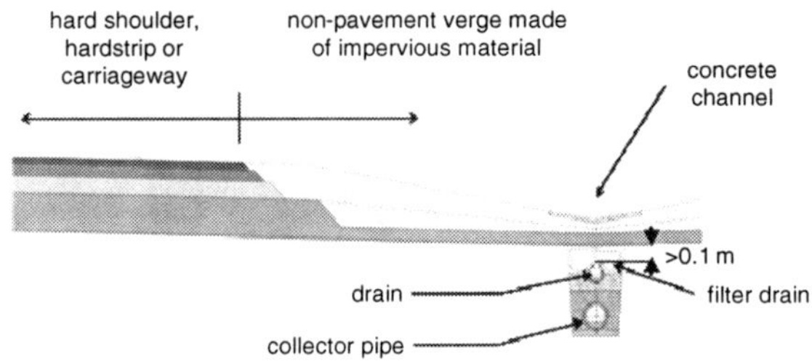

Fig. 13.42 Cutting – Combined surface water and groundwater filter drain and drain pipe

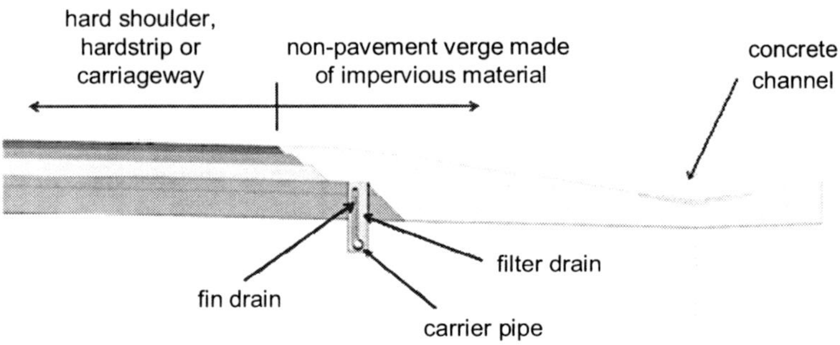

Fig. 13.43 Cuttings – Combined surface water and groundwater filter drain, pavement capping layer with low permeability

in the vicinity of a traffic safety barrier such as appears in a central reservation, Fig. 13.42 shows a typical combined drainage system for both surface runoff and subterranean water collection whilst Fig. 13.43 shows a subterranean fin drain arrangement for pavement sub-drainage.

13.7 Construction and Maintenance of Drainage Systems

13.7.1 Construction

When construction commences it is necessary to be responsive to the geological and geotechnical conditions encountered and not to adhere to those assumed at the design stage. Therefore the in-situ conditions should be carefully inspected throughout the construction process. Also, care should be taken that the construction activities do not have a deleterious effect on drainage.

The planned drainage systems for a project can only be finalized during the work's execution, when the local geotechnical conditions are fully understood. Thus, it is important that an adequate specification is produced for the anticipated type of drainage system and for suitable materials, so that the implementation teams are able to deliver the best solutions.

The many phases which constitute the construction of a road are sometimes delayed, and this can be drainage related, due to:

- Alteration in design flows;
- Obstruction of the surface and underground water flow path, due to earth moving and material placement;
- Possible surface and underground water contamination, due to earth moving, machine cleaning and associated incidents;
- Increase in the soil's compaction in the areas where there is flow to or from an aquifer; and
- Alteration in the hydrological regime, as a consequence of the disturbed soil caused by the construction of the road structure.

Thus, it is necessary to plan the phases of project to adopt preventative measures so that there is:

- Adequate design flow, taking into account the future plans of the drainage area (land use) as well as current needs;
- Optimization of the programming of the earthworks and drainage activities, taking into account the season in which they are to be performed;
- Adoption of a plan to control erosion and soil sedimentation; and
- A work phasing plan, so that the heavy trucks and machinery do not cross the water courses, and do not affect the infiltration and recharge of the aquifer.

Normally, the supplier is obliged to demonstrate the way in which he establishes, maintains and implements a Quality Management System (QMS) to control the construction. In Europe, this quality management system must comply with the requirements of the ISO 9001:2000 standard as well as with any national or European legislation that might be relevant. The system must account for regulations applicable to quality assurance as well as for the QMS and should be based on a Quality Plan assembled for the construction project that contains the procedures, inspection and testing plans, work instructions, audit plan, training and information plan, as well as other plans containing the different specialist activities involved in the project.

13.7.1.1 Testing Plan

The tests to be performed on site, or in parts of the project, are defined in the testing plan. To verify the characteristics and behaviour of the materials to be used, samples must be taken and tests performed as specified in the contract specifications. These specifications typically define the type and frequency of the tests as illustrated in Table 13.4. The codes used are listed at the end of the Table.

Table 13.4 Tests on soils, rock and aggregates

Embankment materials – Soils

Test code	Number of tests	Periodicity and quantity
PSD, LL, PL, C_{OM}, S_{eq}, Cr, PEAA	1 of each	For each excavation and/or at each 25 000 m^3 excavated, or every time there is an alteration in soil nature.
w, ρ_d	3 of each	For profile in each layer

Embankment materials – Rock/Soil fill

Test code	Periodicity and quantity
LA, PSD, PEAA, e, FR, DR	They will be performed in the experimental section and when required by the Quality Controllers if material are heterogenic with a minimum of 1 test for each 50 000 m^3 of constructed embankment.

Capping layer material – Soil

Test code	Number of tests	Periodicity and quantity
PSD, LL, LP, MB, S_{eq}	1 of each	For each 2 500 m^3 or working day
Cr, CBR	1 of each	For each 10 000 m^3
w, ρ_d	3 of each	For each 12.5 m
PLT	1	For each 2.0 km

Granular material

Test code	Number of tests	Periodicity and quantity
PSD, LL, LP, MB, S_{eq}	1 of each	For each 2 500 m^3 or working day
Cr	1	For each 10 000 m^3
LA	2	For each homogenous formation or 1 per day
PEAA, %C	1 of each	For each 10 000 m^3 or working day
w, ρ_d	1 of each	For each 12.5 m or 1 per day
w, PLT	1	In each 2.0 km

Lime/Cement treated soils

Test code	Number of tests	Periodicity and quantity
PSD, LL, PL, Cr, CBR(7d), CBR-i	1 of each	For each working day
w, ρ_d	1 of each	In each 12.5 m
PLT	1	In each 2.0 km
$\sigma_{T\text{lab}}$ (7 & 8d)*	1	For each working day
σ_T (id)*	1	Core boring sample each 200 m

* Only for soil treated with cement.

Test codes and their designation as used Table 13.4:

w	Water content of soil and aggregates	%C	Percentage of crushed and broken broken surfaces in aggregates
C_{OM}	Organic matter content		
Cr	Compaction test	LA	"Los Angeles" test
ρ_d	Dry density in-situ	FR	Coefficient of fragmentability †
LL	Atterberg liquid limit	DR	Coefficient of degradability †
PL	Atterberg plastic limit	e	Void ratio

Table 13.4 (continued)

Test codes and their designation as used Table 13.4:

PSD	Grain/particle size distribution	σ_{Tlab}	Indirect tensile strength (Brasilian) test, laboratory curing
S_{eq}	Sand equivalent		
PEAA	Particle density and water absorption	σ_T (*i*d)	Indirect tensile strength (Brasilian) test, *in-situ* curing
CBR	California Bearing Ratio test		
CBR-i	CBR test in place	(*i*d)	after *i* days of curing
MB	Methylene blue test	(7d)	after 7 days of curing
PLT	Plate loading test	(7 & 8d)	after 7 and 8 days of curing
D<80μm	Fines content	†	of rocky material

13.7.2 Maintenance

It is of great importance that the draining system is working properly, hence regular checks (e.g. Fig 13.44) and maintenance are required. Every drainage system should be designed to ensure that inspection and maintenance operations are possible and accessible. Usually, the cleaning of the drainage system should be done at the end of the summer, but inspections could be intensified in periods of high precipitation. However, at least every 5 years it is fundamental that there is a proper inspection of every part of the drainage system.

The problems that practitioners encounter are manifold. In the WATMOVE questionnaire survey (see www.watmove.org) the following issues were mentioned:

- The drainage system becomes clogged with fine materials,
- Crushed pipes,
- Poor outlet conditions, i.e. outlets have negative slopes,
- Root penetration,
- Generation of ferrous hydroxide and calcium carbonate,
- Insufficient capacity,
- Inadequate water velocity,
- The (plastic) cover of the inspection well at the slope may be damaged (sometimes due to snow clearance of the road).

Fig. 13.44 Sometimes drainage problems can be seen easily and from the surface

An earlier study by Dunnam & Daleiden (1999) revealed similar problems as well as blockages by vegetation and animal nests.

In order for maintenance works to take place and to ensure a long life for the road, it is essential to plan a maintenance programme, based on a series of procedures, measures, actions and practices. The establishment of maintenance actions, of a systematic nature, and emergency actions should be part of that intervention plan.

The systematic plans, i.e. those used in normal conditions, aim to guarantee that the drainage systems remain in a good working condition. These plans will comprise, at least, inspection actions, vigilance and cleaning (which include the removal of sediments), clearance of channels and ditches and the removal of vegetation.

Inspections could be visual or by video surveillance of pipes, depending on pipe diameter. Close-circuit TV cameras are available, mounted on the end of umbilical cables and incorporating lighting, to achieve down-pipe inspections (Dunnam & Daleiden, 1999; Fleckenstein & Allen, 1996; FHWA website).

The intervention plans, for accident/emergency situations, requires fast intervention of maintenance teams, to mitigate the negative impacts on personal safety and environmental contamination. To do so, a sequence of procedures should be established and adapted for different scenarios. These will require prior surveys, covering:

- The boundaries of water areas and environmental compartments;
- aquifer vulnerability;
- sensitivity of each compartment;
- existing drainage systems;
- the assessment of potential hazard sources (including industrial areas); and
- transportation requirements for dangerous substances.

In order for these plans to fully work, one must ensure proper management of the road infrastructure with maintenance programs, specialist human resources and operational and logistical support.

13.7.3 Identifying Rehabilitation Needs

In road rehabilitation, it is important to design the drainage at the same time as the other rehabilitation measures (e.g. strengthening of the structure). The designer should aim to recognise locations where poor drainage is the major cause of road damage. In cold regions the springtime is usually the best time for field studies of the drainage system, since the water level is high, frost damage can be seen, vegetation is low and the bearing capacity and slope stability are lowest. Very wet/soft slopes can indicate a wet structure. There may even be water pressure inside the structure, as in Fig. 13.45 where water is pouring out from the slope of a highway just before a culvert (which is acting as a water barrier).

The appearance of certain vegetation (e.g. rushes and willow trees) on the slopes can also be a sign of excess moisture in the geotechnical structures, appearing typically on low permeability slopes. Video or photos can help to record the conditions. Ground radar surveys may also be used to identify extra moisture on

Fig. 13.45 Water exiting from an embankment slope where it has collected due to a culvert (in the background) acting as a barrier

road structures and in subsoil or on rock surfaces and the level of the groundwater. Bearing capacity measurements can be used to identify weak (moist) locations and rock locations (maybe channelling water) below the road. The designer should analyse available observations, measurements and the requirements for both the road and its surroundings.

13.8 Future Performance

The observation of climate changes, as a consequence of global warming, reveals the aggravation of extreme situations, including alternating torrential rain periods with drought situations. Therefore, it's important to ensure that road drainage systems are calculated for extraordinary phenomenon (both precipitation and flow) associated with a predetermined return period, which includes an allowance for the worsening of weather conditions. Designing to historic weather patterns is likely to mean that there will be an increase in the frequency when elements of the drainage system will not have adequate capacity, with inevitable consequences for the safety of users and, eventually, for the survival of the infrastructure itself.

Today's road drainage elements should be monitored to allow measurement of the return period for which they are designed, adopting in latter life of the drainage system a solution for increasing the capacity in face of the reality found on site. One should also ensure that inside this observation phase (which could take some years), efficient cleaning and maintenance plans are implemented with adequate frequency, so that the drainage elements are in perfect functioning order. For example, after drought seasons the tendency to become clogged with sand is higher.

Future infrastructures should be dimensioned by adopting the revised values and parameters collected through information recovered in rainfall and/or water flow monitoring stations placed near the locations of the particular road scheme.

13.9 Conclusion

A safe and comfortable road requires a great investment in scheme planning, careful design, quality construction and ongoing maintenance. At each stage of the road's life, the hydraulic, geotechnical and pavement performance must be considered alongside the environmental response of the road and its "corridor".

Drainage standards exist to aid design, performance and maintenance, but they are not to be followed as laws, but as a reference and recommendation for a project. As important as the standards are for many facets of a highway's design (including the safety aspects), it is the engineer's experience and good sense that must determine the road scheme planning and its detailed execution at a project level. Greater rigor by consultants and owners concerning the choice of drainage solutions is important, yet they should be given greater flexibility in the project's execution schedule and in construction of the work's drainage system. It is also the consultant and owner who are best placed to determine the appropriate safety implementation.

The wide variety of solutions available to ease the road through the hydraulic environment will necessitate careful study and selection in order that the most economically, socially and environmentally beneficial solution is found. A similar attention will need to be paid to the selection of materials and components. This is partly because of the wide range of geosynthetic materials and composites with properties specifically "tuned" to drainage applications that are now readily available and partly because of the ever-increasing pressure to use marginal, waste and by-product construction materials in place of the conventional aggregates with which designers may be more familiar.

This chapter has, albeit briefly, sought to indicate something of the breadth of solutions and the considerations. It will be apparent that much more could be written and much more detailed design advice and sample calculations could have been presented. However this chapter is already the longest in the book and the book longer than intended! It will suffice for now to advise readers of the wealth of information available in the references listed below and at the end of the previous chapters.

Water is often considered the chief enemy of the pavement engineer. It is one of the materials to which the environmental expert gives prime attention. It is the very basis of the hydrogeologist who seeks to protect it so as to ensure continued pure water supplies. It is right, therefore to make it the subject of this book which, one hopes, will help to ensure that it is treated appropriately by everyone who has a role to play in providing and maintaining roads in our precious environment.

References

Ajdič, M., Brenčič, M. Ločniškar, A., Petkovšek, A., Prestor, J. & Žmavc, J., 1999, "*Guidelines for groundwater protection measures in the areas around highways*", Ljubljana: Company for Highways in Republic of Slovenia (in Slovene).

Bergado, D.T., Manivannanand, R. & Balasubramaniam, A.S., 1996, "Filtration criteria for prefabricated vertical drain geotextile filter jackets in soft Bangkok clay", *Geosynthetics International*, **3**(1), pp. 63–83.

Berntsen, G., & Saarenketo, T., 2005, "*Drainage on low traffic volume roads*", April, Roadscanners Oy, 123pp. Downloadable from www.roadex.org.

Biczysko, S.J., 1985, "Permeable sub-bases in highway pavement construction". 2nd Symposium on Unbound Aggregates in Roads, *Proc. 2nd Symp. on Unbound Aggregates in Roads* (UNBAR2), ed. Jones, R.H., Dept. of Civil Eng., Univ. of Nottingham, 1985, pp. 81–92.

Brattebo, B. & Booth, D., 2002, "Permeable parking lot demonstration project – the six year follow-up", *Washington Water Resource*, Summer, pp 5–10.

Christopher, B.R., 1998, "Performance of drainable pavement systems: a U.S. perspective", *Int. Symp. Subdrainage in Roadway Pavements & Subgrades*, ed. E. J. Hoppe, Grafistaff, Granada, Spain, pp. 205–215.

Christopher, B.R. & Holtz, R.D., 1985, "*Geotextile Engineering Manual*", U.S. Department of Transportation, Federal Highway Administration, Washington, DC, Report No. FHWA-TS-86/203, March, 1044pp.

Carreteras, D.G. 2004, "*Recommendations for the design and construction of sub-surface drainage in roadworks*", (in Spanish: "Recomendaciones para el proyecto y construcción del drenaje subterráneo en obras de carreteras"), Orden Circular 17/2003, Centro de Publicaciones, Ministerio de Fomento, Madrid.

Dawson A.R., 1985, "Water movements in road pavements", *Proc. 2nd Symp. on Unbound Aggregates in Roads* (UNBAR2), ed. Jones, R.H., Dept. of Civil Eng., Univ. of Nottingham, pp. 7–12.

Dawson, A., Hill, A. & Boothroyd, P., 2006, "The Hydraulic and Geo-Environmental Behavior of the Sub-Pavement Environment", *Geotech. Special Publ.* No. 154, Am. Soc. Civil Eng'rs., Ed's. Huang, B., Meier, R., Prozzi, J. & Tutumluer, E., pp 287–295.

Dempsey, B.J., 1988, "*Pavement Drainage System Design*", Module I, Drainage survey and evaluation, in Wisconsin DOT Highway Pavement Drainage Symposium, University of Wisconsin, Madison.

Department of Transport, 1990, "A permeameter for road drainage layers", HA 41/90, *Design Manual for Roads and Bridges*, **4**(2), April, London, 21pp.

Dunnam, Z.W. & Daleiden, J.F., 1999, "*Observations of Highway Edge Drain Systems in Florida*", Austin, Fugro.

Edel, R., 2002 "*Odwodnienie dróg*" (in Polish), Wydawnictwa Komunikacji i Łączności, Warsaw, 320 pp.

EN 13252, 2000, "*Geotextiles and geotextile-related products – Required characteristics for use in drainage systems*", Comité Européen de Normalisation.

Ferguson, B.K., 2005, "*Porous Pavements*", CRC press, Boca Raton, 600pp.

FHWA web-site, Maintenance of Highway Edge drains, available on-line at http://www.fhwa.dot.gov/pavement/concrete/edge.cfm.

FHWA, 1980, "*Highway Sub-Drainage Design*", Report no. FHWA-TS-80-224. US Department of Transportation, Washington, DC.

Fleckenstein, L.J. & Allen, D.L., 1996, "Evaluation of Pavement Edge Drains and Their Effect on Pavement Performance", *Transportation Research Record* 1519: Effectiveness of Subsurface Drainage, pp. 28–35, Washington: Transportation Research Board.

Floss, R. & Berner, U., 1989, "A new method for the determination of horizontal and vertical permeability of cohesionless basecourse materials", *3rd Symposium on Unbound Aggregates in Roads* (UNBAR 3), eds.. Jones, R.H. and Dawson, A.R., Dept. of Civil Eng., Univ. of Nottingham, pp. 78–85.

Forsyth, R.A., Wells, G.K. and Woodstrom, J.H., 1987, "Economic impact of pavement subsurface drainage," *Transportation Research Record* 1121, TRB, National Research Council, Washington, D.C., pp. 77–85.

Hall, K.T. &, Correa, C.E., 2003, "*Effects of Subsurface Drainage on Performance of Asphalt and Concrete Pavements*", National Cooperative Highway Research Program (NCHRP), Report 499, Transportation Research Board, Washington, D.C.

Hermansson, Å., 2004, Laboratory and Field Testing on Rate of Frost Heave versus Heat Extraction. *Cold Regions Science and Technology* **38**(3), pp. 137–151.

Highways Agency, 2006, "*Surface and Sub-Surface Drainage Systems for Highways*", HD 33/06, **4**(3) of Design Manual for Roads and Bridges, 33pp.

Huang, Y.H., 2003, "Pavement Analysis and Design", 2nd. Ed., Prentice Hall, New Jersey, 850pp.

IEP, 2001, "*Manual of highway surface drainage*", (in Portuguese: "Manual de Drenagem em Superficial em Vias de Comunicação"), Instituto das Estradas de Portugal, Almada, Portugal.

Interpave, 2005, "*Permeable Pavements – guide to design, construction and maintenance of concrete block permeable pavements*", Edition 3, 26pp.

Joint Departments, 1995, "*Engineering Use of Geotextiles*", Joint Departments of the Army & Air Force (USA), Technical Manual, TM 5-818-8/AFJMAN 32-1030, July.

Jones, R.H. & Jones, H.A., 1989, "Horizontal permeability of compacted aggregates", *3rd Symp. on Unbound Aggregates in Roads* (UNBAR 3), eds. Jones, R.H. & Dawson, A.R., Dept. of Civil Eng., Univ. of Nottingham, pp. 70–77.

Kasibati, K. and Kolkman, L.L., 2006, "Highway Drainage Systems and Design", in: *The Handbook of Highway Engineering*, ed. Fwa, T. F., Taylor & Francis/CRC Press, Boca Raton, FL., USA, 888pp.

Koerner, K.M., 2005, "*Designing with Geosynthetics*", 5th Edn., Prentice Hall, New Jersey, 816pp.

Mank, J.A.M., Brulot, J. & van de Laak, W.H., 1992, "Incineration slag in road constructions", 2nd Sprint Workshop (organised by the European Commission, DG XIII) on "*Alternative Materials in Road Construction*", Rotterdam, June, 9pp.

Pratt, C., Wilson, S. & Cooper, P., 2002, "*Source control using constructed pervious surfaces – Hydraulic, structural and water quality performance issues*", Construction Industry Research & Information Association (CIRIA), Report C582, January, 152pp.

PrEN 15382, 2005, "*Geosynthetic barriers – Characteristics required for use in transportation infrastructure*", Comité Européen de Normalisation.

RiStWag, 2002, "*Construction Measure Guidelines for Roads in Water Protection Zones*", (in German: "Richtlinien für bautechnische Maßnahmen an Straßen in Wasserschutzgebieten"), FGSV (Forschungsgesellschaft für Straßen- und Verkehrswesen), publication 514, 72pp.

Roy, M. & Sayer, S.K., 1989, "In-situ permeability testing of sub-bases on the M5 Motorway", *3rd Symposium on Unbound Aggregates in Roads* (UNBAR 3), eds. Jones, R.H. and Dawson, A.R., Dept. of Civil Eng., Univ. of Nottingham, pp 86–93.

Stoelhorst, D., 1991, "The use of waste materials in civil engineering: AVI slag can replace gravel in concrete production", in Proc. Int. Conf. Environmental Implications of Construction with Waste Materials, "*Waste Materials in Construction*", Eds. Goumans, J.J.J.M., van der Sloot, H.A. & T.G. Aalbers, Studies in Environmental Science, 48, Elsevier, pp. 71–80.

Użdalewicz Z., 2001, "Geosynthetics in drainage systems. French drainage", (in Polish: "Geosyntetyki w drogownictwie. Drenaż francuski"), *Bezpieczne Drogi*, No. 7.

Vägverket, 2004, "*Technical Standards for Road Construction – Common Assumptions*", (in Swedish: "Allman teknisk beskrivning for vagkonstruktion – Gemensamma förutsättningar – 'ATB Väg'"), Publication 111, Börlange, Sweden, 39pp.

Winkelman, T.J., 2004, "*Open-graded drainage layer performance in Illinois*", Final Report, Physical Research Report No. 147, Illinois Dept. of Transportation, August.

Woods Ballard, B. & Kellagher, R., 2007, "*The SUDS manual*", Construction Industry Research & Information Association (CIRIA), Report C697, March, 600pp.

Annex A
Seasonal Variation in Pavement Design and Analysis – Some National Examples

A.1 Introduction

The primary objective of this annex is to present examples of how moisture condition is taken into account in pavement design and analysis.

The pavement design regarding the influence of water has significantly different objectives in different European countries. While in central and northern Europe the most important question is how to protect the pavement structure from frost action, in southern European countries it is more critical to control excessive water that suddenly penetrates the pavement after heavy rainfall. At the same time identical processes play significantly different roles in different climates. For example, the suction, that has negative effect in locations of freezing and thawing, has positive effect in southern countries by increasing the stiffness, and this influences the decision of which type of material is used in unbound base and sub-base. In northern countries the percentage of fines is limited typically to 5–8%, while the Mediterranean countries allow up to 15% fines.

In many parts of Europe freeze/thaw effects play a crucial role in pavement design. How these effects are taken into account in pavement design varies in complexity from using the frost index, or Stefan or modified Berggren equations (Aldrich & Paynter, 1956) to calculate frost depth to coupled heat/moisture flow models. Most of the European pavement design methods take into account phenomena related to freezing and thawing by using the frost index. This indicator is supposed to account for the "quantity of frost" to which a pavement was subjected for a given period. The frost depth is then calculated based on some empirical equation as a function of frost index. Some pavement design methods also consider the influence of sunshine on this phenomenon.

Many design methods take into account the loss of bearing capacity during thawing. This is typically done by adjusting the modulus of each of the unbound materials by a reduction factor that depends on the frost susceptibility of the material (COST 337, 2002).

The European project AMADEUS (Amadeus, 2000) identified the need for further research of deterioration mechanisms related to freezing and thawing and development of a long-term predictive model for freeze/thaw related deterioration.

The project also identified the need for more information on how the bearing capacity of different types of soils is affected by the drainage conditions.

The World Bank's Highway Development and Management Model HDM-4 (ND Lea International, 1995) models the seasonal performance variation using a simple two season model ("wet" and "dry" seasons). The program uses the following parameters:

- mean monthly precipitation,
- drainage effectiveness,
- surface cracking,
- potholed area,

to calculate the ratio between "wet" and "dry" pavement adjusted structural numbers. Loizos et al. (2002) modified this approach in the Road Infrastructure Management System (RIMS) developed for the Greek government, to enable multiple season analysis, more suitable to a European context, by introducing the environmental function.

In the following, examples of how seasonal variations are handled in design systems are given.

A.2 Finland

In Finland the design and analysis of pavement structures is done by separate consideration of different criteria:

- Frost resistance (structure and subgrade)
- Traffic loading/Resistance to fatigue
- Resistance to settlement (deformation in structural layers and subgrade)
- Rutting due to studded tyres

A.2.1 Frost Design

Design of public roads in Finland still, commonly, uses a semi-empirical method based on acceptable calculated frost heave, which depends on

- road class;
- structural durability;
- how homogenous the subgrade is;
- the freezing index, F_{10}, of the geographical location (583–1416 °C.days); and on
- an empirical factor of frost heave of the subgrade, which depends on the proportion of soil passing the 63 μm and 2 mm sieves and whether it is a wet/dry location.

On homogenous clay soil, the measured frost heave values can also be used.

According to the results of the "Road Structures Research Programme" the control of frost behaviour of the road is divided into two parts: control of frost heave and control of the effects of thaw weakening. The frost heave of a road is estimated using the segregation potential (SP) concept, in which SP is the parameter that describes the frost susceptibility of the subgrade. The total thickness of the road structure is designed on the basis of frost susceptibility of the natural subgrade, and on the thermal conductivity of the used materials. If necessary, the structure is protected using insulation against frost, so that the permitted frost heave, set as the design criterion, is not exceeded. The procedure can be applied in designing new roads and improving old roads in all road classes. The program also calculates settlement profiles based on the investigations and identification of variations in the water content of soil layers along the road line.

A.2.2 Design to Traffic/Resistance to Fatigue

Fatigue design is made using either methods based on Odemark's formula or on analytical calculation methods (e.g. a multi-layer linear elastic design program). In both, the seasonal variation can only be considered by choosing the material properties for each (especially the unbound) layer so as to represent weighted mean values for the whole year. The "worst" thawing and moist period of the year (the so-called spring bearing capacity period) has a very large influence on the mean value using this weighting approach. Calculations are also made using a single temperature (+20 °C).

A.2.3 Rutting

The obligatory use of winter tyres (most of them studded) during four to five months each winter causes most of the rutting in Finnish medium and high volume roads. The design against rutting is based on empirical information between rutting speed and amount of traffic, asphalt type, binder and aggregate properties. No seasonal variation is considered, but it is well known that moist/wet conditions at the road surface will increase the rate of rutting.

At the moment only discussions about the possibility of using several, varying-length, time periods (each associated with individual material parameters – depending on moisture, temperature and density) have been carried out. During the analyses mentioned above the vertical deformation (the elastic recoverable strain) is also checked to to ensure that it remains less than the maximum allowable limit for each layer material.

As a brief summary it can be said that seasonal variation is only taken in to account by choosing parameter values based on the most dominating period, the spring thaw period.

A.3 USA

The new AASHTO Mechanistic-Empirical Pavement Design Guide (MEPDG) (AASHTO, 1998 & 2004) integrates climatic factors, materials properties, and traffic loadings to predict pavement performance. The Enhanced Integrated Climatic Model (EICM) that is used in the new AASHTO MEPDG integrates three models:

- The two-dimensional drainage infiltration model (ID Model), developed at Texas A&M University, that evaluates the destination of the rainfall on the pavement.
- The Climatic-Materials-Structure Model (CMS Model) developed at the University of Illinois. A one dimensional finite difference engine is used to calculate coupled heat-moisture flows in pavement structures and predict pavement temperature.
- The CRREL frost heave and thaw settlement model developed at the United States Army Cold Region Research and Engineering Laboratory (US Army CRREL) which computes temperature and moisture flow at different temperatures and predicts the depth of frost and thaw penetration.

The EICM is based on the Integrated Climatic Model, developed for the Federal Highway Administration in 1989, containing several improvements. It replaced the Gardner equation for the soil-water characteristic curve (SWCC) with the equations proposed by Fredlund and Xing (1994). It also provides better estimates of the saturated permeability and specific gravity of soils, given known soil index properties such as grain-size distribution (percent passing sieve no. 200 sieve (75 μm) and effective grain size with 60% passing by weight) and Plasticity Index (PI). The unsaturated permeability prediction based on the SWCC and proposed by Fredlund et al. is also incorporated in the model (Fredlund et al., 1994).

The model uses actual climatic data (hourly or monthly) and predicts the following parameters throughout the entire pavement/subgrade profile:

- Temperature
- Resilient modulus adjustment factors
- Pore water pressure
- Water content
- Frost and thaw depths
- Frost heave and
- Drainage performance.

The model evaluates the expected changes in moisture condition from the initial or reference condition (generally, near optimum moisture condition and maximum dry density) as the subgrade and unbound materials reach equilibrium moisture condition. The model also evaluates the seasonal changes in moisture condition, and consequently the changes in resilient modulus, M_r. In addition the model calculates the effect of freezing, and thawing and recovery of M_r and uses these M_r values for calculation of critical pavement response parameters and damage at various points within the pavement system.

A.4 Croatia

In Croatia, the design of pavement structures primarily considers the traffic load. However, if the subgrade soil is frost susceptible and if the hydraulic conditions are unfavourable, the originally designed pavement structure should be additionally tested to determine the impact of freezing. If there is a risk of freezing, certain technical measures have to be planned within the pavement structure or under it in order to avoid the risk of freezing or to significantly lower the impact of freezing. In Croatia the pavement testing concerning freezing is carried out in the following manner.

Based on the soil mechanics characteristics, the pavement materials and subgrades are grouped into one of the following four groups according to their freezing susceptibility (in compliance with the national standard HRN U.E1. 012):

G1 – lightly susceptible (gravel) containing 3–10% of particles smaller than 0.02 mm.
G2 – lightly to moderately susceptible (gravel, sand) containing 10–20% of small particles.
G3 – moderately susceptible (gravel, sand, clay with PI higher than 12) containing over 20%, i.e. over 15% of particles smaller than 0.02 mm.
G4 – highly susceptible (dust, very fine dusty sand, clayey dust) containing over 15% of particles smaller than 0.02 mm.

According to the national standard HRN U.C4.016, it is established whether the hydrological circumstances are favourable or not.

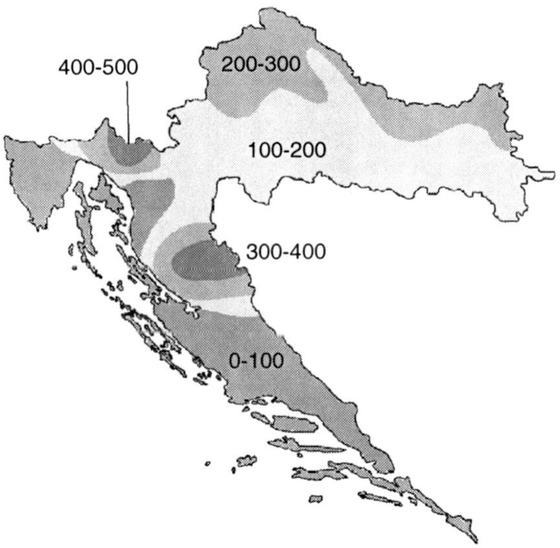

Fig. A.1 Contour map with air freezing index (FI) for the territory of Croatia (Sršen et al. 2004). Reproduced by permission of the Croatian Association of Civil Engineers

Then the freezing depth is determined according to the national standard HRN U. B9. 012. It is required to determine the Freezing Index (FI) for that purpose. The FI value is taken from the freezing index contour map shown in Fig. A.1. That map is derived from an IGH study on Freezing Index determination for national roads in 2003, elaborated in cooperation with the Metrological and Hydrological Service. The study is based on the data on the highest and lowest air temperatures noted in 39 meteorological stations in Croatia in a period of 25 years (1976–2000) (Sršen et al. 2004).

The freezing depth under the pavement surface is read from a diagram based on FI and from the data on pavement thickness (designed or completed), its spatial mass and humidity and from data on the soil type under the pavement.

A.5 Denmark

The Danish design system deals with seasonal variations by adjusting the expected bearing capacity (E-modulus) of each pavement layer. In the design software MMOPP (Mathematical Model of Pavement Performance) (Ullidtz, 1993) the user can choose an advanced design procedure, where the performance of the road is simulated over (for example) 40 years. The program is given constant E-modulus values as material parameters for each pavement layer. These moduli are then varied over the seasons as shown in Table A.1. The constant E-values given as input represent the summer values. In wet seasons the E-moduli of unbound layers are reduced. In frost seasons the values are increased.

Table A.1 Coefficient multiplied to the E–value dependent on season and layer

Season	No. of days/year	Air temp. °C	Asphalt concrete	Unbound base	Sub base	Subgrade
Winter	49	−2	4	4.2	10	20
Winter thaw	10	1	3.7	0.33	10	20
Spring thaw	15	1	3.7	0.67	0.7	0.6
Spring	46	4	3.1	1	0.85	0.8
Summer	143	20	1	1	1	1
Heatwave	10	35	0.3	1	1	1
Autumn	92	7	2.6	1	1	1

A.6 Sweden

When designing pavements in Sweden the different layers are given stiffness varying with the season of the year. The procedure is very similar to the one used in Denmark (see above). In Fig. A.2 the first column holds the thickness of each layer and the other six columns show how the stiffness of each layer is predicted to vary over the year.

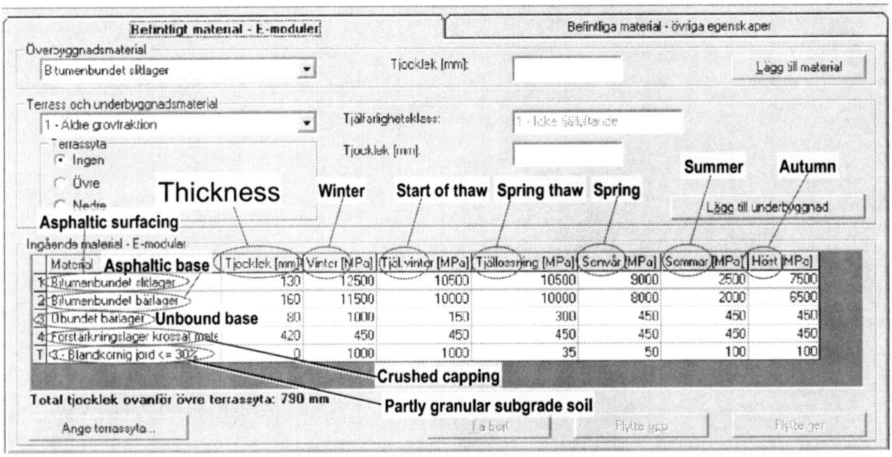

Fig. A.2 Road design dependent on season and layer – Swedish case using PMSObjekt (Vägverket, 2005) showing moduli of pavement layers in 6 "seasons"

A.7 Poland

The Catalogue of Typical Flexible and Semi Rigid Pavement's Construction, was developed for the Polish General Directorate of Public Roads. Typical structures were designed, based on analysis of stresses and deformations in pavement, using multilayer elastic and viscoelastic half space theory. According to this Catalogue and the Roads Design Guidelines (General Directorate of Public Roads, 1995), in pavement design process following factors should be taken into account:

- climatic and ground-water conditions,
- intensity and kind of traffic structure during whole designed life period (20 years),
- values of allowable loads from vehicles (100 kN/axle),
- function of pavement.

Climatic conditions are freezing depth, average annual temperature and temperature differences.

Subgrade bearing capacity groups (G1–G4) depend on type of soil, water conditions and CBR value. The G1 is the best subgrade, mainly sandy soils of CBR $\geq$ 10, G4 is the weakest subgrade, mainly cohesive soils of CBR $<$ 3. Bearing capacity group has an impact on the necessity and kind of subgrade improvement. For typical structures taken from the Catalogue, the bearing capacity of the subgrade to be achieved is as follows. The secondary static modulus, E_2, must be greater or equal to 100 or 120 MPa and the compaction ratio, I_s, must be greater or equal to 1.00 or 1.03 depending on the traffic loading.

Water conditions are evaluated depending on ground water depth (z) from the bottom of the pavement structure. If subgrade drainage is required, a capping layer made from frost non-susceptible materials with a coefficient of permeability, $K \geq 9.3 \times 10^{-5}$ m/s should be used. The capping layer (at least 15 cm thick) should be placed across the whole width of road bed. For the situation where there is unimproved soil under the capping, a "tightness condition" is imposed for the layers:

$$\frac{D_{15}}{d_{85}} \leq 5 \tag{A.1}$$

where:

D_{15} is the dimension of sieve, through which 15% of grains of separating layer or drainage layer will pass and d_{85} is the dimension of sieve, through which 85% of grains of the foundation soil will pass. In situations when the above layer tightness condition cannot be fulfilled, then between these layers a separating layer (of thickness at least 10 cm of suitably graded soil) should be arranged or a non-woven geosynthetic interlayer should be inserted.

In the case of frost susceptible subgrade soils, it is necessary to check if the total thickness of all layers (taken from the Catalogue) and any improved subgrade layer is sufficient to achieve frost resistance. In situations when this condition cannot be fulfilled, then the lowest layer of improved soil should be thickened.

References

AASHTO, 1998, "*Guide*", **2**, Section 2.0, Fundamental Design Parameters

AASHTO, 2004, "*Mechanistic Empirical Pavement Design Guide*", Draft Report, Part 2 – Design Inputs, web site www.trb.org/mepdg.

Aldrich, H.P. & Paynter, H.M., 1956, "Derivation of a Rational Formula for the Prediction of Frost Penetration", *Highway Research Board Bull.*, 135, pp. 145–149.

AMADEUS, 2000, "*Advanced Models for Analytical Design of European Pavement Structures*", Final Report.

COST 337, 2002, "*Unbound Granular Materials for Road Pavements*", Final Report, June, 386pp.

Fredlund D.G., Xing, A. & Huang, S., 1994, "Predicting the Permeability Function for Unsaturated Soils Using the Soil-Water Characteristic Curve", *Canadian Geotechnical J.*, 31(**3**), 1994, pp. 521–532.

Fredlund D.G. & Xing, A., 1994, "Equations for the Soil-Water Characteristic Curve", *Canadian Geotechnical J.*, **31**(4), pp. 533–546.

General Directorate of Public Roads, 1997, "Typical Flexible and Semi-rigid Pavements Structures", Warsaw.

General Directorate of Public Roads, 1995, "Roads Design Guidelines WPD-1, WPD-2 and WPD-3", Warsaw.

Loizos A., Roberts, J. & S. Crank, S., 2002, "Asphalt Pavement Deterioration Models for Mild Climatic Conditions", 9th Int. Conf, Asphalt Pavements, **II**, Copenhagen, Paper 3: 3–4.

ND Lea International, 1995, "Modelling Road User Effects in HDM-4", RETA-5549 REG, Highway Development and Management Research, Final Report, 240pp.

Sršen, M., Kovačić, M., & Kaučić, D., 2004, "Determining soil freezing depth under the pavement structure" (orig. in Croatian), J. Croatian Assoc. Civil Engineers, **56**, pp. 145–154.

Ullidtz, P., 1993, "Mathematical Model of Pavement Performance under Moving Wheel Load", *Transportation Res. Rec.*, **1384**, pp. 94–99.

Vägverket, 2005, "PMS Objekt – a windows-based tool for the analysis and design of pavement constructions" (in Swedish: *PMS Object är ett windowsbaserat verktyg för analys och design av vägöverbyggnader*), Borlänge.

Annex B
Terminology Used for Standard Pavement and Associated Drainage Items

B.1 Introduction

The purpose of this document is to show some standard designs for pavements and associated drainage items and to present the names of these in several languages. The diagrams contained within each section show the general layout only. No detail is included regarding the specifications of the materials or specific dimensions. Many of the figures in this Appendix have been adapted and simplified from drawings made available courtesy of the Highways Agency (Manual of Contract Documents for Highway Works, Volume 3 - Highway Construction Details, Section 1 "Carriageway and Other Details" (March 1998, updated with amendments including November 2005, May 2006 & November 2006)). Where possible translations have been given in German (de), Spanish (es), French (fr), Italian (it), Greek (gr), Polish (po), Portuguese (pt), Serbian (cs), Danish (dk) and Slovenian (si).

B.2 Highway Cross Sections

This section provides general details of the layout of single and dual carriageway roads. Details of the carriageway are given in Section B3, and the carriageway edge and drain arrangements in Section B4.

B.2.1 Single Carriageway

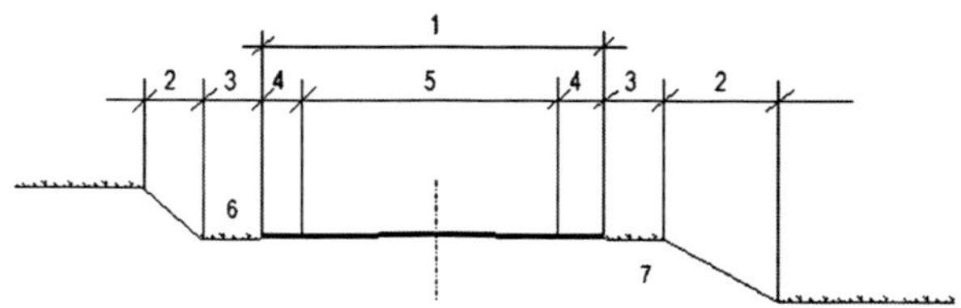

Language	Item no.						
	1	2	3	4	5	6	7
English	pavement	slope/batter	verge	hardstrip	carriageway	in cutting	on embankment
German	Straßenbefestigung	Böschung	Bankett	Randstreifen	Fahrbahn	Einschnitt	Damm
Spanish	pavimento	trasdos de muros	borde	arcén	calzada	desmonte	terraplén
French	plate forme	talus de remblai	accotement	voie d'arrêt	chaussée	berme	remblai
Italian	pavimentazione	scarpata	argine	banchina	carreggiata	in sterro	in rilevato
Greek	Οδόστρωμα	Κλίση πρανών	Έρεισμα	Στερεόν εγκιβωτισμού	Οδός διπλής κατεύθυνσης	Σε Όρυγμα	Σε Επίχωμα
Polish	nawierzchnia	skarpa wykopu/ skarpa nasypu	pobocze gruntowe	pobocze utwardzone/pas awaryjny	jezdnia	w wykopie	w nasypie
Portuguese	pavimento	taludes	Berma não pavimentada	berma pavimentada	faixa de rodagem	em escavação	em aterro
Serbian	kolovoz	kosina useka/nasipa	bankina	ivična traka	vozne trake	usek	nasip
Slovenian	vozišče	brežina vkopa/nasipa	bankina	odstavni pas	vozna pasova	vkop	nasip
Danish	belægning/kørebane	skråning	yderrabat	kantbane	kørespor	afgravning	påfyldning

B.2.2 Dual Carriageway

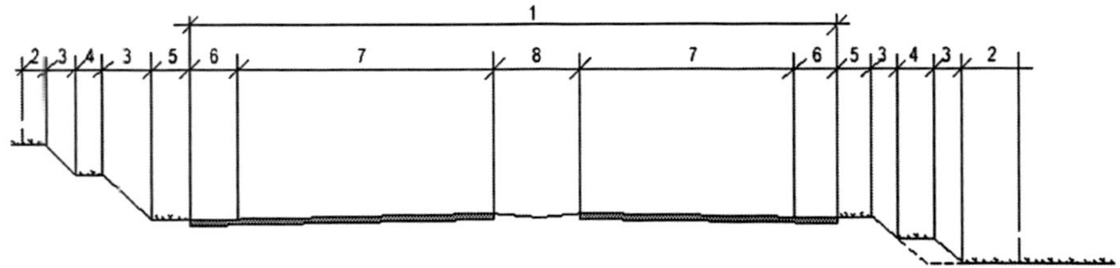

Language	Item no. 1	2	3	4	5	6	7	8
English	highway	berm	slope/batter	bench	verge	hard shoulder	carriageway	central reserve
German	Schnellstraße	Berme	Böschung	Berme	Bankett	Seitenstreifen	Richtungsfahrbahn	Mittelstreifen
Spanish	carretera	berma	trasdos de muros	berma	borde	arcén	calzada	mediana
French	autoroute	berme	talus de remblai	risberme	accotement	voie d'arrêt	chaussée	terre-plein central
Italian	autostrada	berma	scarpata	berma	argine	banchina	carregiata	spartitraffico/margine interno
Greek	Αυτοκινητό δρομος	Μπαγγίνα	Κλίση πρανών	Ερεισμα	Ερεισμα	Ερεισμα	Οδόστρωμα	Κεντρική νησίδα (ασφαλείας)
Polish	autostrada	półka	skarpa wykopu/ skarpa nasypu	półka	pobocze gruntowe	pobocze utwardzone/pas awaryjny	jezdnia	pas rozdziału

(continued)

Language	Item no.							
	1	2	3	4	5	6	7	8
Portuguese	auto-estrada	terreno natural	taludes	banqueta	berma não pavimen-tada	berma pavimentada	faixa de rodagem	separador central
Serbian	autoput	berma	kosina useka/nasipa	berma	bankina	ivična traka	kolovozne trake	centralni (razdelni) pojas
Slovenian	avtocesta	berma	brežina vkopa/nasipa	berma	bankina	odstavni pas	vozni pasovi	vmesni/sredinski pas
Danish	motorvej	terrænkant	skråning	banket	yderrabat	kantbane/nødspor	kørespor	midterrabat

B.3 Pavement Sections

This section provides details of the general layers and sections in pavement construction.

B.3.1 Flexible Pavement (with Verge)/Εύκαμπτο οδόστρωμα

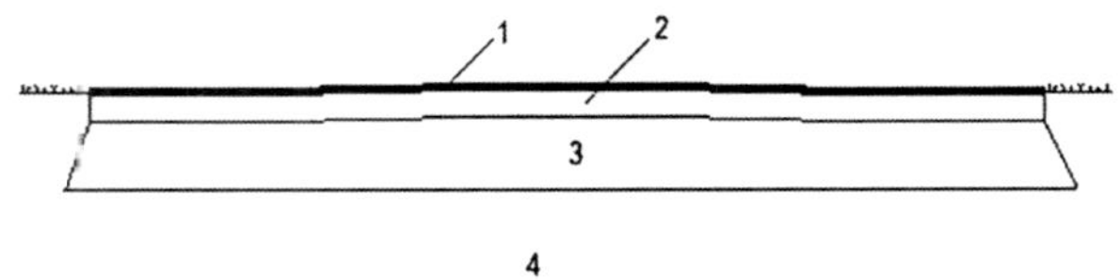

Language	Item no.			
	1	2	3	4
English	surface course[1] (wearing course)	binder course[1] (base course)	base[1] (road base)	subgrade
German	Deckschicht	obere Tragschicht	untere Tragschicht	Untergrund
Spanish	capa de rodadura	capa de base	sub-base	explanada
French	couche de roulement	couche de base	couche de fondation	couche de forme ou sol
Italian	strato d'usura	base	fondazione	sottofondo
Greek	Στρώση κυκλοφορίας	Στρώση βάσεως	Βάση Οδού	Υπέδαφος
Polish	warstwa ścieralna	górna warstwa podbudowy	dolna warstwa podbudowy	podłoże

(continued)

Language	Item no.			
	1	2	3	4
Portuguese	camada de regularização e desgaste	camada de base	camada de sub-base	plataforma de terraplenagem
Serbian	habajući sloj	gornji noseći sloj (podloga)	donji noseći sloj (podloga)	posteljica
Slovenian	vezana obrabna in zaporna plast	vezana zgornja nosilna plast	nevezana nosilna plast	posteljica/temeljna tla
Danish	slidlag	bærelag	bundsikringslag	planum/underbund

[1]Standard terminology has been adopted for pavement layers in EN standards and these are given here. In parentheses are given the traditional terms

B.3.2 Flexible Pavement (with Kerb)

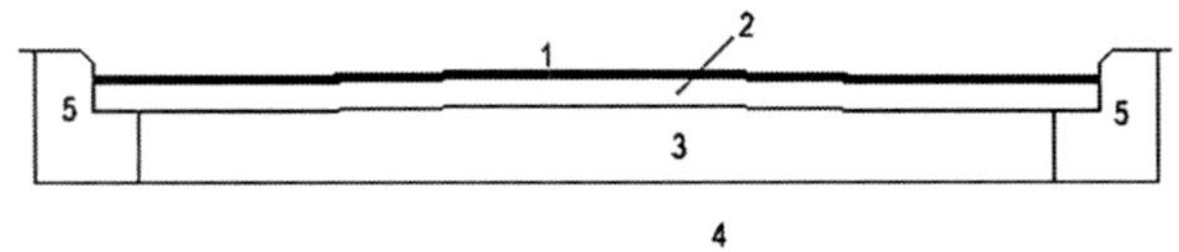

Language	Item no. 1	2	3	4	5
English	wearing course	base course	road base	subgrade	kerb and foundation
German	Decke	obere Tragschicht	untere Tragschicht	Untergrund	Bordstein
Spanish	capa de rodadura	capa de base	sub-base	explanada	bordillo y cimiento
French	couche de roulement	couche de base	couche de fondation	couche de forme ou sol	bordure de trottoir
Italian	strato d'usura	Base	fondazione	sottofondo	cordolo con fondazione
Greek	Στρώση κυκλοφορίας	Στρώση βάσεως	Βάση Οδού	Υπέδαφος	Κράσπεδο και Θεμέλιο
Polish	warstwa ścieralna	górna warstwa podbudowy	dolna warstwa podbudowy	podłoże	krawężnik z fundamentem
Portuguese	camada de regularização e desgaste	camada de base	camada de sub-base	plataforma de terraplenagem	lancil com fundação
Serbian	habajući sloj - zastor	gornji noseći sloj (podloga)	donji noseći sloj (podloga)	posteljica	ivičnjak (sa temeljom)
Slovenian	vezana obrabna in zaporna plast	vezana zgornja nosilna plast	nevezana nosilna plast	posteljica/temeljna tla	robnik
Danish	slidlag	bærelag	bundsikringslag	underbund	kantsten

B.3.3 *Permeable Pavement (SUD)*/Πορώδες Οδόστρωμα

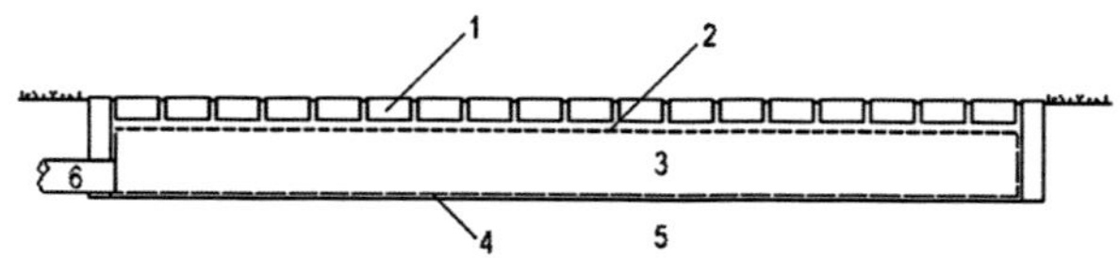

Language	Item no.					
	1	2	3	4	5	6
English	permeable wearing course (in this case paving blocks in permeable filler)	geotextile liner	permeable aggregate	geotextile or impermeable liner	subgrade	drainage pipe (where the sub-grade is insufficiently permeable to permit soakaway operation)
German	durchlässige Deckschicht	Geotextileinsatz	durchlässiger Zuschlagstoff	Geotextileinsatz oder undurchlässiger Einsatz	Untergrund	Entwässerungsleitung
Spanish	capa de rodadura permeable (en este caso adoquinado en arena fina permeable)	capa de geotextil	árido permeable	capa de geotextil o capa impermeable	explanada	tubo de drenaje (cuando la explanada no es suficientemente permeable para permitir operación de drenaje)
French	couche de roulement perméable	recouvrement de géotextile	agrégat perméable	géotextile ou recouvrement imperméable	couche de forme ou sol	tuyau de drainage
Italian	strato permeabile di usura	geosintetico	aggregato permeabile	geosintetico o impermeabilizzante	sottofondo	tubo di drenaggio

(continued)

Language	Item no.					
	1	2	3	4	5	6
Greek	Διαπερατή Στρώση Κυκλοφορίας	Στρώση Γεωϋφάσματος	Διαπερατό Αδρανές Υλικό	Γεωύφασμα ή Αδιαπέρατη Στρώση	Υπέδαφος (στρώση έδρασης)	Σωλήνας Αποστράγγισης
Polish	przepuszczalna warstwa ścieralna	warstwa geotekstyliów	kruszywo prze-puszczalne	warstwa geotekstylna lub warstwa nieprzepuszczalna	podłoże	rura drenarska
Portuguese	pavimento drenante	manta geotextil	agregado permeável	manta geotextil ou camada impermeável	plataforma de terraplenagem	tubo de drenagem/colector
Serbian	porozni zastor	geotekstil	nevezana podloga - porozna	geotekstil ili nepropusna folija	posteljica	drenažna cev
Slovenian	vezana obrabna plast - drenažni asfalt	geotekstil	nevezana nosilna plast	geotekstil ali nepropustna folija	posteljica/temeljna tla	drenažna cev
Danish	vandgennemtrængelig belægning (fx belægningssten)	geotekstil	porøst materiale	geotekstil	underbund	drænrør

SUD (Sustainable Urban Drainage) refers to pavements that are designed to behave more like natural soils (i.e. they allow the infiltration and storage of stormwater). As the SUD pavements have the capacity to hold water and release it slowly, unlike impermeable asphalt and concrete pavements, their use reduces the risk of flash flooding downstream from their discharge points.

B.4 Pavement Edge Details

This Section provides general layouts of the pavement edges. More specifically it provides the general designs used for surface and subsurface drainage. In most cases drainage pipes have outlets into large piped networks collecting surface run-off and sub-surface drainage water. These in turn outlet into ditches, streams, rivers and soakaways depending on which is most accessible given the local environment.

B.4.1 Trench ("French") Drain (Subsurface Drainage Only)

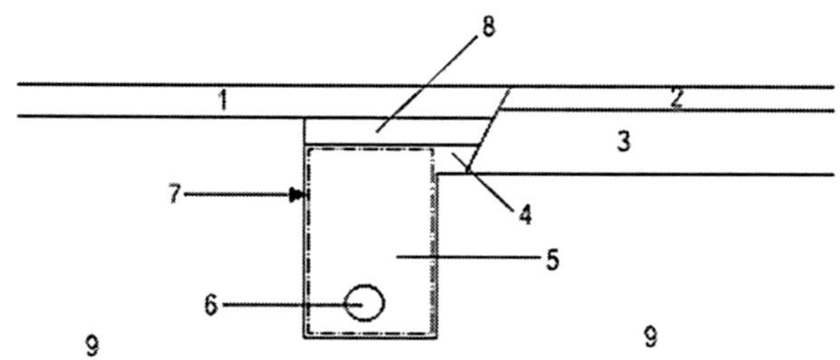

Language	Item no.								
	1	2	3	4	5	6	7	8	9
English	topsoil	wearing course (bound aggregate)	base course (unbound aggregate)	permeable filler	permeable aggregate	porous drainage pipe	geotextile liner	cohesive soil/impermeable filler	subgrade
German	Mutterboden	Deckschicht	obere Tragschicht	durchlässiger Füller	durchlässiger Zuschlagstoff	poröse Entwässerungs-leitung	Geotextileinsatz	undurchlässiger Füller	Untergrund

(continued)

Language	Item no.								
	1	2	3	4	5	6	7	8	9
Spanish	tierra vegetal	capa de rodadura	capa de base	arena fina permeable	árido permeable	tubo de drenaje poroso	capa de geotextile	suelo cohesivo/arena fina impermeable	explanada
French	terre végétale	couche de roulement	couche de base	matériau de remplissage perméable	agrégat perméable	tuyau poreux de drainage	recouvrement de géotextile	matériau de remplissage imperméable	couche de forme ou sol
Italian	terreno vegetale	strato di usura	base	filler permeabile	aggregato permeabile	tubo di drenaggio poroso	geosintetico	filler impermeabile	sottofondo
Greek	Επιφανεια κόέδαφος	Στρώση Κυκλοφορίας (Αδρανές με συνδετικό)	Στρώση Βάσης (Ασύνδετο Αδρανές)	(Διαπερατό λεπτόκκοκο υλικό)	Διαπερατό Αδρανές	Σωλήνας Αποστράγγι σηςμε Πόρους	Στρώση Γεωυφάσ ματοςάσμα	Συνεκτικό Έδαφος/ Αδιαπέραστο ΥλικόΠλ ηρώσεως	Υπέδαφος (στρώσηέ δρασης)
Polish	gleba	warstwa ścieralna	podbudowa	zasypka przepuszczalna	kruszywo drenu	porowata rura drenarska	osłona geotekstylna	grunt spoisty/zasypka nieprze-puszczalna	podłoże
Portuguese	terra vegetal	camada de regularização e desgaste	camada de base	material drenante	material drenante	tubo de dreno poroso	geotextil	enchimento em material impermeável	plataforma de terraple-nagem
Serbian	tlo	zastor (vezani agregat)	podloga	drenažna ispuna	drenažna ispuna (agregat)	porozna drenažna cev	geotekstil	koherentno tlo – nepropusni sloj	posteljica
Slovenian	humus	vezana nosilna in obrabna plast	nevezana nosilna plast	drenažni zasip	drenažni zasip	perforirana drenažna cev	geotekstil	glinasti naboj	posteljica/ temeljna tla
Danish	overfladejord	asfalt	ubundne materialer	filter materiale	filter materiale	dræn med porøse rør	geotekstil	tæt materiale (leret)	underbund

B.4.2 Trench ("French") Drain (Surface and Subsurface Drainage)

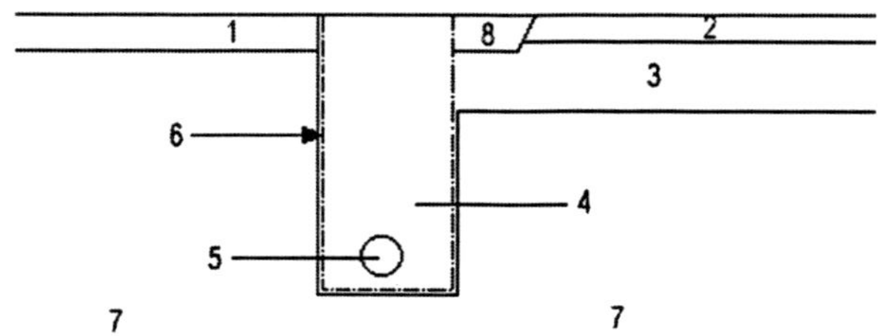

Language	Item no.							
	1	2	3	4	5	6	7	8
English	topsoil	wearing course (bound aggregate)	base course (unbound aggregate)	permeable aggregate	porous drainage pipe	geotextile liner	subgrade	impermeable filler
German	Mutterboden	Deckschicht	obere Tragschicht	durchlässiger Zuschlagstoff	poröse Entwässerungs-leitung	Geotextileinsatz	Untergrund	undurchlässiger Füller
Spanish	tierra vegetal	capa de rodadura	capa de base	árido permeable	tubo de drenaje poroso	capa de geotextil	explanada	arena fina impermeable
French	terre végétale	couche de roulement	couche de base	agrégat perméable	tuyau poreux de drainage	recouvrement de géotextile	couche de forme ou sol	matériau de remplissage imperméable
Italian	terreno vegetale	strato di usura	base	aggregato permeabile	tubo di drenaggio poroso	geosintetico	sottofondo	filler impermeabile

(continued)

Language	Item no.							
	1	2	3	4	5	6	7	8
Greek	Επιφανειακό έδαφος	Στρώση Κυκλοφορίας (Αδρανές με συνδετικό)	Στρώση Βάσης (Ασύνδετο Αδρανές)	Διαπερατό Αδρανές	Σωλήνας Αποστράγγισης με Πόρους	Στρώση Γεωυφάσματ οςάσμα	Υπέδαφος (στρώση έδρασης)	Αδιάβροχο (Αδιαπέραστο) Υλικό Πλήρωσης
Polish	gleba	warstwa ścieralna	podbudowa	kruszywo drenu	porowata rura drenarska	warstwa geotekstylna	podłoże	grunt spoisty/zasypka nieprzepuszczalna
Portuguese	terra vegetal	camada de regularização e desgaste	camada de base	material drenante	tubo de dreno poroso	geotextil	plataforma de terraplenagem	enchimento em material impermeável
Serbian	humus	zastor (vezani agregat)	podloga (nevezani agregat)	drenažna ispuna	porozna drenažna cev	geotekstil	posteljica	nepropusni sloj
Slovenian	humus	vezana nosilna in obrabna plast	nevezana nosilna plast	drenažni zasip	perforirana drenažna cev	geotekstil	posteljica/ temeljna tla	glinasti naboj
Danish	overfladejord	asfalt	ubundne materialer	filter materiale	dræn med porøse rør	geotekstil	underbund	tæt materiale (leret)

B.4.3 Channel Block Drainage

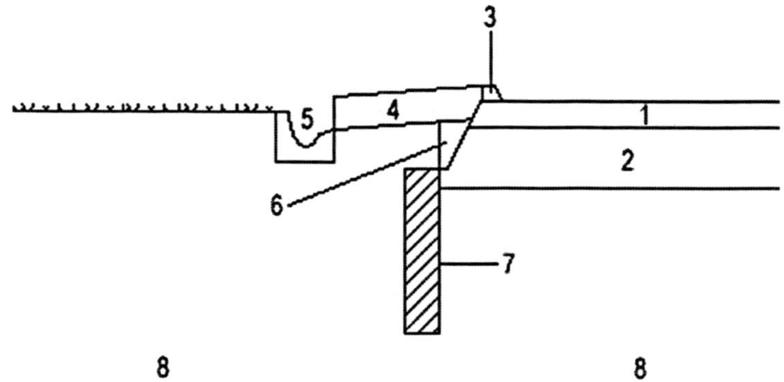

Language	Item no.							
	1	2	3	4	5	6	7	8
English	wearing course (bound aggregate)	base course (unbound aggregate)	kerb	impermeable construction with drainage channels at intervals along the length of the highway	outlet channel	filler	fin drain (see section B5)	subgrade
German	Deckschicht	obere Tragschicht	Bordstein	Undurchlässige Konstruktion mit abschnittsweisen Entwässerungsrinnen entlang der Schnellstraß	Abflusskanal	Füller	Drainage	Untergrund
Spanish	capa de rodadura	capa de base	bordillo	construcción impermeable con canales de drenaje a intervalos a lo largo de la carretera	canal de desagüe	rellenador	diafragma drenante	explanada

(continued)

Language	Item no.							
	1	2	3	4	5	6	7	8
French	couche de roulement	couche de base	bordure de trottoir		caniveau	matériau de remplissage	drain	couche de forme ou sol
Italian	strato di usura	base	cordolo	construzione impermeable con canali drenanti intervallati per tutta la lunghezza dell' autostrada	cunetta	filler	dreno a pinna	sottofondazione
Greek	Στρώση Κυκλοφορίας (Αδρανές με συνδετικό)	Στρώση Βάσης (Ασύνδετο Αδρανές)	Κράσπεδο		Εξωτερικός Αγωγός	Λεπτόκκοκο υλικό Πληρώσεως	Αποστραγγιστικό πλέγμα	Υπέδαφος
Polish	warstwa ścieralna	podbudowa	krawężnik	konstrukcja szczelna z otworami odwadniającymi w odstępach wzdłuż jezdni	kanał odbiorczy	zasypka	dren żebrowy (kompozyt drenażowy)	podłoże
Portuguese	camada de regularização e desgaste	camada de base	lancil	construção impermeável com canais drenantes em intervalos ao longo do comprimento da rodovia	canal de drenagem	enchimento	ecran drenante	plataforma de terraple-nagem
Serbian	zastor (vezani agregat)	podloga (nevezani agregat)	ivičnjak	Nepropusni zastor sa drenažnim kanalima na intervalima duž puta	drenažni kanal	ispuna	drenažni filter (rov)	posteljica
Slovenian	vezana nosilna in obrabna plast	nevezana nosilna plast	robnik		odvodni kanal	zasip	vzdolžno drenažno rebro	posteljica/temeljna tla
Danish	asfalt	ubundne materialer	kantsten	tæt konstruktion med drænkanaler jævnt fordelt langs vejen	drænkanal	tæt materiale	drænbånd	underbund

B.4.4 Gulley Drain

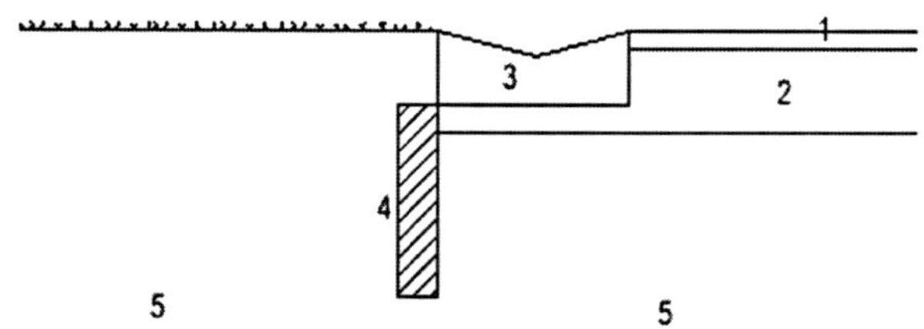

Language	Item no.				
	1	2	3	4	5
English	wearing course (bound aggregate)	base course (unbound aggregate)	concrete channel	fin drain (see section B5)	subgrade
German	Deckschicht	obere Tragschicht	Betongerinne	Drainage	Untergrund
Spanish	capa de rodadura	capa de base	cuneta de hormigón	diafragma drenante	explanada
French	couche de roulement	couche de base	canal de drainage	drain	couche de forme ou sol
Italian	strato di usura	base	cunetta in calcestruzzo	dreno a pinna	sottofondo

(continued)

Language	Item no.				
	1	2	3	4	5
Greek	Στρώση Κυκλοφορίας	Στρώση Βάσης (Ασύνδετο Αδρανές)	Ανοικτός Αγωγός Υδροσυλλογής από σκυρόδεμα	Αποστραγγιστικό πλέγμα	Υπέδαφος
Polish	warstwa ścieralna	podbudowa	ściek betonowy	dren żebrowy (kompozyt drenażowy)	podłoże
Portuguese	camada de regularização e desgaste	camada de base	valeta	ecran drenante	plataforma de terraplenagem
Serbian	zastor (vezani agregat)	podloga (nevezani agregat)	kanaleta (slivnik)	drenažni filter (rov)	posteljica
Slovenian	vezana nosilna in obrabna plast	nevezana nosilna plast	kanaleta	vzdolžno drenažno rebro	posteljica/temeljna tla
Danish	asfalt	ubundne bærelag	beton vandrende	drænbånd	underbund

B.4.5 Kerb And Gully Pot Drainage

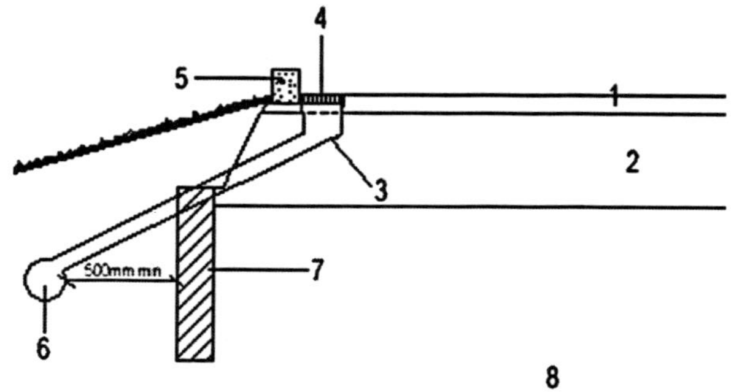

Language	Item no.							
	1	2	3	4	5	6	7	8
English	wearing course (bound aggregate)	base course (unbound aggregate)	gully pot	grate	kerb	drainage pipe	fin drain (see section B5)	subgrade
German	Deckschicht	obere Tragschicht	Strassenablauf	Rost	Bordstein	Entwässerungs-leitung	Drainage	Untergrund
Spanish	capa de rodadura	capa de base	cono sumidero	rejilla	bordillo	tubo de drenaje	diafragma drenante	explanada
French	couche de roulement	couche de base	regard	grille	bordure de trottoir	collecteur	drain	couche de forme ou sol

(continued)

Language	Item no.							
	1	2	3	4	5	6	7	8
Italian	strato di usura	base	pozzetto	griglia	cordolo	tubo di drenaggio	dreno a pinna	sottofondo
Greek	Στρώση Κυκλοφορίας	Στρώση Βάσης (με Ασύνδετα Υλικά)	Δοχείο Υδροσυλλογή	Εσχάρα	Κράσπεδο	Σωλήνας Αποστράγγισης	Αποστραγγιστικό πλέγμα	Υπέδαφος
Polish	warstwa ścieralna	podbudowa	studzienka ściekowa	wpust/kratka ściekowa	krawężnik	kolektor	dren żebrowy (kompozyt drenażowy)	podłoże
Portuguese	camada de regularização e desgaste	camada de base	caixa sumidouro	grelha	lancil	colector	ecran drenante	plataforma de terraplenagem
Serbian	zastor (vezani agregat)	podloga (nevezani agregat)	odvodna cev	rešetka	ivičnjak	drenažna cev	drenažni filter (rov)	posteljica
Slovenian	vezana nosilna in obrabna plast	nevezana nosilna plast	kanalizacijska cev	rešetka	robnik	drenažna cev	vzdolžno drenažno rebro	posteljica/temeljna tla
Danish	asfalt	ubundne materialer	nedløbsbrønd	rist	kantsten	afvandingsrør	drænbånd	underbund

B.4.6 Swales (with Fin Drain)

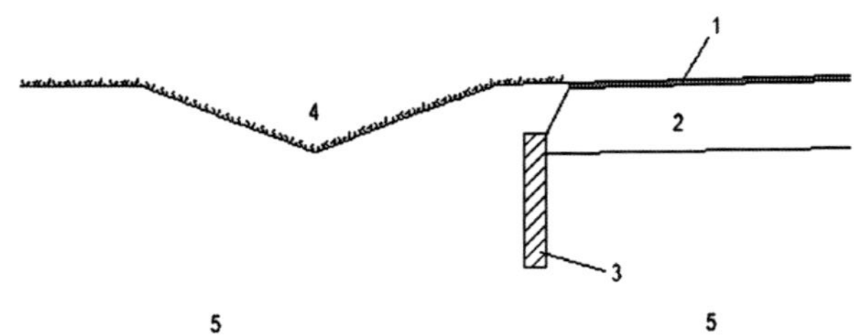

Language	Item no.				
	1	2	3	4	5
English	wearing course (bound aggregate)	base course (unbound aggregate)	fin drain (see section B5)	swale	subgrade
German	Deckschicht	obere Tragschicht	Drainage	Mulde	Untergrund
Spanish	capa de rodadura	capa de base	diafragma drenante	acequia	explanada
French	couche de roulement	couche de base	drain	fossé	couche de forme ou sol
Italian	strato di usura	base	dreno a pinna	avvallamento	sottofondo
Greek	Στρώση Κυκλοφορίας	Στρώση Βάσης (με Ασύνδετα Υλικά)	Αποστραγγιστικό πλέγμα	Αποστραγγιστική Τάφρος	Υπέδαφος
Polish	warstwa ścieralna	podbudowa	dren żebrowy (kompozyt drenażowy)	mulda	podłoże
Portuguese	Camada de regularização e desgaste	camada de base	ecran drenante	valeta larga	plataforma de terraple-nagem
Serbian	zastor (vezani agregat)	podloga (nevezani agregat)	drenažni filter (rov)	odvodni jarak	posteljica
Slovenian	vezana nosilna in obrabna plast	nevezana nosilna plast	vzdolžno drenažno rebro	odvodni jarek	posteljica/temeljna tla
Danish	asfalt	ubundne materialer	drænbånd	trug	underbund

B.4.7 Ditch and Swale (with French Drain)

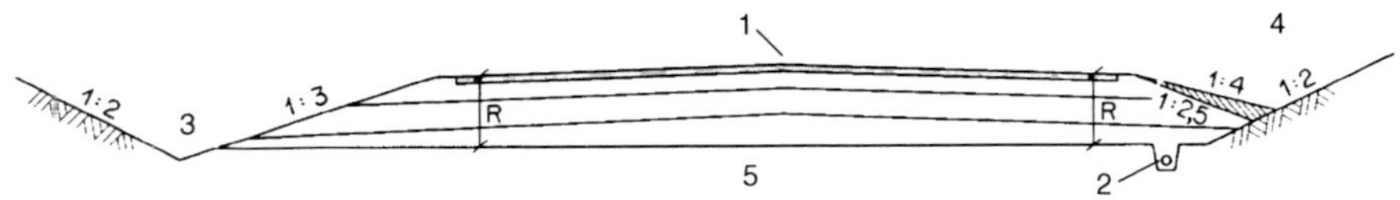

Language	Item no.				
	1	2	3	4	5
English	wearing course (bound aggregate)	trench ("French") drain	ditch	swale	subgrade
German	Deckschicht	Rohrdrainage	Straßengraben	Mulde	Untergrund
Spanish	capa de rodadura	zanja de drenaje	cuneta	acequia	explanada
French	couche de roulement	tranchée drainante	fossé	fossé	couche de forme ou sol
Italian	strato di usura	dreno	fosso	avvallamento	sottofondo
Greek	Στρώση Κυκλοφορίας			Αποστραγγιστική Τάφρος	Υπέδαφος
Polish	warstwa ścieralna	dren francuski	rów	mulda	podłoże
Portuguese	camada de regularização e desgaste	dreno francês	valeta larga	valeta larga	plataforma de terraplenagem
Serbian	zastor (vezani agregat)	drenazni rov	kanal	odvodni jarak	posteljica
Slovenian	vezana nosilna in obrabna plast	nevezana nosilna plast	vzdolžni drenažni jarek	jarek	odvodni jarek
Danish	asfalt	dræn	grøft	trug	underbund

B.5 Trench ("French") and Fin Drains

This section provides general designs of french and fin drains used to drain the subsurface section of highways.

B.5.1 Trench ("French") Drains

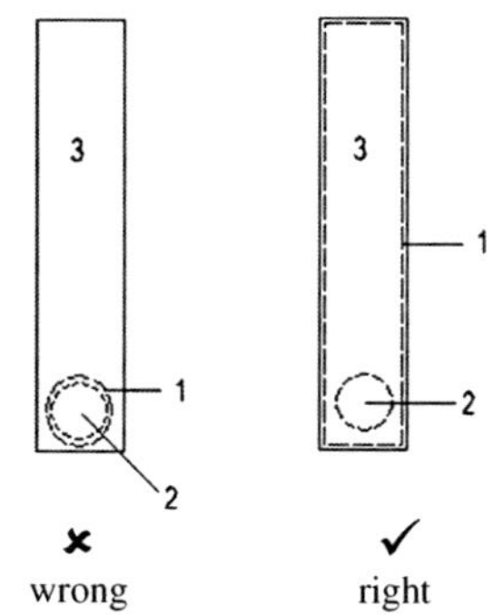

Language	Item no.		
	1	2	3
English	geotextile liner	slotted drainage pipe	permeable aggregate filler
German	Geotextileinsatz	poröse Entwässerungsleitung	durchlässiger Zuschlagstoff
Spanish	capa de geotextil	tubo de drenaje ranurado	arena fina permeable
French	recouvrement de géotextile	tuyau poreux de drainage	agrégat perméable
Italian	geosintetico	tubo di drenaggio poroso	aggregato permeabile
Greek	Στρώση Γεωυφάσματος άσμα	Σωλήνας Αποστράγγισης με Σχισμές	Διαπερατή Στρώση Πληρώσεως
Polish	warstwa geotekstylna	porowata rura drenarska	kruszywo wypełniające dren
Portuguese	geotextil	tubo de drenagem poroso	material drenante
Serbian	geotekstil	porozna drenažna cev	drenažna ispuna
Slovenian	geotekstil	perforirana drenažna cev	drenažni zasip
Danish	geotekstil	slidset afvandingsrør	filtergrus

B.5.2 Fin Drain Designs

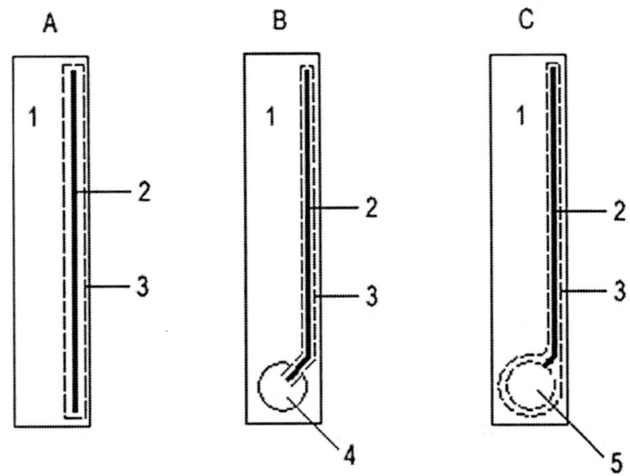

Language	Item no.				
	1	2	3	4	5
English	permeable aggregate filler	core	geotextile liner	drainage pipe	slotted drainage pipe
German	durchlässiger Zuschlagstoff	Kern	Geotextileinsatz	Entwässerung-sleitung	poröse Entwässerungsleitung
Spanish	arena fina permeable	núcleo	capa de geotextil	tubo de drenaje	tubo de drenaje ranurado
French	agrégat perméable	structure interne	recouvrement de géotextile	tuyau de drainage	tuyau poreux de drainage

Language	Item no.				
	1	2	3	4	5
Italian	aggregato permeabile	struttura interna	geosintetico	tubo di drenaggio	tubo di drenaggio poroso
Greek	Διαπερατή Στρώση Πληρώσεως	Πυρήνας	Στρώση Γεωυφάσματος άσμα	Σωλήνας Αποστράγγισης	Σωλήνας Αποστράγγισης με Σχισμές
Polish	kruszywo wypełniające dren	rdzeń	warstwa geotekstylna	rura drenarska	porowata rura drenarska
Portuguese	material drenante	estrutura interna	geotextil	tubo de drenagem	tubo de drenagem poroso
Serbian	drenažna ispuna	jezgro (ispuna)	geotekstil	drenažna cev	porozna drenažna cev
Slovenian	drenažni zasip	jedro	geotekstil	drenažna cev	perforirana drenažna cev
Danish	filtermateriale	kerne	geotekstil	afvandingsrør	slidset afvandingsrør

The designs of fin drains vary dependant upon the flow capacity that is required. The geosynthetic has the capacity to transport a certain flow (i.e. it has a certain transmissivity) - see diagram A, above. Where this is insufficient a pipe can be added to increase the outflow capacity, diagrams B and C. The dimension of the pipe varies depending on the additional capacity that is required.

B.5.3 Core Designs for Fin Drains

Core designs other than those shown here are also used. The purpose of the core is to support the geotextile membrane without hindering the flow of water within the thickness of the geosynthetic.

The figure, above, shows a plan view of a geocomposite highway edge drain, comprising a core (3/4/5) between two geotextile sheets (1/2). Some cores have one permeable side (e.g. geotextile layer 1) and one impermeable side (e.g. in place of geotextile layer 2) where water is to be collected from one side of the fin drain and not to be drained from the other, but these are seldom employed in highway construction except where an impermeable face (e.g. part of a retaining wall) abuts the pavement. The core may be formed in a number of ways, some of which are illustrated here: 3=plastic plates with pillars to separate faces; 4=concertina waffle structure, 5=dimpled plastic sheet. Nets/grids and stiff, entangled polymeric strands may also be used to provide a permeable core.

Language	Item no.	
	1/2	3/4/5
English	geotextile	core
German	Geotextileinsatz	Kern
Spanish	capa de geotextil	núcleo
French	recouvrement de géotextile	structure intérieur
Italian	geosintetico	struttura interna
Greek	Στρώση Γεωυφάσματος άσμα	Πυρήνας
Polish	warstwa geotekstylna	rdzeń
Portuguese	geotextil	estrutura interna
Serbian	geotekstil	jezgro (ispuna)
Slovenian	geotekstil	jedro
Danish	geotekstil	Kerne

B.6 Water Disposal

This section provides general designs of the final elements the water runs through before going to existing water bodies.

B.6.1 Retention Ponds/Λεκάνες συγκράτησης

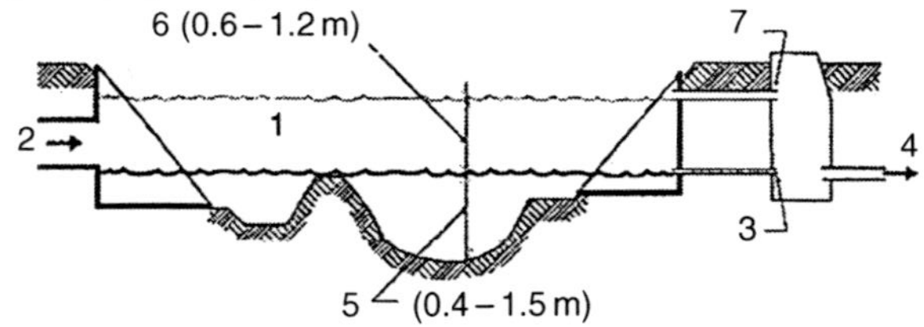

Language	Item no. 1	2	3	4	5	6	7
English	retention pond	inlet	throttle pipe	outlet	minimum level	operating depth	overflow pipe
German	Regenrückhaltebecken	Zulauf	Drosselrohr	Ablauf	tiefstes Absenkziel	Stauhöhe	Überlaufrohr
Spanish	cubeta de retención	entrada	tubería de regulación de nivel	desagüe	nivel mínimo	profundidad de trabajo	tubería de desbordamiento
French	bassin de rétention	arrivée	valve pointeau	sortie	niveau le plus bas	niveau de fonctionnement	tuyau de trop plein
Italian	bacino di ritenzione	ingresso	condotta di strozzamento	uscita	livello minimo	profondita' di esercizio	condotta di sfioro
Greek	Λεκάνη συγκράτησης	Είσοδος	Σωλήνας αποστράγγισης	Έξοδος	Ελάχιστο επίπεδο	Βάθος λειτουργίας	Σωλήνας υπερχείλισης

Language	Item no.						
	1	2	3	4	5	6	7
Polish	zbiornik retencyjny	wlot	rura dławiąca	odpływ	poziom minimalny	poziom roboczy	rura przelewowa
Portuguese	Bacia de retenção	Entrada	Válvula de borboleta	Saída	Nível mínimo	Profundidade de operação	Descarregador de superfície
Serbian	Retenzioni bazen	Dovodna cev	Regulaciona cev/ventil	Odvodna cev	Minimalni nivo	Operativna dubina	Prelivna cev
Slovenian	zadrževalni bazen	dovodna cev	regulacijska cev/ventil	odvodna cev	najnižji/minimalni nivo	obratovalna globina	prelivna cev
Danish	regnvands/forsinkelses bassin	indløb	drosselledning	udløb	minimal vanddybde	stuvningshøjde	overløbsledning

B.6.2 *Soakaways*/Αγωγοί απαγωγής ομβρίων

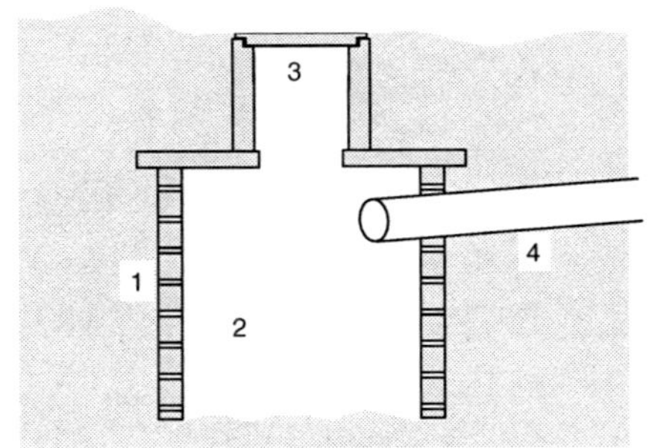

Language	Item no. 1	2	3	4
English	porous wall	void	access cover	inlet pipe
German	durchlässige Wand	Hohlraum	Schachtabdeckung	Zulaufrohr
Spanish	pared porosa	hueco	recinto cerrado de acceso	tubería de entrada
French	mur poreux	vide	accès supérieur	conduite d'arrivée
Italian	muro poroso	vuoto	copertura di accesso	condotta d'ingresso
Greek	Διαπερατά τοιχώματα	Κενό	Κάλυμμα επίσκεψης	Σωλήνας εισόδου
Polish	ściana porowata	komora	właz	rura wlotowa
Portuguese	Parede porosa	Vazios	Tampa de acesso	Colector de entrada
Serbian	Porozni zid	Otvor sahta	Poklopac sahta	Dovodna cev
Slovenian	perforirana stena	jašek	pokrov jaška	dovodna cev
Danish	vandgennemtrængelige sider	hulrum	dæksel	indløbsrør

Annex C
Glossary of Words and Abbreviations

Italicised words indicate a separate entry in the glossary

AAD	a core bituminous binder (asphalt cement) in the US SHRP system
AADT	annual average daily traffic
AAS	atomic absorption spectrometry
AASHTO	American Association of State Highway & Transportation Officials
absorption	*sorption* to the inside of a solid
AC	*asphalt* concrete (see *asphalt*)
acute response	a short-term response, usually undesirable
adsorption	*sorption* to the surface of a solid
advection	mass movement of a substance carried in a fluid flow
AEV	*air entry value*
air entry value	the pressure of air necessary to induce air flow through a water-*saturated* porous medium
alkalinity	acid-neutralizing capacity
alternative material	a material that is an alternative to those traditionally used in construction. typically it may be an industrial *by product*, a *waste*, a *recyclate*, a treated material, etc.
amphoteric	capable of functioning either as an acid or as a base
analyte	a species of chemical or micro-organism that is to be analysed
analytical model	material *constitutive* relationship suitable for use in stress-strain analysis
aquaplaning	see *hydroplaning*
aquiclude	a saturated ground stratum, or strata, of low permeability which can only yield inappreciable quantities of water

aquifer	a permeable water-bearing layer of rock or sediment that is capable of yielding useable amounts of water
"AREA"	area of a deflection basin, scaled to the central deflection
arthropod	segmented invertebrate of the phylum *Arthropoda* (e.g. insect)
asphalt	mixture of aggregate and fine particles held together in a *bitumen*-based binder. Known as asphaltic concrete in many countries
axis translation method	system for measuring the *suction* of a soil or aggregate specimen using a high air entry value porous stone
background value	the *concentration* of a chemical species, micro-organism or suspended particulate that is found under normal conditions prior to any man-made effect
base	the main structural layer of a pavement
base-line	the value of *contaminants*' concentration, *contaminant* loads, water flows, etc. before any remedial or disruptive activity has taken place that would change values
BCI	base curvature index – often defined as the deflection at a point 600 mm from the centre of an area loaded by *FWD*, less the deflection at a point 900 mm from the same centre
BDI	base damage index – defined as the deflection at a point 300 mm from the centre of an area loaded by *FWD*, less the deflection at a point 600 mm from the same centre
bearing capacity	the limiting ability of a pavement to carry traffic
Benkelman beam	device to measure the *deflection* of a pavement's surface due to the passage of a heavy vehicle
bioaccumulation	the accumulation by a biological organism of a substance that is not an essential component or nutrient of that organism (or is not at the levels at which they are accumulated)
biomagnification	process in which the *concentration* of a substance (such as a *pollutant*) increases from one food-chain level to the next
biota	the animal and plant life of a region or period

bitumen	highly viscous *hydrocarbons* derived from petroleum and used as a binding agent, e.g. in road surfacings
bituminous geosynthetic barrier	factory-assembled structure of *geosynthetic* materials containing a bituminous layer in the form of a sheet which, thereby, acts as a barrier
by-product	a material generated in a process incidental to the main purpose of that process
California bearing ratio	test giving an index value of a soil or aggregate layer's strength
Californian drain	parallel and closely-spaced drainage tubes installed in the ground
capillarity	phenomenon that is associated with the surface tension acting between two *immiscible* fluids in a porous medium by which the medium is able to pull a fluid into its pores
capillary break	a layer of construction comprised of a material with large pores that cannot sustain a high suction and, thereby, prevents capillarity effects near the surface from exerting a suction beneath that layer
capillary pressure	the difference in pressure across the interface of two *immiscible* fluids
capillary zone	the soil area just above the *phreatic surface* where water rises due to *capillarity*
capping	a *subgrade* improvement layer lying immediately under the *pavement*
catchpit	a pit in the ground, without a drain, which is capable of collecting and containing the solids from passing water flows in a drainage system
CBR	*California bearing ratio*
CCP	constant confining pressure
CEN	Comité Européen de Normalisation (the European Committee for Standardization)
centrifuge	a device to spin an object at high speeds thereby developing a high, radial acceleration, usually to speed-up physical processes normally driven by gravitational acceleration
chelate	heterocyclic compound having a central metallic ion attached by covalent bonds to two or more non-metallic atoms in the same molecule
chronic response	a long-term and ongoing response, usually undesirable

clay geosynthetic barrier	factory-assembled structure of *geosynthetic* materials containing hydrophilic clay in the form of a sheet which, thereby, acts as a barrier
clogging	the action when fine particles block open pores in an otherwise porous material thereby significantly reducing the *hydraulic conductivity*
compaction	*mechanical* process by which the constituent particles of a *particulate solid* are forced closer together and air displaced from the pores (cf. *consolidation*)
compaction ratio	the ratio of the difference between actual and loose densities to the difference between dense and loose densities ('dense' and 'loose' are defined by reference to standard tests, not the maximum and minimum values achievable)
compartment	see *environmental compartment*
compensation	*mitigation* method in which the environmental quality of some new or substitute habitat or water body is restored, created or enhanced in place of that which has, or is likely to, become degraded
complex	a chemical compound consisting of a central atom and *ligands* (consisting of a group, molecule or ion) tied to the central atom with at least one co-ordination bond
concentration	the amount of a substance expressed as a proportion of the water, air or solid in which it resides. Usually expressed as mass of substance per volume of liquid or as mass of substance per mass of solid
concrete	a material formed of aggregate bound with Portland cement. Also known as Portland cement concrete.
conductivity	a measure of a material's ability to conduct energy, particularly electrical or thermal
confined aquifer	an *aquifer* that is bounded above and below by *aquicludes*
consolidation	process by which the constituent particles of a *particulate solid* are drawn closer together due to water drainage from the pores (cf. *compaction*)
constitutive model	a formulaic description of the stress-strain relationship of a material
contaminant	a chemical, micro-organism or particulate component that is normally absent from the *environmental compartment* in which it resides, or which is present at an elevated *concentration*
convection	energy transfer by a bulk macroscopic motion of fluid (liquid and gas) particles from a hot region to a cool region

coupled problems	a computational problem comprising two or more physical components that exhibit two-way (mutual) interaction such that the response of the system to some forcing function modifies the forcing function and/or the system's future response, often indirectly
crossfall	the gradient of a layer of a road or embankment construction measured at 90° to the direction of the road
cryo-suction	the effect by which water is pulled towards a *freezing front*
curviameter	monitoring tool that measures a *deflection* of the pavement thereby assessing the road's condition
cyclic triaxial test	see *repeated load triaxial test*
d.o.f.	degree of freedom
de-bonding	the separation of *bitumen* from the stones to which it formerly adhered
deflection	transient movement of an object, or part of the object, due to the application of a stress or force (c.f. *deformation*)
deformation	generally understood as a movement of an object, or part of the object, due to the application of a stress or force. More precisely used to mean *strain*
desorption	the release of chemicals that are *sorbed* onto or inside a solid to a surrounding fluid
dielectric	the relationship between an applied electric field and the moment of a dipole that it induces
diffusion	the three-dimensional redistribution of a substance dissolved or suspended in a fluid from areas of high concentration to areas of lower concentration due only to the energy of the molecules of the *contaminant* and water
direct pollution	*pollution* resulting from spilling of a harmful substance directly into an *environmental compartment*
dispersion	the three-dimensional redistribution of a substance dissolved or suspended in a fluid that is caused by local variations in movement of that fluid usually resulting from *inhomogeneity* in the *pathway*
dissolution	the action of dissolving a substance into a fluid
drainability	degree to which a porous medium can drain when it is subject to no external stresses. Inverse of *retention*
drainage spur	aggregate-filled drainage trenches installed in a slope face to both drain and buttress the slope
dynaflect	an electro-magnetic system for measuring the dynamic *deflection* of a surface or structure caused by an oscillatory load

dynamic plate bearing test	*plate bearing test* to assess soil or aggregate layer *modulus* that uses impact loading to induce a brief loading pulse
ecotoxicity	the ability to have an adverse effect on an *environmental compartment* and/or *biota*
eddy	rotational flow within a water current causing energy losses
effective porosity	a measure of *porosity* that excludes isolated pores and pore volume made unavailable to the process of concern
effective stress	a computed stress representing the difference between an applied external stress and the pore fluid pressure. See *pore water pressure*
elasto-plastic equivalent model	material *constitutive* relationship that describes resilient and plastic strains, due to an applied stress, in one formulation
elasto-plasticity	stress-strain behaviour incorporating both elastic (recoverable) and plastic (non-recoverable) strains
EMC	*event mean concentration*
EN	European Norm (Standard)
environmental compartment	a sub-section of the environment that is rather uniform in nature e.g. air, topsoil, subsoil or a water body
environmental regulator	the legally established authority for setting and policing compliance with environmental laws
EP Tox	extraction procedure toxicity test
ETR	*evapo-transpiration*
evapo-transpiration	discharge of water to the atmosphere via evaporation from lakes, river, seas, and plant transpiration
event mean concentration	the average concentration of a *contaminant* over a certain period weighted for the flow
extraction method	a method to separate compounds, often based on their relative solubility
falling weight deflectometer	automated *plate bearing test* to assess pavement layer *modulus*, that uses a dropping weight to induce a high-stress pulse similar to that applied by a vehicle
fatigue	the progressive and localised damage that occurs when a material is subjected to repeated application of a stress or a strain
FI	*freezing index*
filter drain	sub-surface drain to both collect and convey water. It is formed by excavating a trench and backfilling it with natural or broken granular material. Within it may be laid a porous or perforated pipe. Usually *geotextile* wrapped

fin drain	sub-surface drain in which water is collected via a *geo-composite*
finite element	a computational technique solving a partial differential equation system; applied e.g. for computing the mechanical response of a solid object, or the seepage in a porous media, etc.
first flush	a rapid, initial change in water quality (*concentration* or load) that occurs during the first interaction between water and the medium through or over which it is travelling, e.g. at the start of rain-induced flow after a dry period
flow disruption	the opposite of *hydraulic transparency*
freezing index	the product of the multiplication of the number of days of continuous *frost* by the number of degrees less than 0 °C
freezing front	the location at which freezing is taking place in a soil or aggregate's pores due to *frost*. The material above the freezing front is already frozen
French drain	see *filter* drain
frost	A climatic condition less than 0 °C
frost heave	the rise of the ground surface due to *frost* action. Water is drawn into a soil profile by so-called *cryo-suction* and then freezes
frost susceptible	liable to suffer *frost heave* and spring *thaw* problems
FWD	*falling weight deflectometer*
GBR	*geosynthetic* barrier
GBR-B	*bituminous geosynthetic barrier*
GBR-C	*clay geosynthetic barrier*
GBR-P	*polymeric geosynthetic barrier*
GCO	*geocomposite*
geocomposite	manufactured, multi-layered product using at least one *geosynthetic* product among the components
geosynthetic	manufactured product, at least one of whose components is made from a synthetic or natural polymer, in the form of a sheet, a strip or a three-dimensional structure, and used in contact with soil and/or other materials
geosynthetic barrier	manufactured low-permeability *geosynthetic*-based or *geocomposite*-based product with the purpose of reducing or preventing fluid flow across its thickness. See *clay geosynthetic barrier, bituminous geosynthetic barrier* or *polymeric geosynthetic barrier*

geotextile	planar, permeable, polymeric (synthetic or natural) textile material – which may be nonwoven, knitted or woven - used in contact with soil and/or other materials
greenfield	an area with low development, except for some agricultural use, in which soil and water quality values can be used as *background values*
grip	a shallow ditch connecting a road's edge to a roadside ditch
groundwater	water in the pore space below the *phreatic surface*. Sometimes used more generally to describe water found in any soil pore
groundwater vulnerability	the *sensitivity* of *groundwater* quality to an imposed contaminant load, which is determined by the intrinsic characteristics of the *aquifer* (see *vulnerability*)
gulley	drainage pit covered by an open metal grating located at the edge of a road
gutter	surface longitudinal drainage at the edge of the pavement to convey *runoff*
HAEV	high *air entry value*
hazard	in a general sense it is a situation which poses a level of threat to life, health, property or environment. More technically, it is the attribute that is the consequence of the probability of an adverse event and the degree of harm that can happen if this event occurs
HCB	hexachlorobenzene
HCV	heavy commercial vehicles (as a proportion of AADT)
heavy metal	metal with a density $> 6\,Mg/m^3$ in elemental form. Typically are only semi-mobile in the natural environment
hinterland	the near-road environment
hydraulic conductivity	see *permeability*
hydraulic gradient	the gradient of water head, being the head difference between two points divided by the length of the flow path between those points
hydraulic transparency	property of an object having no influence on the ability (rate, time, route, etc.) of water to pass
hydrocarbon	compound containing only hydrogen and carbon
hydrodynamics	the principles of fluids dynamics applied to water
hydrograph	graph of the water level, or rate of flow, of a body of water as a function of time
hydrometer	device to measure the density of a liquid

hydroplaning	action of a vehicle sliding on a film of water between the road surface and the vehicle's tyres. Also known as *aquaplaning*
hypolimnion	the layer of water below the *thermocline* in a standing water body (such as a lake or reservoir), i.e. the water layer with a steep temperature gradient
hysteresis	describes the property of a material or system which exhibits a delay between the making of a change, and the response to, or effect of, that change.
ICP	inductively coupled plasma
IEC	ion exchange chromatography
immiscible	the inability of two fluids to become mixed
impact mitigation	activity to reduce or compensate for a deleterious impact of an action
incompatibility	exists when there is a chemical interaction between a ground fluid and a soil that causes physical changes to the soil through which it is flowing. Also used more generally to indicate the impossibility of successfully blending two substances
indirect pollution	*pollution* resulting from the introduction of a harmful substance into an *environmental compartment* in an indirect manner (i.e. via some pathway)
inductance	the property of an electric circuit by which an electromotive force is induced as the result of a changing magnetic flux
infiltration	the movement of water or fluid through a sub-aerial surface
infiltrometer	a device to assess *infiltration*
insular saturation	air isolated in larger pores, so that flow of that air is not possible
interceptor drain	sub-surface drain installed to prevent sub-horizontal movement of water from the *hinterland* to the pavement and embankment
invariant	see *stress invariant*
ion selective electrode	an electrical sensor capable of selectively measuring a particular ionic concentration in water containing several different dissolved ions
irrotational flow	fluid flow without *eddy* patterns
ISE	*ion selective electrode*
isotherm	graphical curve describing changes in a physical system at a constant temperature

isotropy	having the same properties in all directions
karst	limestone landscape characterised by sinks, ravines, caves and underground streams created as a result of the partial *dissolution* of the rock by water
Lacroix deflectograph	automated version of *Benkelman beam* that is deployed repeatedly along a pavement by a moving lorry
laminar flow	water flow that is non-*turbulent*
latent heat	the amount of energy in the form of heat released or *absorbed* by a substance during a change of phase
LC	(a) lethal *concentration*
	(b) liquid chromatography
leachant	a fluid that causes *leaching*
leachate	a fluid resulting from *leaching*
leaching	the process by which *contaminant*s move from the solid fraction of a soil or construction material to immediately adjacent water
ligand	ion or molecule attached to a metal atom by covalent bonding in which both electrons are supplied by one atom
lightweight falling weight deflectometer	see *dynamic plate bearing test*
load of contaminant	the mass of *contaminant* being imported to an ecosystem *compartment*, usually summed over a long period such as one year
longitudinal drainage	drains that carry water parallel to the direction of a road
lysimeter	an instrument for monitoring the water that *percolates* through a certain depth of soil
lysimeter, seepage	water collected passively
lysimeter, suction	water collected by applying a *suction* to draw the water to the collection point. Also known as *tension lysimeter*
lysimeter, tension	see *suction lysimeter*
mastic	mixture of *bitumen* and dust/fine aggregate
mechanical	concerning the response to forces, stresses, *deformations* and/or *deflections*
mechanical behaviour	the strain or *deformation* response of a material to applied loading
Meyer's bottle	device for collecting a water specimen at a particular depth in a *water body* by opening an inlet port at that depth
microflora	microbial life

mitigation	action taken to reduce the impact of some other event or action. See *compensation*
modulus	a value of *stiffness*
moisture	a synonym for water, often used when the water doesn't *saturate* the medium in which it resides
moisture content	see *water content*
moisture sensitivity	ease with which a material's property changes in the presence of water
monolith	the form of a solid that exists in large crystalline, or otherwise adhering, blocks
NAPL	non aqueous phase liquid
Natura 2000	an ecological network of nature protection areas in the EU, designed to assure the long-term survival of Europe's most seriously threatened habitats and species
OGDL	open graded drainage layer
open drainage	drainage systems open to the atmosphere
optimum water content	the *water content* that will allow a material to reach the highest packing of its solids under a standardised (e.g. proctor) compactive effort (see *compaction*)
osmotic control method	system for measuring or controlling the *suction/water content* of a soil or aggregate specimen using a semi-permeable membrane and a concentrated solution of *PEG* to induce pore *suction*
PAH	polyaromatic *hydrocarbon*
particle size distribution	the mass distribution of particle-size classes of a *particulate solid*
particulate solid	an assemblage of particles of various sizes and shapes that are held together by gravitational, *suction* or weak cementation forces or by a combination of these. There are pore spaces between the particles
pathway	the route, or potential route, that a *contaminant* takes between its *source* and its *receptor*
pavement	any layer(s) added to the natural ground surface
pavement design	procedure to dimension road pavement layer thicknesses and to select constituent materials
PCB	polychlorinated biphenyl
PCC	Portland cement concrete (see *concrete*)
PCE	passenger car equivalent (a measure of number of vehicles using a road)
PEG	polyethylene glycol

pellicular water	thin skin of *adsorbed* water
pendular water	water held at particle contact points by surface tension forces
percolation	water movement downwards through *unsaturated soil*
Percostation	an installation of devices that measures the *dielectric* value and other properties of an adjacent material
permanent deformation	element of *deformation* that is not recoverable on unloading
permanent deformation model	*constitutive* relationship to describe the development of plastic strain as a function of the magnitude of repeatedly applied transient stress for a material and of the number of applications of the stress
permeability	a measure of the ease with which water flows through a *particulate solid*. Unless otherwise indicated, the solid must be *saturated*. Also known as *hydraulic conductivity*
permeameter	a device to measure *permeability* of a soil or aggregate
pervious	relatively permeable
pH	a measure of acidity or *alkalinity* derived from the *concentration* of hydrogen ions in a fluid
phreatic	subject to a pressure (applies to *groundwater*)
phreatic surface	the contour of atmospheric pressure in water in the ground
phytoremediation	the use of vegetation in the treatment of *pollutant*s in ground or surface water, e.g. in road runoff
piping	process by which seeping water erodes material from the seepage path thereby substantially increasing the volumetric rate of water movement
PI	plasticity index
plasticity theory based model	material *constitutive* relationship that incorporates the estimation of irrecoverable strains
plate bearing test	in-situ, quasi-static, test to assess soil or aggregate layer *modulus* using a circular plate to apply loading to the layer's surface
pollutant	a substance existing at sufficient *concentration* such that its effects are harmful to human health, other living organisms, or the environment
pollution	the introduction into the environment of a *pollutant*
polymeric geosynthetic barrier	factory-assembled structure of *geosynthetic* materials containing a low-permeability synthetic sheet which, thereby, acts as a barrier

polyvalent metal	having a valency > 1
pore suction	a pressure less than atmospheric in a soil or aggregate pore
pore water pressure	the pressure of water that is acting in a soil or aggregate pore
porosity	relative volume of pores in a soil or aggregate
porous asphalt	*asphalt* that is designed to allow water flow within itself
precipitation	(a) rainfall (b) loss of dissolved chemical to its solid form which then falls to the bottom of the water in which it was dissolved
proctor compaction	a standardised compactive effort (see *compaction*)
pumping	(a) moving water by use of a pump (b) water movement through a soil or aggregate induced by repeatedly stressing that soil or aggregate
radiation	the direct transfer of energy through a medium or vacuum
ravelling	progressive *stripping*. The term *unravelling* is often used having the same meaning
receptor	the ultimate destination, e.g. *environmental compartment*, of a substance (often a *contaminant*), that is being transported through the environment. Often, intermediate destinations, e.g. *groundwater*, are defined as the receptor. Also known as *target*
recyclate	a material resulting from the action of re-cycling
redox potential	the tendency of a chemical species to acquire electrons and thereby be reduced
remediation	*pollution mitigation* method applied after a *pollution* event
repeated load triaxial test	laboratory test for soil or aggregate that applies pulsed stresses simulative of traffic-induced stress pulses
resilient behaviour	behaviour which considers the element of *strain* or *deformation* that is recoverable on unloading. Mainly used to describe the response to repeated loading after stabilisation of stress–strain cycles
resilient behaviour model	*constitutive* relationship to describe the transient stress to resilient strain relationship of a material
resilient modulus	ratio of the transient stress applied to the resilient strain caused (see *modulus*)
resonant frequency	a frequency that matches that at which an object tends to oscillate with maximum amplitude
retention	see *soil-water retention*
retention pond/lagoon	an area of ground or a construction that can, temporarily, contain runoff so as to improve the water quality and/or reduce the peaks of the *hydrograph*

risk	a) the probability that a particular adverse event will occur b) the product of the likelihood of such an event **and** the severity of the outcome of the impact
RLT	*repeated load triaxial*
RLTT	*repeated load triaxial* test
road equipment	equipment installed in and adjacent to a road, especially to improve the safety and comfort of a road. Includes lighting, fences/barriers, vertical signs, pavement markings, traffic signals, sign posts, litter bins, roadside seats, etc.
road furniture	as *road equipment* but excluding pavement markings
ROAM	a computational tool for the study of the interaction between moisture damage and *mechanical* response to loading
runoff	rainfall water that runs across and off a surface (especially from a road's surface)
rutting	channels in a road surface due to plastic *deformation* of the road construction
Sabkha soils	salt-encrusted soil occurring in hot arid or semi-arid regions
saturated	state of a physical system that contains as much of another substance as is possible at a particular temperature or pressure
saturated soil	soil where the pores are completely *saturated* with water
SCI	surface curvature index – often defined as the deflection under the centre of the *FWD*, less the deflection at a point 300mm from the centre of the area loaded by *FWD*
semi-confined aquifer	an *aquifer* that has an *aquiclude* beneath it, but not above it
sensitivity (general)	the relationship of the change of a response to the corresponding change of a stimulus
sensitivity (of an *environmental compartment*)	(a) the susceptibility of an *environmental compartment* to quality-degrading processes and/or (b) the amount, density and quality of *biota* and substances in an *environmental compartment*, and/or of other highly-valued environmental features. (**Not** related to *moisture sensitivity*)

sequential extraction method	an *extraction method* that obtains multiple samples over a period of time from the same source
shakedown	steady state (non-plastic or plastic) response to repeated loading, obtained after a large number of stress–strain cycles
SHRP	Strategic Highway Research Program (USA)
site mean concentration	A characteristic *contaminant concentration* for a specific site
SMC	*site mean concentration*
soakaway	drainage pit from which water can *infiltrate* into the surrounding ground
soil mechanics	the study of the *mechanical* response of soils
soil skeleton	the geometrical arrangement of the solid particles in a soil
soil water	water stored in soil
soil-water retention	the ability of a soil to keep water in the intergranular pores due to *pore suction*. The inverse of *drainability*. Retention is described by the *SWCC*
soil-water characteristic curve	the relationship between the degree of *saturation* in a partially *saturated soil* and the *pore suction*. Despite the name, it is not entirely a characteristic of the soil being influenced by several factors (e.g. temperature).
solid mechanics	the study of the *mechanical* response of solids
sorbate	the substance that will be *adsorbed/absorbed* by a *sorbent*
sorbent	the material that will *adsorb/absorb*
sorption	the process by which substances move to a solid from immediately adjacent water
source	the point of origin of a substance (often a *contaminant*, or potentially a *contaminant*)
SP	segregation potential
speciation	distribution among the various forms of a metal in a solution
specific surface area	the surface area of all the solid particles expressed as a proportion of the total volume of the assembly of those particles
spectroscopy	study of spectra, especially the experimental observation of optical spectra
SPLP	synthetic precipitation leaching procedure
spring thaw	the unfreezing of frozen ground from the surface, downwards, in spring

stiffness	the resistance of a body to *deformation* caused by an applied stress
stiffness matrix	computational step in calculating stresses in a system from strains, or vice versa
strain	ratio of elongation to length at infinitely small scale. The derivative of a displacement field versus coordinates
stress invariant	a measure of a quantity at a point that is independent of the orientation of measurement
stripping	separation of particles in a bound material by the action of water pressure pulses
sub-base	the pavement layer that supports the *base*. Forms a temporary structural layer during construction
subgrade	the natural or imported soil or rock on which a pavement is constructed
subgrade improvement	a treatment applied to a subgrade soil to improve the *mechanical* properties of that layer (see *capping*)
suction	a pressure less than atmospheric
suction or moisture control	techniques for controlling the pore *suction* or *water content* with a porous material
sump	a low-point in a drainage system designed to allow sedimentation of solids in the flowing water in the drainage system
surface water	water collecting on the ground (in rivers, lakes, streams, ponds, etc.)
swale	surface, unlined, often vegetated, longitudinal ditch near the edge of pavement to convey, store and allow *infiltration* of *runoff*.
SWCC	*soil-water characteristic curve*
target	an alternative name for *receptor*
TCLP	toxicity characteristic leaching procedure
TDR	*time domain reflectometry*
thawing	process by which a frozen solid becomes unfrozen, often leaving excess water in the solid. (See *spring thaw*)
thermal capacity	the ability of a material to store or release heat
time domain reflectometry	technique in which a signal is reflected from the ends of a broken electrical circuit (usually buried in the ground), the character of the reflection being of use in determining the *dielectric* conditions at the point of burial
tipping bucket	device to measure water flow. It tips once for each passage of a certain water volume

TMD	total maximum daily load
tortuosity	the ratio between the actual flow path of a water particle moving through the pore space and the linear distance between the starting and ending points of the path
transverse drainage	drains that carry water across the direction of a road
triaxial test	test in which radial and axial stresses are independently applied to a cylindrical specimen
trigger values	*concentrations* of key indicators, above or below which there is a *risk* of an adverse effect. When the value is reached some pre-determined course of action is instituted
turbulent flow	a flow regime characterized by chaotic, stochastic property changes
two phase porous material	a porous material comprising both solid and liquid
UGM	*unbound granular material*
unbound granular material	*particulate solid* comprised of individual stone (or other) particles held together only by their interlock, *compaction* and by any water-induced *suctions* in the pores
underground wetland	*wetland* ecosystem in a porous medium with a dry surface
unsaturated soil	soil layer where the pores are partly *saturated* with water
UV-VIS	molecule absorption spectrometry in the ultra violet - visible frequency ranges
vadose zone	the zone above the *phreatic* surface
van Dorn bottle	device for collecting a water specimen at a particular depth in a *water body* by closing the ends of a tube when at the desired depth
vapour phase method	system for controlling the suction of a soil or aggregate specimen using air of a known humidity to induce pore suction by evaporation
VCP	variable confining pressure
viscosity	a measure of the resistance of a fluid to deform under the application of shear stress
volumetric water content	volume of water expressed as a proportion of the total volume of material plus water

vulnerability	(a) the degree to which a site or *biota* is susceptible to, and unable to cope with, injury or harm, which is determined by its intrinsic characteristics. Such vulnerability is relatively static and mostly beyond human control (see *groundwater vulnerability*) (b) the tendency or likelihood for an environmentally degrading substance to reach a specified position after introduction at some other location or locations
wash-off	*leachate* produced by water flowing across a solid surface, i.e. polluted *runoff*
waste	any discarded material. In the EU the legal definition usually includes *by-products*
water body	a lake, river, stream, pond or groundwater
water content	mass of water expressed as a percentage of mass of solids that contain the water. Also termed *moisture content*. Strictly, should be termed gravimetric water content to distinguish from *volumetric water content*
water framework directive	high-level legal document (2000/60/EC) that establishes the principles for the protection and remediation of water in EU member states
water table	upper level of *groundwater* (therefore coincident with *phreatic surface*)
water quality	a numeric or qualitative assessment of the concentration or load of *contaminant*s in a water body or water sample, usually as compared to some standard value or *background value*
water source protection zone	an area where special regulations apply in order to safeguard a water supply or other water resource
wetland	an environment at the interface between a terrestrial and an aquatic ecosystem
WFD	*water framework directive*

Annex D
List of Symbols

This list excludes most of the material model parameters given in Chapter 9 and not subsequently used. Dimensions are included for many symbols, especially where readers may be uncertain, using the convention: L = Length, M = Mass, T = Time, C = Temperature, I = Charge, − = no units.

Chapter	Symbol	Usage
2/5	α	material parameter
4	α	thermal diffusivity [L^2T^{-1}]
11	α_L	eigenvalue, (sometimes used without subscript for clarity)
6	α_l	longitudinal dynamic dispersivity [L]
9	α_m	effective stress parameter
6	α_t	transversal dynamic dispersivity [L]
9	β	material elastic stiffness parameter
6	β_t	material parameter that is a function of temperature (only)
9	γ	coefficient of anisotropy
9	γ_h	suction volumetric change index (an indicator of the sensitivity of volume change to change in matric suction)
3	γ_w	unit weight of water [$ML^{-2}T^{-2}$]
11	ε	total strain
9	$\boldsymbol{\varepsilon}^e$	elastic strain of the solid skeleton,
10	ε_c	initial strain due to application of a hydrostatic stress
9/11	ε_{ii}	a normal (direct) strain component of the strain tensor, $\boldsymbol{\varepsilon}_{kl}$ (subscripts indicate direction)
9/11	ε_{ij}	a component of the strain tensor, $\boldsymbol{\varepsilon}_{kl}$ (subscripts indicate direction)
9	ε_{ij}^{em}	elastic mechanical strain
9	$\dot{\varepsilon}_{ij}^{e}$	elastic strain increment
9	$\dot{\varepsilon}_{ij}^{p}$	plastic strain increment

9	ε_{ij}^{pm}	mechanical plastic strain, associated with the mechanical yield surface
9	$\boldsymbol{\varepsilon}_{kl}$	the strain tensor
9/10	ε_q	deviatoric strain
9/10	ε_v	volumetric strain $= \varepsilon_1 + \varepsilon_2 + \varepsilon_3$
9	ε_v^{eh}	reversible hydric strain
9	ε_v^{ph}	hydric plastic strain, associated with the hydric yield surface
10	$\varepsilon_{vertical}^{p}$	vertical plastic strain
11	$\varepsilon_{xx}, \varepsilon_{yy}$	see ε_{ii}
9	$\varepsilon_1,\ \varepsilon_2,\ \varepsilon_3$	the principal strains
11	η	isoparametric coordinate
2	Θ	normalized water content (= 1 when saturated and 0 at θ_r)
2/3/5/9	θ	volumetric water content
2	θ_α	volumetric air content
2	θ_r	the residual (or the irreducible) water content
9	κ	proportionality coefficient which describes the hydric behaviour
3	κ_r	relative permittivity, *also known as* relative dielectric constant [–]
2	λ	material parameter pore size distribution index
4/11	λ	thermal conductivity (W/m°C) [$MLT^{-3}C$]
4	$\boldsymbol{\lambda}$	matrix of thermal conductivities
11	μ	dynamic viscosity [$ML^{-1}T^{-1}$]
3	μ_r	relative magnetic permeability [–]
9	ν	Poisson's ratio
6	ξ	an empirical dispersivity measurement [L]
11	ξ	isoparametric co-ordinate
2/11	ρ	density [ML^{-3}]
6/2/10/13	ρ_d	dry density of a soil material [ML^{-3}]
3/2/6	ρ_w	density of water [ML^{-3}]
9/11	$\boldsymbol{\sigma}$	(Cauchy's) total stress tensor
1/9	σ	total stress [$ML^{-1}T^{-2}$]
1/9	σ'	effective stress [$ML^{-1}T^{-2}$]
9/11	$\boldsymbol{\sigma}'$	effective stress tensor
9	$\overline{\sigma}$	net stress defined as σ-u_a [$ML^{-1}T^{-2}$]
10	σ_c	hydrostatic confining stress [$ML^{-1}\ T^{-2}$]
2	σ_i	interfacial energy, surface tension [$ML^{-1}T^{-2}$]
9	σ_{ii}	a normal stress component of the stress tensor, $\boldsymbol{\sigma}_{ij}$ [$ML^{-1}T^{-2}$]
9	$\boldsymbol{\sigma}_{ij}$	the stress tensor
9/11	$\sigma_{ij},\ \sigma_{kl}$	a component of the stress tensor, $\boldsymbol{\sigma}_{ij},\ \boldsymbol{\sigma}_{kl}$ [$ML^{-1}T^{-2}$]
9/10	$\sigma_1,\ \sigma_2,\ \sigma_3$	principal total stresses; σ_3 = cell pressure within triaxial equipment [$ML^{-1}T^{-2}$]

10	$\bar{\sigma}_1,\ \bar{\sigma}_2,\ \bar{\sigma}_3$	the principal net stresses [$ML^{-1}T^{-2}$]
9	τ	shear stress applied to a particle assembly [$ML^{-1}T^{-2}$]
11	$\boldsymbol{\tau}$	variable indicating relative time within a time step
9	ϕ	friction angle at critical state [°]
9	χ	empirical parameter for effective stress calculation $\chi = 1$ for saturated soils and $\chi = 0$ for dry soils)
2	Ψ	capillary suction head (also known as matric potential) [L]
2	Ψ_b	the air entry value [L]
6	ω	coefficient related to the tortuosity [–]
1/3/5/9/13	A	cross sectional area of the specimen [L^2]
10	A_{1c}	final, incrementally accumulated, plastic strain according to an asymptotic strain development framework
3/5	a	cross sectional area [L^2]
11	$\mathbf{B}$	matrix of derivatives of the shape functions, N
13	B	width of section of pavement network [L]
11	B_{Lj}	a member of the matrix $\mathbf{B}$
10	$B_1,\ B_2$	material constants
13	b	width of piece of surface/distance between drains [L]
6/7/11	C	the concentration in a liquid [ML^{-3}]
9	$\mathbf{C}^e$	drained compliance matrix
2	C_c	coefficient of gradation
3	C_e	capacitance [$T^4I^2M^{-1}L^{-2}$]
2	C_u	uniformity coefficient
2	c	material constant
4/11	c	thermal capacity, or specific heat (J/(kg.°C) or J/(m^3.°C)) [$L^2T^{-2}C^{-1}$ or $ML^{-1}T^2C^{-1}$]
9	c	plastic behaviour: apparent cohesion constant [$ML^{-1}T^{-2}$]
11	$\bar{c}$	mean thermal capacity
3	c_e	velocity of the electromagnetic pulse [LT^{-1}]
3	c_v	coefficient of consolidation
3	c_0	velocity of light [LT^{-1}]
9/11	$\mathbf{D}$	diffusion – dispersion tensor
6	$\mathrm{D}(\theta)$	soil-moisture dispersion coefficient
6	D^*	effective diffusion coefficient [L^2T^{-1}]
6	$D^*(\theta)$	effective soil diffusion coefficient [L^2T^{-1}]
6	D_d	diffusion coefficient [L^2T^{-1}]
6/11	D_h	hydrodynamic dispersion coefficient, (combined effects of diffusion & dispersion)

6	D_l	hydrodynamic dispersion coefficient parallel to the principal direction of flow [L^2T^{-1}]
6	D_t	hydrodynamic dispersion coefficient perpendicular to the principal direction of flow [L^2T^{-1}]
10	$D_0, D_1 \ldots D_i$	surface deflection of a pavement measured by sensors 0, 1 . . . i from the centre of an applied loading
2/13/Annex	D_{10}, D_{30}, D_{60}, etc.	particle size of which 10, 30, 60, etc. (%) of particles are smaller
5	$d(\theta)$	water-induced damage parameter
9/11	$\mathbf{E}$	the constitutive stress-strain tensor. Superscripts indicate as follows: e = classical elastic; ep = elasto-plastic. Also used to indicate the stress-strain (stiffness) matrix.
9/10/Annex	E	layer or material stiffness modulus (material stiffness modulus is usually denoted M_r) [$ML^{-1}T^{-2}$]
3/Annex	E_1, E_2	inferred layer modulus from first, second loading cycle of a plate load test
9/11	E_{ijkl}	member of tensor $\mathbf{E}$
11	E_{KL}	member of matrix $\mathbf{E}$
2/13	e	void ratio
6	F	mass flux of solute [$ML^{-2}T^{-1}$]
11	F_L^{int}, F_L^{ext}	nodal internal/external forces or fluxes
9	F_n	force applied on a porous medium (constituted by a solid matrix and pores) [MLT^{-2}]
Annex	F_{10}	10 year maximum value of freezing index [CT]
13	F_{10}, F_{30}, F_{60}, etc.	as D_{10}, D_{30}, D_{60}, etc., except applied to particle sizes of a filter soil/material [L]
11	$f, f_i, f_{i,adv}, f_{diff}, etc.$	flux of fluid or heat or concentration (indices indicate nature and type of flux)
3	f	resonant frequency [T^{-1}]
9	f_n	normal force between particles [MLT^{-2}]
6	f_R	retardation factor
9	f_s	shear force between particles [MLT^{-2}]
2	G	deep percolation or groundwater recharge [LT^{-1}]
9	G	shear modulus [$ML^{-1}T^{-2}$]
5	G^*	bitrumen binder stiffness [$ML^{-1}T^{-2}$]
9	G_a	material elastic stiffness parameter [$ML^{-1}T^{-2}$]
2/11	g	acceleration due to gravity [LT^{-2}]
11	H	enthalpy [ML^2T^{-2}]

1/2/3/6	$h,\ h_1,\ h_2$	head (i.e. pressure expressed in terms of the height of water that would produce it), head at position 1 and position 2 [L]
2	$\nabla \mathbf{h}$	differential matrix of head, h
9	$\mathbf{I}$	second order identity tensor
Annex	I_s	compaction ratio, being E_2
2/5/13	IR	infiltration (volume/time/area) [LT^{-1}]
3	i	hydraulic gradient [–]
5	i_c	infiltration through one crack in units of volume/time/length of crack [L^2T^{-1}]
6	J	rate coefficient [L^2M^{-1}]
1/2/3/6/11/ 13/Annex	K	permeability coefficient (also known as hydraulic conductivity) of a saturated porous material [LT^{-1}]
9	K	the bulk modulus [$ML^{-1}T^{-2}$]
2/11	$\mathbf{K}$	permeability matrix
9	K_a	material elastic stiffness parameter [$ML^{-1}T^{-2}$]
9	K_m	material parameter
2	$K_w(\theta)$	permeability coefficient for a partially saturated porous material [LT^{-1}]
2	$K_1,\ K_2$	material constants of permeability for Eqn. 2.18 [LT^{-1} & L^2T^{-2} respectively]
2	k	relative permeability
6/11	k_d	partition factor, [L^3M^{-1}]
9	$k_1,\ k_2,\ k_2$	material elastic stiffness parameters
3/11/13	L	specimen length, or element dimension, or cell size [L]
3	L_e	inductance [$ML^2T^{-2}I^{-2}$]
1/3/5/11/13	l	length or displacement [L]
6	$M(t)$	pollutant mass on the pavement surface [ML^{-2}]
9/10	M_r	resilient modulus [$ML^{-1}T^{-2}$]
9	$M_{r\mathrm{OPT}}$	value of M_r, resilient modulus, at optimum water content [$ML^{-1}T^{-2}$]
6	M_0	pollutant mass on the pavement surface at the start of a storm [ML^{-2}]
9/11/13	m	counter
2	m_s	mass of solids [M]
3	m_v	coefficient of volume compressibility [LT^2M^{-1}]
2	m_w	mass of water [M]
2	N	material parameter
10	N	number of load applications
11	N_L	shape functions (for node L)
2/4/10	n	porosity
9	n	material elastic stiffness parameter

6	n_e	effective porosity (no units)
11	O^2	second order, infinitesimal, terms in a series equation
13	O_{90}	effective opening size of a geosynthetic
2/13	P	precipitation (rainfall) [LT^{-1}]
11	Pe	Peclet's number [–]
11	P_j	volume force, member of vector $\underline{P}$
13	P_m	precipitation (rainfall) in hour m [LT^{-1}]
9/10	p	mean normal stress $= (\sigma_1 + \sigma_2 + \sigma_3)/3$ [$ML^{-1}T^{-2}$]
11	p	arbitary pressure term (could represent pore pressure, concentration or temperature). A subscript may be used to indicate the time or point at which it is evaluated
9	$\bar{p}$	mean normal net stress [$ML^{-1}T^{-2}$]
9	p^*	the mean stress adjusted for anisotropic usage [$ML^{-1}T^{-2}$]
9	p_a	the reference pressure (100 kPa), approximately equal to atmospheric pressure [$ML^{-1}T^{-2}$]
11	Q	sink term [LT^{-1}]
6	Q_T	maximum sorption capacity of sorbent at temperature, T
1/2/3/13	q	rate of discharge (volumetric flux) [L^3T^{-1}]
9/10	q	the deviatoric stress [$ML^{-1}T^{-2}$]
4	$\mathbf{q}$	heat flow vector (W/m^2) [MT^{-3}]
9	q^*	deviatoric stress adjusted for anisotropy [$ML^{-1}T^{-2}$]
13	q_E	rate of discharge from slopes [L^3T^{-1}]
13	q_L	rate of pipe discharge [L^3T^{-1}]
13	q_P	rate of discharge from pavement [L^3T^{-1}]
2/6	R	surface runoff [LT^{-1}]
11	$\mathbf{R}$	storage matrix
11	R_{KL}	member of storage matrix, $\mathbf{R}$
11	r	storage parameter
3/10	$r,\ r_i,\ r_1,\ r_2$	radius, radial distance at point i, 1 and 2 [L]
2/11	S	water storage [L] (or heat storage, Chapter 11)
6	S	mass of sorbate sorbed per mass of sorbent (typically in units of mg/kg) [–]
2	S_a	degree of air saturation
13	S_{eq}	sand equivalent
5	S_{md}	bitumen-stone tensile bond strength [$ML^{-1}T^{-2}$]
2/4/10/11	S_r	degree of saturation
2/	S_{rr}	irreducible water saturation
5	S_0	dry adhesive strength [$ML^{-1}T^{-2}$]

9	S_θ	gradient of the soil desorptive curve (the rate of change of the logarithm of s with the logarithm of θ)
2/9/10	s	matric suction pressure defined as the difference between pore air and pore water pressures, u_a-u [$ML^{-1}T^{-2}$]; also known as capillary pressure
4/6/11	T	temperature (°C) [C]
3/4/5/6/11	$t,\ t_1,\ t_2$	time, time at positions, or times, 1 and 2 [T]
11	t_0	start time [T]
11	U	amplification factor
1/2/9/11	u	pore fluid pressure (the fluid is usually water) [$ML^{-1}T^{-2}$]
2/9	u_a	pore air pressure [$ML^{-1}T^{-2}$]
3/11	$u_i,\ u_1,\ u_2$	pore fluid pressure in cell or positions or times i, 1 and 2 [$ML^{-1}T^{-2}$]
11	u_L	pore fluid pressure at node L [$ML^{-1}T^{-2}$]
9	u_m	pore fluid pressure for fluid m [$ML^{-1}T^{-2}$]
2/7/11/13	V	volume (total) [L^3]
2	V_s	volume of the solid skeleton [L^3]
2	V_v	volume of voids [L^3]
2/3	V_w	volume of water [L^3]
2/6	v	linear velocity of water (or other fluid) [LT^{-1}]
2	$\mathbf{v}$	velocity matrix [LT^{-1}]
2	W	weight [MLT^{-2}]
3	W_d	dry weight of soil [MLT^{-2}]
3	W_b	moist (bulk) weight of soil [MLT^{-2}]
10	w_{OPT}	optimum water content (gravimetric)
2/3/9/10/11/13	w	water content (gravimetric)
2/6/11	x	co-ordinate (by convention, horizontal or sub-horizontal); distance or displacement in the direction of that co-ordinate [L]
11	x_L	displacements of node L [L]
11	$\underline{l}$	a generalised field, being a set of co-ordinates, distances or displacements
2/6	y	co-ordinate (by convention, horizontal or sub-horizontal); distance in the direction of that co-ordinate [L]
11	Z	represents a set of state variables, describing a material's history
2/6/11/13	z	co-ordinate (by convention, vertical); distance in the direction of that co-ordinate [L]
13	z_w	depth to water table [L]

Operator Symbols and Other Conventions

Δ	"change in value of" at large scale
δ	"change in value of" at infinitesimally small scale
∇	differential operator
Σ	"sum of values of"
δ_{ij}	Kronecker "delta" operator (has the value 1 if $i = j$, otherwise has the value 0)
∂	partial derivative operator
∂_i	an individual partial derivative operator = ${}^{\partial}/_{\partial x_i}$
$fn\,(\ldots)$	function of (. . .)
$\underline{M}^T$, $\mathbf{M}^\mathrm{T}$	transpose of vector *M*, or of matrix **M**
$X_{(i)}$	subscript in parentheses indicates iteration number
BOLD	bold character indicates the matrix or tensor form
$\dot{}$	a dot over a variable indicates "rate of" that quantity
$\_$	underlined character indicates the vector quantity

COST

COST is the the acronym for European COoperation in the field of Scientific and Technical Research. It is the oldest and widest European intergovernmental network for cooperation in research. Established by the Ministerial Conference in November 1971, COST is presently used by the scientific communities of 35 European countries to cooperate in common research projects supported by national funds.

The funds provided by COST - less than 1% of the total value of the projects - support the COST cooperation networks (COST Actions) through which, with EUR 30 million per year, more than 30.000 European scientists are involved in research having a total value which exceeds EUR 2 billion per year. This is the financial worth of the European added value which COST achieves.

A "bottom up approach" (the initiative of launching a COST Action comes from the European scientists themselves), "à la carte participation" (only countries interested in the Action participate), "equality of access" (participation is open also to the scientific communities of countries not belonging to the European Union) and "flexible structure" (easy implementation and light management of the research initiatives) are the main characteristics of COST.

As precursor of advanced multidisciplinary research, COST has a very important role for the realisation of the European Research Area (ERA) anticipating and complementing the activities of the Framework Programmes, constituting a "bridge" towards the scientific communities of emerging countries, increasing the mobility of researchers across Europe and fostering the establishment of "Networks of Excellence" in many key scientific domains such as: Biomedicine and Molecular Biosciences; Food and Agriculture; Forests, their Products and Services; Materials, Physical and Nanosciences; Chemistry and Molecular Sciences and Technologies; Earth System Science and Environmental Management; Information and Communication Technologies; Transport and Urban Development; Individuals, Societies, Cultures and Health. It covers basic and more applied research and also addresses issues of pre-normative nature or of societal importance.

Index

E

Q

Printed in the United States
152085LV00002B/73/P